Pest and Vector Management in the Tropics

Pest and Vector Management in the Tropics

with particular reference to insects, ticks, mites and snails

Anthony Youdeowei

Department of Agricultural Biology, University of Ibadan, Nigeria

Mike W. Service

Department of Medical Entomology Liverpool School of Tropical Medicine UK

1st Reprint in India, 2015

Published by:

United Book Prints
4806/24, Ansari Road,
Daryaganj,
New Delhi - 110 002
Tel.: 011-41511055
E-mail: unitedbookprints@gmail.com

Print: 2015

ISBN: 978-93-83692-11-8

Printed in India

Contents

Preface

Insect pests and disease vectors are a major threat to increased agricultural production and to the health and well-being of the human populations who live in tropical developing countries. The need to reduce, and if possible eliminate altogether, the ravages of these insect pests and disease vectors has attracted the attention of scientists, technologists, administrators, United Nations agencies, national governments and international aid agencies for many decades. Early efforts to control arthropod pests and disease vectors relied primarily on the application of broad-spectrum pesticides. Unfortunately this approach often caused undesirable side effects, including environmental pollution, the development of insecticide resistance and sometimes sudden increases in old, or even new and worse, pests. Research efforts were therefore directed towards seeking more rational, long term and environmentally acceptable methods for combating pests and vectors but which at the same time were efficient. This led to the concept and philosophy of 'integrated pest and vector management'. The main advances in this field have been made in the developed temperate countries, but even here efficient management has been achieved for only a few crops in some countries. In the tropics, however, especially in the developing countries where arthropod pests and disease vectors are of major concern, comparatively less work has been done. Moreover, the limited information which is available is published in technical journals scattered over the world. Relatively little attempt has been made to synthesize research results to produce management protocols for the more important tropical pest and disease vectors. Although broad generalizations and principles have been produced by organizations such as the Food and Agriculture Organization (FAO) of the United Nations and the World Health Organization (WHO) for certain pests and vectors, one problem remaining is that in many cases pest management programmes have to be developed for particular areas.

Another problem is that insufficient attention has been paid to the co-ordinated training of personnel in integrated pest and vector management techniques which are relevant to the particular situations prevailing in the tropics. Admittedly, regional, national and sometimes international institutions have sometimes played a part in developing such training programmes, but this has often been on an uncoordinated basis. With this background among other factors in mind, the International Centre of Insect Physiology and Ecology (ICIPE) and the United Nations Environment Programme (UNEP), both in Nairobi, Kenya, jointly established in1977 the ICIPE/UNEP International Group Training Course in Ecologically Sound Pest and Vector Management Systems. Its main objective was to train young scientists from tropical developing countries in the 'components essential for ecologically sound insect pest and vector management'. After five years organizing and evaluating this training course, it became abundantly clear that there was a need to document existing control strategies for selected important pests and vectors. Many of the strategies relied almost entirely on pesticides which, for ecological, biological and economic reasons, may have only limited future application. It was necessary to emphasize that environmentally acceptable approaches need to be developed as well as ways of effectively integrating the variety of existing methods for the more rational management of insect pest and vector populations. It is recognized that at present, not all agricultural pests and vectors can be controlled effectively without resorting to insecticides; this is particularly obvious with certain disease vectors. Although there have been increasing efforts to reduce insecticidal usage and improve non-insecticidal control strategies, we feel that there is still considerable scope for improving their implementation. It is important to be realistic and take into consideration the cost-effectiveness of such management programmes in relation to the resources of developing countries, especially in Africa.

This book attempts to describe the types of integrated approaches that are being, or can be used against important pests and vectors. The examples discussed here concentrate on the African continent, but various pest situations encountered in Central and South America and Asia are also included. Particular attention has been focused on the following features:

(a) Attempts have been made to combine, in one volume, the management of crop pests and disease vectors in recognition of the close link which often exists between agricultural and health problems. Also, pest control officers and scientists frequently encounter both problems in the mixed farming situations found in many tropical developing countries.

(b) Environmental changes taking place both now and in the future, combined with the politics and economics of insect pest and vector control, and the issue of choice that exists in a few cases between their eradication and management are discussed.

(c) The book has a strong ecological bias and stresses the need to understand pest and vector population dynamics for successful management. The importance of correct sampling procedures, surveys, evaluation of management strategies, relevant training of personnel, and government policies are emphasized.

(d) Attention is focused on problems, such as the socio-economic aspects of pest and vector management. Whenever possible an integrated approach is advocated. Ways of formulating the best method for any one pest or vector, which may involve several different control methods, are discussed.

(e) Special note is made of the dangers already existing, and those likely to be created, in connection with development schemes, especially those involving large-scale irrigation projects and increased urbanization – development schemes which usually entail considerable changes in the local ecology and environment. The ways in which such man-made and drastic changes to the environment affect insect pest and vector populations are described.

This book is intended to provide up-to-date information and an easily readable text for agricultural and health extension workers, as well as for scientists involved in pest and vector management in Ministries of Agriculture and Health. It should also be of value to agricultural and medical undergraduates, postgraduate students, pest and vector management consultants, administrators and policy-makers and planners, all of whom, it is hoped, will find areas of special interest in the book.

Nairobi, Kenya
July 1981

Anthony Youdeowei
Mike W. Service

Acknowledgements

In writing this book we have drawn extensively on the copious materials which have been used and featured at the ICIPE/UNEP International Group Training Course on components essential for ecologically sound pest and vector management systems, held in Nairobi between 1977 and 1981. We acknowledge the foresight of Professor Thomas R. Odhiambo, Director of ICIPE, and Professor Reuben Olembo, Dr Letitia Obeng and Dr O.U. Alozie all of UNEP, and others who originated and initially planned this course. We are extremely grateful to the International Centre of Insect Physiology and Ecology (ICIPE) and the United Nations Environment Programme (UNEP) both in Nairobi, for the opportunity to participate as scientific co-ordinator (AY) and as lecturer (MWS) throughout this annual training course.

This book has emerged from detailed discussions with many of our colleagues, in particular David Baldry, Ray Kumar, Frank Walsh, David Rogers and the late Reuben Abasa, to all of whom we are deeply grateful for suggestions and useful comments. We wish to thank personally the following for providing certain material: Professor G.S. Nelson (Fig. 1.5); Professor I. McIntyre (Fig. 1.7); Dr R.H.L. Disney (Fig. 4.22); Dr H.S. Omar (Fig. 4.23); Dr V.A. Dyck (Fig. 6.14); Dr E.A. Heinrichs (Fig. 6.15); Dr L.A. Falcon (Table 6.1) and Dr. R.W. Crosskey (Figs 7.5 and 8.6).

We are also indebted to the following for permission to reproduce photographs, illustrations and tables: The World Health Organization; the Food and Agriculture Organization of the United Nations; the Director of the International Centre of Insect Physiology and Ecology, Nairobi, Kenya; Professor P.H. Vercammen-Grandjean; the Director of the National Institute of Virology, Pune, India; The G.W. Hopper Foundation, Commonwealth Scientific and Industrial Research Organization, Australia; CIBA-GEIGY; American Geographical Society; International Rice Research Institute; National Cereals Research Institute. The editors of the Nairobi Times (Kenya), Canadian Entomologist, Journal of Animal Ecology, Rivista Saude Publica (S. Paulo), Tropenmedizen und Parasitologie; the publishers Academic Press, Applied Science, Blackwell Scientific Publications, Hafner, Oxford University Press, S. Karger, Wiley, Longman Group Ltd., Servicio de Biblioteca e Documentacao and University Press (formerly OUP Nigeria).

List of contributors

Abasa, R.O. *formerly* Department of Zoology, University of Nairobi, P.O. Box 30197, Nairobi, Kenya.

Amin, Mutamed, A. Department of Community Medicine, Faculty of Medicine, University of Khartoum, Khartoum, Sudan.

Baldry, D.A.T. UNDP/FAO, Tsetse Applied Research and Training Project, P.O. Box 30563, Lusaka, Zambia.

de Lima, C.P.F. Ministry of Agriculture, National Agricultural Research Laboratories, P.O. Box 30028, Nairobi, Kenya.

Ezueh, M.O. National Cereals Research Institute, Umudike, P.M.B. 1006, Umuahia, Nigeria.

Forattini, O.P. Universidade de Sao Paulo, Faculdade de Saude Publica, Sao Paulo, Brazil.

Kumar, R. Zoology Department, University of Ghana, P.O. Box 67, Legon, Ghana.

Nadchatram, M. Institute for Medical Research, Jalan Pahang, Kuala Lumpur 02-14, Malaysia.

Robert, M.J. Department of Medical Entomology, Liverpool School of Tropical Medicine, Liverpool L3 5QA, UK.

Rogers, D.J. Department of Zoology, University of Oxford, Oxford OX1 3PS, England.

Rosenfield, P. Special Programme for Research and Training in Tropical Diseases, World Health Organization, 1211 Geneva 27, Switzerland.

Salama, H.S. National Research Centre, Laboratories of Pest and Plant Protection, Sh. El-Tahrir, Dokki, Cairo, Egypt.

Saxena, R.C. International Rice Research Institute, P.O. Box 933, Manila, Philippines.

Tatchell, R. FAO/DANIDA, Tick and Tick-borne Disease Control Project, P.O. Box 913, Khartoum, Sudan.

Thompson, D.J. Zoology Department, University of Liverpool, Liverpool L69 3BX, UK.

Walker, P.T. Centre for Overseas Pest Research (London), Porton Down, Salisbury, Wiltshire, 5PA 0JQ, UK.

Walsh, J.F. World Health Organization, Onchocerciasis. Control Programme; B.P. 549, Ouagadougou, Upper Volta.

1 Overview of pest and vector problems in the tropics

Man in the tropical environment

This book is concerned with the management of pests and vectors in the tropics, with particular reference to the developing countries. An overview of the pest and vector problems in the tropics will provide a basis for defining the problems and placing them in correct perspective.

In his attempt to live and survive on earth, man manipulates his environment to provide his needs. In practical terms, this implies interaction with his physical environment, and with the other living organisms sharing the ecosystem. It is in the interaction with living organisms that competition and conflicts arise. Man has always suffered extensively from the depredations of other animal species which affect him both directly, by attacking his person and inflicting mechanical injury or transmitting disease, and indirectly, by causing damage to his crops, animals or other products and materials which he needs. Even early man in his primitive cave is believed to have suffered from small insect pests such as lice, fleas and bugs. As the human population increases and man's demand for natural resources increases both in magnitude and diversity, the ecological problems become more and more complex and serious. It is now quite clear that ecological problems present the most serious and pressing dilemmas which will confront man in the coming decades (see Ch. 8).

The tropics are characterized by warm climates and in the humid areas conditions favour the rapid development and multiplication of plant and animal life throughout the year. Insects, for example, abound in the warm humid tropics, and Thomas R. Odhiambo, Director of the International Centre of Insect Physiology and Ecology (ICIPE) in Nairobi, Kenya, contends that 'insects are the most important element of our tropical natural resource'.

Man, especially in the tropics, is always in competition with insects for available natural resources; competition for food being the most important and most urgent. Two of the greatest challenges of today are the provision of an adequate supply of food, and improving the health of the rapidly increasing human population. The rate of food production increase in tropical developing countries is less than 1.0 per cent annually, while in most countries the population is growing at a rate of 2.0–2.5 per cent, or more. There is thus a serious gap between food supply and food demand and the gulf between starvation and adequate feeding continues to widen. The sahelian countries in West Africa with their drought problems, and countries within the arid and semi-arid tropical regions, are among the worst affected. Insect pests compound this problem by attacking agricultural crops in the field and in storage, causing considerable losses of agricultural produce.

Problems with agricultural pests

Field pests

The losses of major agricultural crops from field pests are high and are estimated at an average of 35 per cent. Table 1.1 gives data for losses of major world crops due to insects and other pests. It is estimated that at least 20 per cent of all major crops are lost even before harvest in the field.

Table 1.1 Percentage losses caused by pests of the world's major crops (data from *World Food and Nutrition Study*, Vol. 1, 1977)

Crop	Losses (%)			
	Insects	Diseases	Weeds	Total
Rice	26.7	8.9	10.8	46.4
Wheat	5.0	9.1	9.8	23.9
Maize	12.4	9.4	13.0	34.8
Sorghum/Millet	9.6	10.6	17.8	38.0
Potatoes	6.5	21.8	4.0	32.3
Cassava	7.7	16.6	9.2	33.5
Sweet potatoes	8.9	5.0	11.7	25.5
Tomatoes	7.5	11.6	5.4	24.5
Soyabeans	4.5	11.1	13.5	29.1
Peanuts	17.1	11.5	11.8	40.4
Palm Oil	11.6	7.4	9.6	28.6
Copra	14.7	19.3	10.0	44.0
Cotton seed	11.0	9.1	4.5	24.6
Bananas	5.2	23.0	3.0	31.3
Citrus	8.3	9.5	3.8	21.6

Table 1.2 Regional percentage crop losses caused by pests (data from H. H. Cramer, 1967)

Region	Losses (%)			
	Insects	Diseases	Weeds	Total
Europe	5	13	7	25
North and Central America	9	11	8	28
South America*	10	15	8	33
Africa*	13	13	16	42
Asia*	21	11	11	43

* Regions with high crop losses.

Average regional percentage crop losses due to pests is given in Table 1.2. The data clearly show that crop losses caused by insect pests are highest in the tropics, particularly in South America, Africa and Asia. In many instances losses occur at different stages of crop development in the field (Figs. 1.1, 1.2, 1.3). For example, A. Raheja demonstrated for cowpeas in

Fig. 1.1 A cassava farm in southern Nigeria completely damaged by the grasshopper, *Zonocerus variegatus*. (Anthony Youdeowei)

Fig. 1.2 *Zonocerus variegatus*, one of the most important pests of agricultural crops in West Africa. Adults are scraping the bark of cassava plants after complete defoliation of the plants, in Nigeria. (Anthony Youdeowei)

Fig. 1.3 Cocoa pods in Indonesia damaged by capsids, *Helopeltis* sp. (Anthony Youdeowei)

northern Nigeria that major losses occur during the flowering and pod-forming growth periods (Table 1.3). Clearly it can be seen that without adequate crop protection it would be impossible to grow cowpeas in northern Nigeria successfully. In fact reasonable yields of cowpeas are obtained only with regular application of pesticides to the crop in the field.

For cash crops like coffee, oil palm, cocoa and cotton, losses due to insect pests mean considerable financial deficits. In Nigeria, losses due to pest and disease attack on cocoa ranged from 104 000 tonnes in 1975/76 to 192 200 tonnes in 1979/80. At the 1975/76 producer price of grade I cocoa, these losses represent financial deficits of 69–130 million Nigerian Naira (i.e. $US121–228 million).

Table 1.3 Estimated losses due to insect pests in yield of cowpeas at Samaru, northern Nigeria (data from A. Raheja, 1980)

Growth period	Nature of damage	Loss of potential yield (%)			Causative agents
		1971	1972	1973	
Establishment	Reduced establishment	2	3	2	Soil arthropods, *Ootheca* spp.
Pre-flowering	Foliage damage	9	7	9	*Ootheca* (spp.), *Alcidodes leucogrammus*
Flowering	Reduced flower production	46	52	48	*Maruca testulalis*, coreids, thrips, aphids
Pod formation	Flower shedding and pod loss	26	25	21	*M. testulalis*, coreids
Ripening	Seed loss	9	7	17	Coreids and other insects
Totals		92	94	97	

Stored products

The section on stored product pests in Chapter 6 outlines the various factors which cause losses of stored products. Insects constitute a major factor in crop losses in storage. Various estimates of the magnitude of these losses have been made for different parts of the tropics. In Sierra Leone, about 30 per cent of stored rice is lost to insect pests; 25–45 per cent of the maize stored in Ghana is lost to insect pests while farmers in northern Nigeria expect 50–60 per cent of their stored cowpeas to be attacked by insects after only six months of storage. Similar levels of losses to various stored food and other products occur in east and southern Africa and in other parts of the tropical world.

Problems with disease vectors

Many of man's most debilitating diseases are transmitted by arthropod vectors. Mosquitoes spread malaria, filariasis and yellow fever; tsetse flies transmit sleeping sickness; simuliid blackflies transmit river blindness; body lice transmit typhus; triatomine bugs transmit Chagas' disease; *Leptotrombidium* mites spread scrub typhus; and soft ticks carry tick-borne relapsing fever (Table 1.4). Some of the present-day problems encountered in the control of the major vector-borne diseases are briefly discussed in this section. It should be remembered that vector management is just one aspect of the management of vector-borne diseases; other components include attacks on the parasite causing the disease, education and environmental approaches.

Table 1.4 Some major vector-borne diseases transmitted to man

Disease	Vector	Distribution
Malaria (*Plasmodium* spp.)	*Anopheles* mosquitoes	Most tropics and subtropics see fig. 1.4
Filariasis (*Wuchereria bancrofti*, *Brugia malayi*)	Anopheline & culicine mosquitoes	Most tropics
Yellow fever	Mostly *Aedes* mosquitoes e.g. *Aedes aegypti*; also *Haemagogus* spp. in Americas	Africa, Central and South America
Dengue including haemorrhagic form	*Ae. aegypti*, *Ae. albopictus*	Most tropics; haemorrhagic form in south-east Asia
Encephalitis viruses	Many culicine mosquitoes, and also ixodid (hard) ticks	Temperate and tropical areas, e.g. USA, Europe, Asia
River blindness (*Onchocerca volvulus*)	Simuliid blackflies	Africa, Yemen, Central and South America
Sleeping sickness (*Trypanosoma* spp.)	Tsetse flies	Africa
Chagas' disease (*Trypanosoma cruzi*)	Triatomine bugs	Central and South America
Leishmaniasis (*Leishmania* spp.)	Phlebotomid sandflies	Tropics and subtropics
Epidemic typhus (*Rickettsia prowazeki*)	Body lice	Temperate and tropical areas
Endemic typhus (*Rickettsia mooseri*)	Rodent fleas	Temperate and tropical areas
Scrub-typhus (*Rickettsia tsutsugamushi*)	Leptotrombiculid mites	Asia

Endemic relapsing fever (*Borrelia duttoni*)	*Ornithodoros* ticks	Tropics, subtropics and temperate areas
Epidemic relapsing fever (*Borrelia recurrentis*)	Body lice	Tropics and temperate areas
Plague (*Yersinia pestis*)	Rodent fleas	Mainly tropics
Loiasis (*Loa loa*)	*Chrysops* flies	West Africa
Schistosomiasis* (*Schistosoma* spp.)	Aquatic snails	Mainly Africa, south-east Asia, South America

* Not strictly a vector-borne disease, see p. 10.

Malaria

During the optimism of the DDT era in the late 1940s it was thought that many of the important vector-borne diseases would at last be conquered. There were some early successes. For example, DDT powder dusted onto people to kill body lice prevented typhus outbreaks; this same insecticide also aided the eradication of malaria from the southern USA and Europe. Successes such as these led to the declaration by the World Health Assembly in 1955 that worldwide malaria eradication was technically feasible, except in Africa south of the Sahara. However, there was a sense of urgency in accomplishing this, as in 1950 two malaria vectors had already developed DDT resistance. The strategy proposed was that of spraying the interior surfaces of houses with deposits of DDT which would remain effective in killing indoor-resting mosquitoes for about six months (see p. 275).

In India this approach resulted in reducing the number of malaria cases from an estimated 75 million in 1947 before the campaign started, to only 100 000 in 1965. This was a significant achievement and malaria eradication seemed to be within sight, but thereafter there were serious deteriorations and epidemics of malaria culminating in an estimated 7–10 million cases in 1977. What caused the breakdown of the eradication campaign? There were many reasons, one of which was the evolution of resistance in the malaria vectors; but more importantly the resurgence of malaria was probably caused by a lack of trained manpower, poor entomological surveillance, lack of funds and administrative problems and complacency. People no longer considered malaria a serious disease and therefore tended to neglect some of the simple protective measures formerly undertaken. Another problem was that there was a migration of people into areas that were formerly very malarious, so when there was a resurgence of the disease such people were very liable to get malaria. Although the number of malaria cases increased dramatically due to poor vector management, mortality decreased due to improved case detection and treatment – that is better disease management.

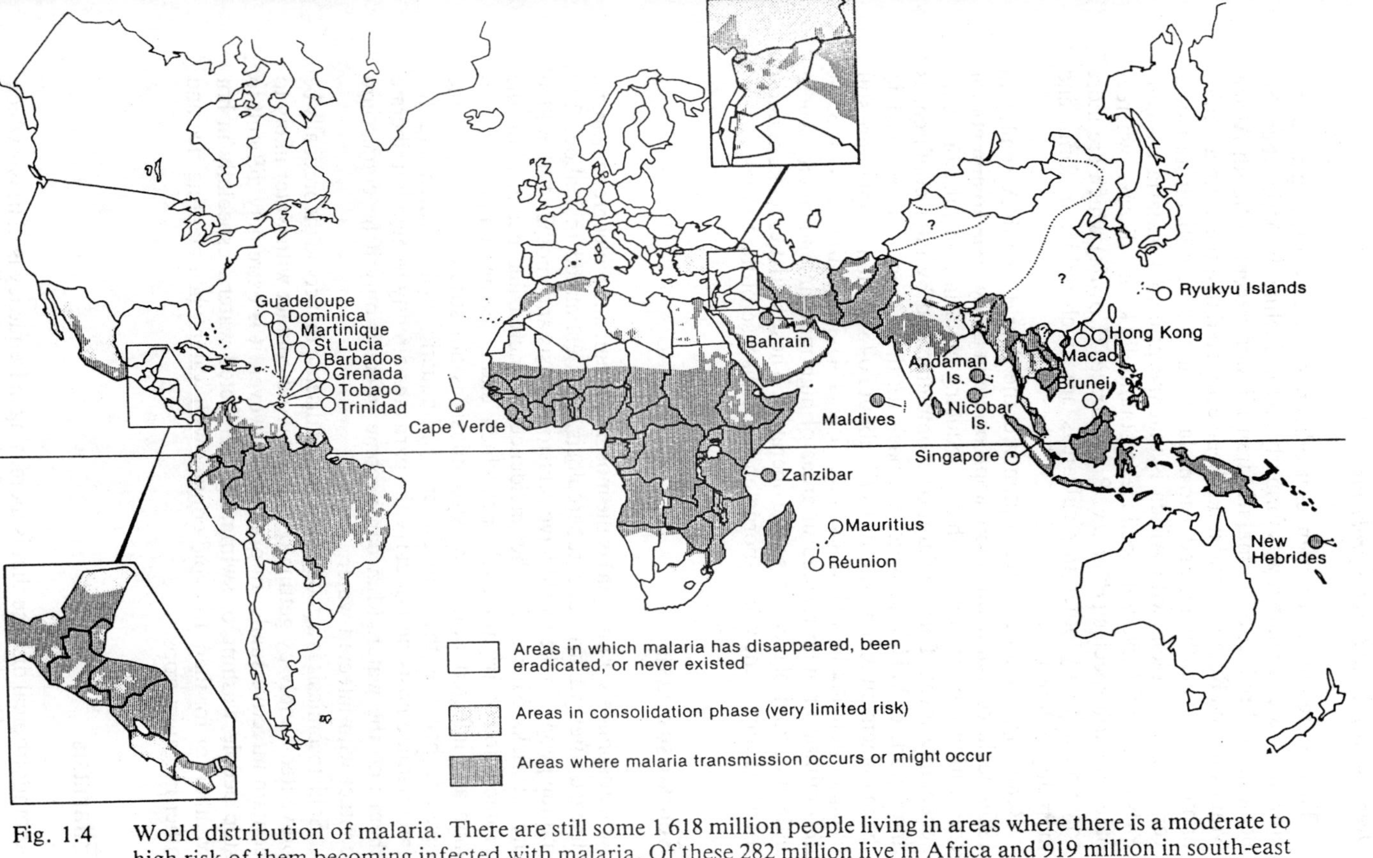

Fig. 1.4 World distribution of malaria. There are still some 1 618 million people living in areas where there is a moderate to high risk of them becoming infected with malaria. Of these 282 million live in Africa and 919 million in south-east Asia. (Courtesy of the World Health Organization)

There were similar resurgences in other parts of the world and malaria is now rife in areas where it was formerly on the decline. In Africa there has been no real reduction in malaria transmission except in South Africa, Mauritius and Réunion, and it has been estimated that there are a million or more infant deaths a year due to malaria, but this is a difficult statistic to verify. The problem with Africa is that the principal malaria parasite, *Plasmodium falciparum*, produces a potentially killing disease, whereas other malaria species (*P. vivax*, *P. malariae*, *P. ovale*) which are more common in many other tropical areas are much milder in form, and mortality is lower.

Realizing it had been over-optimistic, the World Health Assembly in 1969 declared that it was more appropriate in many areas to aim not for eradication but for control, that is reducing malaria transmission to an acceptable level. This declaration, complemented by later conferences and reports of WHO, provides the basis for the present-day approach to the management of malaria. In 1977, about 2048 million people were still living in potentially malarious areas (Fig. 1.4), it is estimated that 352–480 million of them resided in areas of high malaria endemicity where little or no control measures were operating. These statistics make malaria the most important vector-borne endemic disease of the tropics.

Schistosomiasis

Schistosomiasis (bilharzia) is the most important and widespread tropical disease after malaria and tuberculosis. It is estimated that about 200 million people in the world have schistosomiasis and another 400 million are at risk. Moreover, the incidence is increasing because of the proliferation of irrigation schemes which lead to greater contact between man and infected water. Schistosomiasis is not strictly a vector-borne disease, because the aquatic or semi-aquatic snails which are the intermediate hosts do not carry the parasites directly to man but liberate them into the water. Man can become infected only if he comes into contact with infected water.

Schistosomiasis is a good example of a disease that could theoretically be prevented simply by getting people to avoid contact with water likely to contain infective snails. In practice, however, it is extremely difficult to stop people washing or swimming in infected waters, especially in hot weather, or for them to avoid contact with water when planting rice and other irrigated crops.

Filariasis

Another tropical disease that is increasing in frequency is filariasis which is transmitted by mosquitoes, and which can lead to gross deformities such

Fig. 1.5 Many people who become infected with filarial worms of the parasite *Wuchereria bancrofti* (giving rise to the disease bancroftian filariasis) show few or no symptoms. But sometimes in severe cases gross deformities occur leading to the condition shown here and called elephantiasis. (Courtesy of G.S. Nelson)

as elephantiasis (Fig. 1.5). There are probably some 400 million people living potentially under the threat of contracting mosquito-borne filariasis (due mainly to two parasites, *Wuchereria bancrofti* and *Brugia malayi*), which is a higher number than a hundred years ago when it was first discovered that a mosquito was the vector! One reason for the upsurge of filariasis is the rapid population growth that has occurred in Africa, India and south-east Asia. In some countries attempts to give people a piped water supply and latrines have created numerous collections of polluted waters due to leaking pipes and poorly constructed latrines, and such collections of water provide ideal habitats for an important vector of bancroftian filariasis, *Culex quinquefasciatus* (= *C. fatigans*).

Much of the population growth occurring in tropical developing countries has taken place in the towns and facilities are often inadequate to cope with the spread of overcrowded slums (Fig. 1.6), polluted ditches and other unsanitary conditions, with the result that numerous collections of dirty water accumulate in and around the towns – again leading to larger populations of the mosquito vector of filariasis.

Fig. 1.6 Typical slum habitations in the Americas. Houses are built of a mixture of mud, corrugated tin and even cardboard. Needless to say unhygienic conditions prevail and are conducive for the transmission of various diseases. (M.W. Service)

Unlike malaria it is not essential to interrupt filariasis transmission completely because it is an accumulative disease and low-intensity infections have little effect on general health. A substantial reduction in the number of mosquito bites will greatly alleviate suffering from filariasis.

Improved sanitation, hygiene and living conditions, combined with applications of insecticides to polluted urban waters, could do much to reduce populations of *C. quinquefasciatus*. Should it succeed, the United Nations water and sanitation programme of the 1980s might, through environmental management, provide a permanent solution for most urban, and many rural, affected areas.

African trypanosomiasis

Thirty-five African countries are affected by tsetse flies which are vectors of different forms of trypanosomes including those responsible for sleeping sickness (human trypanosomiasis). Animal trypanosomiasis (nagana) is responsible for the under-exploitation of about 7 million km^2 of land. An estimated 35 million people risk getting human trypanosomiasis, and about 10 000 new cases are reported annually, but there are likely to be many more unreported cases. The main importance of the tsetse fly, however, is as a vector of animal trypanosomiasis, often called nagana, which is responsible for huge economic losses in the cattle industry. Infected cattle belonging to susceptible breeds become thin, emaciated and often die (Fig. 1.7). In many areas of Africa, especially in the west, heavy tsetse fly infestations can reduce animal growth rates by 40–50

Fig. 1.7 A West African Zebu cow sick with animal trypanosomiasis (nagana) and showing typical emaciated condition. (Courtesy of I. McIntyre)

per cent, and milk yields may be diminished to about 0.5–1 litre per animal per day. It has been estimated that the present number of 35 million cattle in Africa could be increased to 160 million animals if nagana could be eliminated. Because of tsetse flies much of the increase in cattle numbers that has occurred has been restricted to marginal grazing areas in dry northern savanna regions where flies rarely exist. Efficient insecticidal operations in certain areas of West Africa have greatly reduced the number of tsetse flies, but grazing pressure has increased at a faster rate than the elimination of flies from such areas. This can lead to overgrazing and the possibility of desertification. However, improving cattle husbandry techniques should lead to increased meat production without increasing the area of grazing or number of cattle.

American trypanosomiasis

In Central and South America about 10 million people are infected with a different form of trypanosomiasis causing Chagas' disease, which is transmitted by large blood-sucking reduviid (triatomine) bugs. It is not basically a killing disease but runs a chronic course of infection leading to incurable heart damage and sometimes death in people in their mid-forties. The disease exists as a zoonosis (see p. 25) in wild animals such as rodents and marsupials, and is contracted by people living in poorly constructed houses in which the vectors shelter in cracks and crevices in the mud walls and thatched roofs. Privileged town dwellers run little risk of contracting Chagas' disease, which is essentially a disease of poor rural families.

Effective drugs are not available to combat the disease and so control is directed against the vectors. For example, huts can be sprayed with residual insecticides, or made of cement or bricks and corrugated roofs which make them inhospitable to the bugs. Both insecticidal spraying and rehousing cost large amounts of money and are therefore not implemented on a wide scale.

Yellow fever and dengue

Another typical zoonosis is the viral disease yellow fever for which monkeys are the natural reservoir. In the forest and rural areas various *Aedes* mosquitoes transmit the disease to monkeys and to man, but in the so-called urban cycle of transmission there is no zoonotic involvement and *Aedes aegypti* is the only vector. Despite an efficient vaccine yellow fever still sporadically occurs in Africa, especially in the west, where there have been several recrudescences in recent years. The largest recent outbreak was in Ethiopia. It also occurs in Central and South America, where there may be epidemics of many thousands of cases and hundreds of deaths. Dengue is another virus transmitted by *Aedes* mosquitoes especially *Ae.*

aegypti. It is usually a mild, febrile prostrating disease with a very low mortality rate. During the 1950s and 1960s, however, a virulent form, termed haemorrhagic dengue, flared up in south-east Asia and the western Pacific area, causing many deaths especially among young children where the mortality rate can be as high as 20 per cent. In 1973 there were more than 14 000 cases of haemorrhagic dengue in Vietnam and 1000 people died; in Fiji there were over 150 000 cases in 1975. Classical dengue has also been increasing in the Caribbean and in other areas of Central America where there are epidemics almost yearly. One of the larger South American epidemics involved at least 1.5 million people in Colombia in 1972. The main domestic vector, *Ae. aegypti*, is a mosquito that breeds in water-storage jars, discarded tin cans, and other assorted containers. It is common in slums and towns where in many areas it has been increasing in recent years, due to the progressive disregard for simple individual sanitation measures, enacted, sometimes by enforcement, in the past. Dengue is mainly an urban disease in south-east Asia, although not in the Caribbean and probably not in the south Pacific islands. In some areas of the tropical Americas attempts have been made to eradicate *Ae. aegypti* from urban areas by DDT spraying, but this has been hampered by the vector developing resistance to DDT, and then to other insecticides. It seems wise not to use too extensively alternative insecticides that are still effective to try to eradicate the vector when dengue is not being transmitted, because they are likely to promote resistance. They would then be useless in suppressing vector populations in emergencies such as dengue epidemics.

Onchocerciasis

About 20 million people in Africa and Central and South America suffer from river blindness (onchocerciasis), a disease transmitted by simuliid blackflies which breed in streams and large rivers. If untreated the disease can cause impaired vision leading ultimately to total blindness. Onchocerciasis is an accumulative disease which is more intense in older people and is aggrevated by chronic exposure. The incidence is commonest in people living near rivers. In the Volta River basin of West Africa about 1 million people out of a population of 10 million are afflicted, and of these some 70 000 are totally or partially blind (Fig. 1.8). Generally when 60 per cent or more of any community are infected and the blindness rate rises above 5 per cent, this becomes an unacceptable burden for the community and the population declines due to emigration. In the Volta River basin alone about 65 000 km^2 of fertile land has been abandoned due to onchocerciasis. People migrate to the less fertile plateau areas where they cause soil erosion and exhaustion, especially when they abandon the quasi-shifting agriculture that is normally practised, in favour of continual cropping to eke out a living.

Fig. 1.8 A frightful picture of blind people being lead from one village to another in the savanna areas of West Africa. Blindness is caused by the parasitic filarial worm *Onchocerca volvulus*. (Courtesy of the World Health Organization)

Disease control

A vigorous vaccination campaign has eradicated smallpox from the world, but unfortunately there are no efficient vaccines against malaria, filariasis, river blindness, sleeping sickness and schistosomiasis. Moreover, even if efficient vaccines were available, eradication of these diseases would be much more difficult because they are vector-borne. For these and other reasons, there are no plans for eradicating these diseases. An added complication is that some also have animal reservoirs (e.g. Rhodesian form of sleeping sickness and forms of leishmaniasis). The immensity of the logistics of control operations is frightening. In Africa for example some 300 million people would have to be vaccinated for protection against malaria. An efficient vaccine against yellow fever, one of the most virulent diseases of man, gives protection for at least 10 years, but even if everyone in potentially dangerous areas could be vaccinated, yellow fever would not be eradicated because of the remaining reservoir of infection in monkeys, which would serve as a source of renewed transmission. Yellow fever therefore will remain a threat for many years.

Unfortunately many of the most effective drugs for the treatment of vector-borne diseases are too toxic to allow them to be used in mass control campaigns. For example, most drugs used to treat sleeping sickness can cause severe side reactions and even death. Similarly, the drug DEC (diethylcarbamazine) kills the microfilariae of *Onchocerca volvulus* (causing river blindness) and *Wuchereria bancrofti* (causing bancroftian filariasis), but in the case of river blindness the presence of enormous numbers of dead microfilariae in the skin stimulates severe and alarming side reactions. Consequently, although drugs such as DEC can be used under strict medical supervision to treat individual cases, they cannot safely be used at the community level against river blindness. Side reactions may also occur when using DEC to treat bancroftian filariasis but they are not so severe and the drug has sometimes been administered at a community level. Anti-malarial drugs are much safer, and toxicity is usually not a problem, but there remain administrative difficulties in ensuring that everyone in a community receives radical treatment; if untreated cases remain they will serve as reservoirs for renewed malaria transmission. Elimination of *Plasmodium falciparum* by drugs is easier than with other malaria parasites because of the short treatment regime required for radical treatment. There may also be logistic and economic restraints in purchasing sufficient drugs for wide-scale distribution. A combination of chemotherapy and vector control measures (= management) may in certain situations prove the best strategy for malaria control. Another problem, as with the use of insecticides, is the appearance of drug resistance. Chloroquin is one of the better curative drugs, but resistant strains of the malaria parasite, *Plasmodium falciparum,* have already appeared in Central and South America, south-east Asia, Bangladesh, India and more recently in the western Pacific and East Africa, from where they are likely to spread to other parts of Africa.

Vector management

Because of the complexities of the epidemiology of most major vector-borne diseases and the difficulties of effectively preventing transmission with drugs or vaccines, vector control methods have in the past provided the main strategy for reducing transmission. The actual methods used to reduce vector populations have changed several times since their introduction in the early 1900s. There is an urgent need to improve further the existing approaches and wherever possible to develop new ones for the control of these tropical diseases which are inter-disciplinary, taking into consideration the best of several methods for minimizing transmission. Such a system of 'integrated disease management' should no longer be regarded in isolation from other aspects of welfare, health, social and economic considerations (see Ch. 9).

The aim of most of the following chapters is to outline the basic concepts and principles of the management of agricultural pests and disease vectors, and to provide some guidelines for their implementation.

Recommended reading

Bruce-Chwatt, L.J. (1980) *Essential Malariology*, William Heinemann, Medical Books, London, 354 pp.

Chiarappa, L. (1971) *Crop Loss Assessment Methods : FAO Manual on the Evaluation and Prevention of Losses by Pests, Diseases and Weeds,* FAO and Commonwealth Agricultural Bureaux 276 pp. and supplement **I** 1973; supplement **II** 1977.

Cramer, H.H. (1967) *Plant Protection and World Crop Production.* Pflanzenschutz-nachrichten, Bayer **20**, 524 pp.

Farid, M.A. (1980) 'The malaria programme - from euphoria to anarchy' *World Health Forum*, **1**, 8–33.

Gillett, J.D. (1971) *Mosquitos.* Weidenfeld and Nicolson, London, 274 pp.

Harrison, G. (1978) *Mosquitoes, Malaria & Man*, John Murray, London, 314 pp.

Kranz, J., Schmutterer H., and Koch W. (1977) *Diseases, Pests and Weeds of Tropical Crops*, Parey, Berlin, 666 pp.

Muller, R. (1975) *Worms and Disease. A Manual of Medical Helminthology,* William Heineman, London, 161 pp.

Nash T.A.M. (1969) *Africa's Bane. The Tsetse Fly*, Collins, London, 224 pp.

Ordish G. (1952) *Untaken Harvest : Man's Loss of Crops from Pests, Weeds and Disease*, Constable, London, 172 pp.

Wood, C. (ed.) (1978) *Tropical Medicine from Romance to Reality,* Academic Press, London, 286 pp.

World Health Organization (1979) 'WHO Expert Committee on Malaria'. 17th Report. *Technical Report Series* No. **640**, 71 pp.

World Health Organization (1979) 'Parasitic Zoonoses'. Report of a WHO Expert Committee with the Participation of FAO. *Technical Report Series* No. **637**, 107 pp.

World Health Organization (1979) 'The African Trypanosomiases'. Report of a Joint WHO Expert Committee and FAO Expert Committee. *Technical Report Series*. No. **635**, 96 pp.

World Health Organization (1981) *Sixth Report on the World Health Situation 1973–1977. Part I Global Analysis,* 290 pp., *Part II. Review by Country and Area.* 412 pp., Geneva, Switzerland.

2 Concepts of pests and vectors and their management

Before we consider the ways of reducing damage to crops growing in the fields and food stored in granaries by agricultural arthropod pests, and the nuisance caused to man by obnoxious insects, ticks and mites, as well as the role of arthropods as vectors of diseases to both plants and animals, we should try to formulate some ideas as to what constitutes a pest and a vector.

Types of pests

Definitions

Although most people understand what is generally implied when an organism is called a pest, a universally accepted definition of the term has been difficult to achieve over the years; therefore several definitions for this term exist in the literature and in the minds of people all over the world. Some examples of the definitions will illustrate the variety of the underlying concepts.

1. According to Robert van den Bosch a pest is a 'species that, because of its great numbers, behavior, or feeding habits, is able to inflict substantial harm on man or his valued resources'. This definition places emphasis on the abundance, behaviour and habits of organisms in relation to human interests, and is not restricted to either animals or plant alone.
2. A.H. Bunting's definition of a pest is 'any organism that competes with man at both the primary and secondary stages of production. They include insects, birds and mammals. From man's point of view, these and the vectors that transmit some of the harmful organisms are pests.' The central theme of this definition is the ability of an organism, whether plant or animal, to successfully compete with man or to transmit a disease pathogen causing ill-health in the host; there is no

indication that this definition takes account of the numbers of the offending organisms.

3. 'All types of biological factors that reduce crop income : insects, weeds, diseases, nematodes' are referred to as pests by E.A. Carlson. Although this definition refers to plant and animal life, it is clear that economic considerations are the main criteria for designating an organism as a pest.
4. In his definition L.R. Clark considers economical, social and nuisance factors as criteria for determining the pest status of an organism; he therefore defined pests as 'those injurious or nuisance species, the control of which is felt necessary either for economic or social reasons'.
5. Paul DeBach regards insects as pests 'when they are sufficiently numerous to cause economic losses'; the two keys concepts here are number of individuals and the levels of economic losses which they cause.
6. According to E.H. Glass, a pest is 'any organism harmful to man, these include insects, mites, nematodes, weeds, and pathogenic organisms i.e. viruses, mycoplasma, bacteria, rickettsiae, fungi and some vertebrates'. This definition is all embracing, drawing attention in particular to the harmful effects of organisms on man as the central theme.
7. The planning group for Science and Technology of the Organization for Economic Co-operation and Development (OECD), in 1977 provided the following definition of a pest which now seems to be widely accepted by scientists. They defined a pest as 'any form of plant or animal life or any pathogenic agent injurious or potentially injurious to plants, plant products, livestock or man.' According to this definition, pests include insects and other arthropods, vertebrates, nematodes, weeds, and mirco-organisms e.g. fungi, bacteria and viruses. The definition also focuses attention on those organisms which are *potentially* injurious.

Although these and many other definitions of pests seem to emphasize different aspects of the abundance, behaviour and effects of some organisms on others, two important aspects are common to all the definitions of pests. Firstly, all definitions recognize that an organism is a pest if it has an adverse effect on man and his interests. Secondly, most definitions consider a variety of both plant and animal life as pests; in other words the definitions of a pest do not relate only to animal life.

In this book, apart from the inclusion of snail hosts of schistosomiasis, we have restricted our discussions to insect, or more accurately arthropod, pests and in this connection we accept the OECD definition and define an insect (or arthropod) pest 'as any species which is injurious or potentially injurious to plants, plant and animal products, livestock or man'. Thus any insect (or arthropod) species which feeds on and damages cultivated plants, attacks plant and animal products in the field or in storage, causes a nuisance or transmits pathogenic organisms to plants, man or to domestic animals and livestock, is regarded as a pest.

The term vector can then be defined specifically as 'a pest species which transmit a pathogenic organism from one situation or living organism to another'. Thus both agricultural insects such as certain planthoppers, as well as medically important insects like mosquitoes, are vectors because they can transmit diseases to plants and man respectively. However, it is traditional practice to distinguish between crop *pests*, such as those that affect agricultural and forestry products, and those arthropods that are important in the field of medical and veterinary sciences, these being called *vectors*. So, although our definition of the term 'insect pest' based on the OECD definition, is broad, we prefer for practical purposes to maintain the distinction between arthropod pests which damage crop plants, and vectors which attack man and other animals.

Categories of agricultural pests

The degree of damage or harm caused by insect pests varies considerably with the particular species of insect concerned, the frequency of occurrence and density of the pest population, the prevailing circumstances and the nature of the host. An insect species may not be important at low population densities but when the numbers increase under certain conditions they can cause great harm and become pests. A typical example is grasshoppers. At low densities these insects cause very little inconvenience. When populations are high, however, they cause considerable crop destruction by the large amounts of vegetation they consume. Again an organism may be an important pest under certain circumstances while in others it is of no importance. Cocoa mirids are a good example. To the cereal grower these insects are of no consequence because they do not feed on cereals and are therefore not looked upon as pests, but to the cocoa farmer mirids are one of the more serious destructive pests because of the high level of damage they inflict on the cocoa crop. *Dysdercus* spp. are important pests of cotton in tropical Africa but they do not attack rice and so they are of no interest or concern to rice farmers. Depending on the frequency of occurrence, the behaviour and level of damage caused, insect pests can be classified into the following major categories.

Key or major pests

These are insect pests which occur perennially and cause serious and persistent economic damage in an ecosystem in the absence of effective control measures. The variegated grasshopper, *Zonocerus variegatus*, is a key pest of cassava, vegetables, citrus and a wide range of cultivated plants in West Africa. The brown planthopper, *Nilaparvata lugens,* is a key pest of rice in south-east Asia and the cocoa-pod borer, *Acrocercops cramerella*, is a key or major pest of cocoa in Indonesia, the Philippines and Malaysia. Whiteflies are major pests of cotton in the Sudan. In any given agro-ecosystem, one or two insects are usually recognized as key or major pests because their damage is high and leads to economic losses. Some major pests cause economic damage at low population densities; these are called 'low density pests', e.g. cocoa mirids in West Africa. Other pests like locusts and grasshoppers usually occur in very dense populations and are the main target of pest management operations.

Minor pests

Some insects cause economic damage only under certain circumstances in their local environment. Under normal conditions their populations are low and the damage which they cause is insignificant.

The cocoa-pod husk miner, *Marmara* sp., is a minor pest of cocoa in Nigeria and Ghana since the damage caused to cocoa pods is normally uneconomic. Minor pests are not usually the focus of pest control activities.

Occasional pests

In certain years or in some localities some insects build up their populations and achieve pest status. Many lepidopterous defoliators occur at irregular intervals and cause heavy economic damage to crops. These are described as occasional pests.

Potential pests

These are insect species which have the potential to develop into major pest status depending upon the circumstances. The giant looper caterpillar, *Ascotis selenaria reciprocaria*, developed into a major pest of coffee estates in Kenya following indiscriminate and uncontrolled use of pesticide in the agro-ecosystem (see p. 163).

Migrant pests

Locusts, aphids and armyworms (*Spodoptera* spp.) move from one zone to another zone where they cause damage to crops. These are a special group of key pests which are classified as migrant pests. Their control normally involves international co-operation between the member countries affected. For example, the African migratory locust, *Locusta migratoria migratorioides*, is jointly tackled by the West African countries which form the OICMA organization (Organisation Internationale contre le Criquet Migrateur Africain).

Plant vectors

A few agricultural pests are important because they transmit diseases to plants : tobacco whitefly (*Bemisia tabaci*) which is a vector of tobacco mosaic and cotton leaf curl viruses; aphids which transmit some 50 plant viruses; various planthoppers which can spread important viral diseases amongst rice plants; and mealybugs which can cause economic chaos to cocoa plantations by transmitting swollen shoot virus. An important consideration with plant pests which are vectors is that a very low population density can cause the spread of diseases, consequently more efficient control than is required for non-vector pests is often necessary.

Types of vectors

The word vector is defined as a carrier of disease or infection. *Anopheles* mosquitoes and fleas are vectors because they transmit malaria and plague respectively. The diseases they carry are known as 'vector-borne diseases' (Table 1.4). While it is agreed that mosquitoes, tsetse flies, fleas, bedbugs, houseflies, etc. can all be regarded as pests, the term vector is retained in this book (see p. 21) for those pests that are potential vectors of disease to man or other animals.

Types of disease transmission

Insects such as houseflies and cockroaches may ingest or pick up on their bodies pathogenic viruses, bacteria, protozoa and helminth eggs when they frequent human or animal excreta, decaying garbage and festering wounds. These pathogens can be transmitted to man when the insects defecate, vomit or crawl over man's food or frequent sores. This simple and direct transference of disease organisms involves neither multiplication nor any form of development of the pathogens in the vector, and is termed *mechanical transmission*. The pathogens are not dependent on them for transmission because they are commonly spread to man by more direct methods, such as people handling food with contaminated hands or by infected water. In contrast to this rather casual method of disease transmission more sophisticated relationships have evolved involving biological development and/or multiplication of the pathogens in the vectors; this is referred to as *cyclical transmission*.

For example, the filarial worms causing mosquito-borne filariasis (*Wuchereria bancrofti* and *Brugia malayi*) and river blindness or onchocerciasis (*Onchocerca volvulus*), undergo essential biological development, but not multiplication, in their mosquito and simuliid hosts. During blood-feeding on an infected person microfilariae are ingested with the blood-meal and although many are destroyed or excreted a few survive and rapidly burrow across the stomach wall into the insect's haemocoele. From here they travel to the thoracic flight muscles of the vector and during the next seven days or so undergo morphological changes, turning into infective (third-stage) larvae which migrate through the vector's head to its mouthparts. At this stage the vector becomes infective and during feeding on a person may deposit on the skin one or two larvae which frequently die, but occasionally manage to survive and penetrate the skin. This method is clearly more complicated than mechanical transmission, but is nevertheless not very efficient because a person usually has to be bitten by large numbers of vectors before any filarial larvae succeed in entering the skin and causing filariasis.

A more efficient method exists in the triatomine bugs of the tropical Americas which feed on people infected with *Trypanosoma cruzi*, the causative agent of Chagas' disease (South American trypanosomiasis). The ingested parasites undergo cyclical development and also multiply within the bug's stomach to produce infective forms which migrate to the hind-gut. Some weeks later these infective forms are excreted in liquid faeces while the vector feeds on a person and, if scratched into skin abrasions or contacting the eyes, may eventually cause Chagas' disease.

An even more effective transmission mechanism occurs with mosquitoes infected with yellow fever, dengue, and other arboviruses which not only multiply enormously in the vector but infect the salivary glands. Consequently, when the mosquito bites and saliva is pumped into the bite to prevent the blood clotting in the insect's mouthparts, viruses are injected directly into a new host. Malaria parasites have the most sophisticated method of vector transmission, because not only do they multiply within the *Anopheles* mosquito, they also have a sexual reproductive cycle. This results in the accumulation in the salivary glands of thousands of infective sporozoites which are inoculated into a new host when the mosquito feeds. A similar cycle without, as far as we know, a sexual stage, is shown by some forms of African trypanosomes in their tsetse fly vectors.

Whenever a vector–parasite relationship involves cyclical transmission there must clearly be an interval between the time a vector picks up pathogens during feeding and the time when the pathogens have completed multiplication and/or biological development in the vector and can be transmitted to a new host. This period is called the *extrinsic incubation period*, the duration of which depends on the pathogens and vectors involved and temperature; it is commonly about one to three weeks. In *Anopheles* mosquitoes the extrinsic incubation period of malaria parasites is about two weeks; as long as three to four weeks in tsetse flies for the trypanosomes responsible for sleeping sickness; and only about a week for relapsing fever rickettsiae in body lice. A long extrinsic incubation period is disadvantageous to the parasite because it reduces the likelihood of the vector being alive at the end of it to transmit the disease.

Theoretically, control sometimes need not eliminate vector populations to reduce disease transmission, but can be effective by decreasing the vectors' longevity. For example, if residual deposits of DDT sprayed onto the interior surfaces of houses reduces the average life of *Anopheles* vectors to less than about ten days, the extrinsic incubation time of malarial parasites cannot be completed and malaria transmission will be prevented.

Trans-stadial and transovarian transmission

In some diseases, mainly those involving viruses, spirochaetes (small mobile parasites akin to bacteria but with spirally twisted bodies) and rickettsiae (intracellular parasites rather like small bacteria), pathogens acquired by various stages in the life-cycle of a vector can be passed onto later developmental stages. For example, the immature stages of ticks (termed larvae and nymphs), as well as adults, take blood-meals, and if a larva or nymph picks up pathogens these can be passed to later stages in the tick's life-cycle. Thus the larval stages of a hard (ixodid) tick can become infected with pathogenic viruses by feeding on infected hosts and this infection can be passed to the nymphal stages and adults, both of which can transmit the disease to a new host when they take a blood-meal. This is termed trans-stadial transmission. Sometimes the process goes further and involves hereditary transmission. For instance, the larvae of certain leptotrombiculid mites in Asia become infected with scrub-typhus (*Rickettsia tsutsugamushi*) when they feed on infected hosts. The rickettsiae are trans-stadially passed onto the nymphs and adults, but as neither take blood meals they cannot infect a new host. However, the infected adult female mites lay rickettsiae-infected eggs. The next generation of larvae hatching from these eggs are consequently already infected and can transmit scrub-typhus to new hosts. This is referred to as transovarian transmission. Both trans-stadial and transovarian transmission are more commonly encountered in ticks and mites than in other arthropods.

Zoonosis

The term zoonosis (plural zoonoses) is usually applied to a disease that is naturally transmitted between vertebrate animals and man. Some zoonotic diseases are transmitted directly without implicating any vectors, such as rabies from dogs, brucellosis from cattle and pork and beef tapeworms from livestock. Many of the more familiar zoonotic diseases, however, are transmitted cyclically by arthropod vectors. Examples are yellow fever from monkeys, the Rhodesian form of sleeping sickness from game animals and cattle, scrub-typhus and leishmaniasis from rodents, and Chagas' disease from rodents and marsupials. Whenever there is a zoonotic relationship the epidemiology of disease transmission becomes more complex and this usually increases the difficulties of breaking the transmission cycle and establishing control.

Concepts of pest management in agriculture

Because of the harmful effects of insect pests on man and his interests, there is a strong desire to eliminate or drastically lessen such effects. In the past four decades, chemical pesticides have been mainly used against pest organisms and these chemicals have made significant contributions to increasing crop yields and reducing illness caused by vector-borne diseases. Unfortunately the short-term dramatic results obtained, the ease of use, the low cost and availability of many pesticides, led to over-use and abuse, especially of the broad-spectrum insecticides like DDT. Well known undesirable side effects resulting from over-reliance on chemical pesticides for pest control include: environmental pollution; creation of new pests; development of resistance to pesticides; the rapid resurgence of pests; and high pesticide residues in food and other products.

As chemical pesticide technology was further developed, the demand for higher quality products, such as fruits, without insect-caused blemishes, increased and this resulted in the vicious cycle of increased pesticide use to meet increased public demand for food and fibre. This problem is more acute in the developed countries, notably the USA, Britain and Japan where the bulk of pesticides are used, than in developing tropical countries, with their comparatively lower pesticide consumption. So instead of solving problems, the extensive widescale use of pesticides has created sometimes new and difficult problems. It soon became clear that chemical pesticides alone could not solve most pest problems. There was therefore an urgent need to re-examine traditional cultural methods for pest control, and to make greater use of naturally occurring mortality factors and to consciously utilize them against pests. In addition new methods which are ecologically sound, long lasting and cause very little disturbance to the ecosystem must be sought. This need deliberately to combine the use of chemical pesticides with other methods of control, together with public concern about the undesirable side effects of pesticide use, stimulated the development of the integrated pest control concept. 'Pest Management' and 'Integrated Control' are often used synonymously or interchangeably, although there are some differences between these terms. *Integrated control* is a pest control strategy which utilizes biological and ecological knowledge, pest monitoring, action criteria, various materials and techniques, together with natural pest population regulatory factors for the management of pests. It is the same as *integrated pest management*, which Robert van den Bosch describes as scientific control. The term *pest management* is broader than integrated pest control; it is defined as the manipulation of pest or potential pest populations by means of all pest control tactics in order to reduce their damage well below the economic threshold or render them harmless. It may involve planting schedules, crop husbandry techniques and cultural practices, use of pest-resistant crop varieties, use of natural enemies, as well as physical, mechanical and chemical control methods. In its simplest terms good pest management is 'sensibly-applied entomological science', together with agronomic and cultural practices. In spite of whatever differences there may be between these terms, there is little point in distinguishing between them. The farmer or pest control specialist (i.e. pest manager) should integrate all control tactics available to him and adopt pest management concepts which will enable him to achieve environmentally and ecologically sound pest control.

The Institue of Biological Control (IOBC) started sponsoring biological control activities in the tropics in 1958, while the Food and Agriculture Organization of the United Nations (FAO) with headquarters in Rome, started activities in Integrated Pest Control in 1963 and the Director-General established the FAO Panel of Experts on Integrated Pest Control in 1966. This expert panel defined integrated pest control as 'a pest management system that in the context of the associated environment and the population dynamics of the pest species, utilizes all suitable techniques and methods in as compatible a manner as possible and maintains the pest population at levels below those causing economic injury'.

The activities of FAO in integrated pest control are extensive. Membership of the Panel of Experts is comprised of about fifty-four specialists from all over the world, appointed in their personal capacity to serve for four years. The panel is now called the FAO/UNEP Panel of Experts on Integrated Pest Control in order to incorporate the input from UNEP (United Nations Environment Programme, with headquarters in Nairobi). The purpose of this expert panel is:

(a) To advise the Director-Generals of FAO and UNEP on policies and programmes in relation to the integrated approach in pest control;
(b) To review principles and standardize procedures and techniques;
(c) To promote joint research programmes for the integrated control of major pests which affect several countries;
(d) To collect and distribute data relevant to integrated control research and development.

The panel meets in sessions to review progress made and the status of knowledge of integrated pest control strategies. It has developed and published guidelines for the integrated control of pests of some crops such as rice and cotton within the context of an FAO/UNEP Co-operative Global Programme for the Development and Application of Integrated Pest Control in Agriculture.

Concepts of vector management

In the euphoria of the early days of the DDT era it was believed that this new man-made 'wonder' insecticide would bring under control many of man's most important vector-borne diseases. It seemed, for example, that global malaria eradication (except for tropical Africa) was technically feasible (see p. 8) and there was also an expectation for a great reduction in the transmission of other scourges of mankind such as sleeping sickness, typhus and river blindness. This optimism was followed by disillusionment when the immense difficulties posed by disease eradication were eventually recognized. The epidemiology of certain diseases, such as malaria, had been over-simplified and control measures that had worked well in one area were extrapolated to other areas and other situations without any realization that the complex ecology of disease transmission would almost certainly vary from place to place and time to time. Another practice was the policy of regarding each disease more or less in isolation from other diseases. For example, in malaria control campaigns there was often little or no attention devoted to reducing the spread of other diseases, even mosquito-borne ones like filariasis, that might be afflicting the people. The insular approach was sometimes unavoidable, but not always so, due to lack of resources to devote to the other diseases.

It has long been appreciated that a low infestation rate of agricultural pests may be economically acceptable, and control of such pests has been designed to keep them at or below a predetermined level – the economic threshold. In contrast, however, it has sometimes been assumed that the economically acceptable level for insect vectors of human diseases is virtually zero. This effectively puts an infinite value on the price of human life. Humane though such an approach undoubtedly is, it is economically unfeasible. There has emerged a change of philosophy and an admission that, at least for the foreseeable future, it will be practically impossible to eradicate certain diseases, through sheer lack of funds if nothing else. Accordingly the emphasis is no longer on eradication of malaria, river blindness, or sleeping sickness, but on reducing transmission to an acceptable level. Thus, for example, the massive control programme in the Volta River basin of West Africa (see p. 284) is not expected to eradicate river blindness (onchocerciasis), but to reduce the populations of vectors (simuliid blackflies) to such a low level that although some transmission will persist, the threat of blindness will be so reduced as to allow communities to return to the fertile valleys. Similarly, an efficient malaria control programme should aim at reducing transmission to a point where malaria is no longer a major health problem, and the risks of contracting the disease are acceptable. The word 'acceptable' is two-edged, but the dilemma must be faced. Resources are finite while the needs of expanding populations are unlimited.

Although environmental management has often proved valuable in reducing vector populations, biological control measures have unfortunately rarely, if ever, given worthwhile reductions of important vectors, except on a very localized scale. It is doubtful whether such an approach has ever significantly decreased the transmission of a major disease. This contrasts with the situation with agricultural pests, where biological control measures have sometimes unequivocably substantially reduced or even eradicated important pests. Past difficulties and failures should not prevent vigorous attempts to develop and introduce biological methods into an integrated vector management programme, but they nevertheless serve as a warning that less empirical procedures are needed if such methods are to have a chance of success.

Conclusions

Pest and vector management should be conceived as a component of a broader approach to increase crop production and improve the health of the community. It should be a flexible process that allows modification to be built into the system to take into account increased undertstanding of the population dynamics of the pests and vectors as well as environmental and ecological changes. In pest and vector management programmes it is desirable to minimize (= optimize) the use of insecticides, avoid their misuse and over-use, and not to ban them altogether. Hopefully this approach will avoid the so-called 'pesticide treadmill' whereby the development of resistant insects outpaces the introduction of new insecticides and the impending crisis of all pests and vectors becoming resistant to all useful insecticides.

Recommended reading

Glass, E.H. (1975) 'Integrated pest management : rationale, potential needs and implementation'. *Entomological Society of America*, Publication **75**, 141 pp.

Jones, D. Price and Solomon, M.E. (eds) (1974) *Biology in Pest and Disease Control*, Blackwell Scientific Publications, Oxford, 398 pp.

Kilgore, W.W. and Doutt, R.L. (eds) (1969) *Pest Control, Biological, Physical, and Selected Chemical Methods*, Academic Press, New York, London, 477 pp.

Knipling E.F. (ed) (1979) *The Basic Principles of Insect Population Suppression and Management*, Agricultural Handbook No. **512**, United States Department of Agriculture, Washington DC, 659 pp.

Norton, G.A. and Holling C.S. (eds) (1979) *Pest Management* IIASA Proceedings Series, Pergamon Press, Oxford, 349 pp.

Ridgway R.L. and Vinson S.B. (eds) (1976) *Biological Control by Augmentation of Natural Enemies. (Insect and Mite Control with Parasites and Predators)*, Plenum Press, New York, 480 pp.

Smith, R.F. and Reynolds, H.T. (1966) 'Principles, Definitions and Scope of Integrated Pest Control'. *Proceedings of the FAO Symposium on Integrated Pest Control,* 1, 11–17.

3 Role of insecticides in integrated pest and vector management

Successful pest and vector control depends on the application of appropriate strategies and tactics. In pest and vector control three main strategies may be adopted:

(a) The preventive or eradication approach.Here measures are taken to prevent the pests or vectors from becoming important, for example when insecticides are applied routinely as a prophylactic.

(b) A containment or corrective approach. In this strategy, measures are applied to reduce pest and vector populations to levels where their damage is well below economic proportions.

(c) Complete reliance on natural forces. Here there is no human intervention to control pest organisms; natural forces of weather and harmful, naturally occurring organisms are allowed to effect control. This strategy is, however, of very limited practical value on its own.

According to Ray Smith 'integrated pest control philosophy depends upon the containment strategy. It relies upon an analysis of the biological and economic systems of both pest control and crop production.' This analysis embraces all significant factors acting on the complex of real and potential pests, their natural enemies and the interactions of such factors among themselves and with other processes in crop production. Basically this means that a clear understanding of the interactions of the complex of factors operating in the ecosystem is fundamental to successful integrated pest and vector management.

Pesticides are toxic chemicals. They offer a quick and immediate solution to pest and vector problems. They are highly effective and economical to use and can quickly be marshalled, especially when pests become important in large areas, to arrest damage by pests and vectors immediately. Pesticides are therefore a powerful tool in pest and vector management and will continue to play a major role in integrated pest and vector management. The critical question is how best to use pesticides so that their maximum benefits are obtained, while at the same time their adverse effects are minimized. This issue is examined in this chapter.

Pesticide use in the tropics

Following the development of synthetic insecticides during the Second World War, these chemicals became increasingly used for insect pest and vector control in developing tropical countries. The amount used, however, is still far below the quantities used in the developed countries such as the USA, Japan and Britain, where most of the pesticides are manufactured. For example, the USA is the world's largest pesticide consumer, utilizing 35–45 per cent of total world production, western Europe consumes about 23 per cent, eastern Europe 15 per cent and Japan about 8 per cent. The developing countries altogether consume only about 7–8 per cent of total world production.

Many synthetic insecticides such as malathion, diazinon and the pyrethroids (natural and synthetic), fungicides such as Dithane Perenox and Zineb, and herbicides like Gramoxone (paraquat) are readily available for purchase and use in tropical developing countries, and their use has continued to increase gradually. Analysis of data from 38 developing countries shows that of all pesticides, insecticides are the most used pest and vector control chemicals; in 1973 the proportion of pesticides used in these 38 developing countries were 66.2 per cent insecticides, 30.4 per cent fungicides and 3.4 per cent herbicides. The actual 1979 data for Liberia are as follows: insecticides 78.2 per cent, fungicides 15.9 per cent and herbicides 5.9 per cent. The main uses of insecticides in the tropics are for the control of agricultural pests such as locusts (desert locust, migratory locust, red locust) armyworm caterpillars, cocoa, coffee and cotton pests, and to a much lesser extent for the control of insect vectors (mosquitoes, tsetse flies, blackflies, triatomine bugs, ticks etc.) of human and animal diseases.

In agriculture, insecticides are used primarily in the control of pests of export crops such as cotton, cocoa, coffee, tea and sisal. Apart from cowpeas and rice which receive a comparatively high insecticide treatment, the other food crops such as maize, yams, cassava and vegetables, receive very limited insecticide treatment. Relatively small quantities of insecticides are used by private individuals to kill vectors. In many tropical countries the greatest sale of pesticides in the vector market is to the farmers for the control of cattle ticks and other veterinary pests. Although there are often instructions accompanying the sale of chemicals to kill ectoparasites of livestock clearly telling the farmer that cattle or sheep dipped or sprayed should not be slaughtered and sold for consumption until sufficient time has elapsed for the animals to excrete most of the absorbed chemicals, such advice is frequently ignored.

Actual on-farm consumption of pesticides in tropical developing countries is limited by the high costs of approved chemicals, application equipment, the lack of stocks of spare parts and satisfactory servicing facilities for spraying machinery, and, in some countries, a poor and inefficient extension service scheme for peasant farmers. Therefore levels of crop production are still well below potential yields since insect pests and diseases are poorly controlled and crop losses are therefore still high.

For important export crops some governments have encouraged farmers to use pesticides to increase crop production by the establishment of subsidy schemes. In Ghana and Nigeria, cocoa pesticides (insecticides and fungicides) are highly subsidized. For example, during the 1977/78 crop year, cocoa pesticide subsidies in Nigeria cost the government 2.6 million Naira ($US4.7 million). In Uganda an 83–100 per cent subsidy for insecticides used on cotton in 1960–63, caused a tremendous increase in the use of insecticides for controlling cotton pests and consequently in cotton production. The subsidy scheme for cotton insecticides was so successful that the government of Uganda contemplated a similar subsidy scheme for cotton fertilizers.

In addition to providing subsidies for pesticides, the governments of some countries establish a subsidy scheme for spraying machinery. In Ghana, mistblowers for spraying HCH (= BHC) to control cocoa pests were highly subsidized and this encouraged cocoa farmers to use them.

It must be pointed out that pesticides do not produce increases in crop yields, rather they prevent losses. For example, it has been calculated on the basis of projected grain supplies in Nigeria that total losses of all grain due to pests and disease organisms were of the order of 4.5 million tonnes in 1976/77 and will be some 8.3 million tonnes by 1984/85, representing an average annual loss of 6.2 million tonnes. The use of pesticides for pest and disease control on grain will prevent these losses. Some tropical crops like cowpeas cannot be grown successfully without pesticide application, and the cost of pesticide use must be such that the entire exercise leads to yields which will give a profitable margin to the farmers.

Pesticide safety in the tropics

According to the World Health Organization about 500 000 people in the world are accidentally poisoned annually by insecticides and the fatality rate is about 1 per cent; but this figure has been disputed, some consider it grossly exaggerated while others estimate that over 20 000 deaths a year are caused by insecticides! Whatever the true figure, there is little doubt that most cases of poisoning occur in the developing countries and involve agricultural insecticides used by farmers having just a few hectares. For example, about 3000 poisonings with about 10 per cent fatality rate, mainly due to parathion, are reported each year from Central America. In 1972, there were 2861 cases of malathion poisoning reported in El Salvador, 30 people died. In the Philippines, in the province of Nueva Ecija alone, 33–44 deaths a year were reported from 1973–76, of agricultural workers from accidental insecticidal poisoning. Surveys showed that not a single farmer was aware that insecticides could enter the body through the skin. There is clearly an urgent need for the registration and restriction of the sale of pesticides in developing countries. All too often insecticides are bought by people who, through lack of education, fail to understand their toxicity. It is difficult for people to appreciate that insecticides which on initial contact may produce no immediate adverse effects, can be hazardous. It is not uncommon to see spraymen mixing insecticide solutions with their hands and transporting the liquid in pails on their heads. Agricultural preparations of malathion have been applied to the body to control lice, with drastic results, and more rarely insecticides have been drunk to get rid of intestinal worms. Insecticides are commonly repacked and sold in medicine and beer bottles without any indication as to what they are or how they should be used; moreover they may have been diluted by unscrupulous traders.

DDT has a truly remarkable safety record. When, for environmental reasons, it has been banned or its usefulness is restricted due to resistance problems, alternative insecticides which are usually more toxic, are used with a carelessness that did not matter so much with DDT, and this has often resulted in poisoning. In 1976, some 2000 malaria workers in Pakistan spraying malathion developed poisoning symptoms, and 5 died, largely due to their failure to observe correct safety procedures. It was subsequently discovered that the malathion contained unexpectedly high levels of isomalathion, a poisonous impurity produced during manufacturing which can increase during storage.

Pollution

The publication of Rachel Carson's book *Silent Spring* in 1962 generated a surfeit of publications and propaganda either condemning or exonerating the use of organic pesticides. The real issues of the pesticide dilemma became obscured in exaggerated claims and counterclaims, and emotive and irrational statements. A more rational and balanced approach has now emerged from this controversial subject.

There is no doubt that widespread and indiscriminate pesticide applications in both developing and developed countries have sometimes caused contamination of the environment and some disastrous ecological damage. This is especially true with agricultural pesticides, which account for 97 per cent of all pesticide residues.

Aerial spraying of African riverine and lake-shore vegetation to kill tsetse flies has undoubtedly killed non-target invertebrates, birds, reptiles, amphibia, fish, small – and sometimes large – mammals. Tsetse flies can often be drastically reduced in numbers by 4–6 aerial sprayings with insecticides such as endosulfan and deltamethrin (=decamethrin), after which there is often no need for further spraying. Applied in this way these insecticides are not very persistent in the environment and may be degraded in about 4–8 weeks. This allows sprayed areas to be gradually reinvaded and recolonized by much of the original fauna – but unfortunately also by tsetse flies. After an area has been freed of tsetse flies it may, however, be occupied by people, the trees cut down for firewood and vegetation removed to facilitate farming. This causes more irreversible damage to the habitat than insecticides.

In some parts of Africa, DDT emulsions have been repeatedly sprayed onto rivers to kill simuliid vectors of onchocerciasis, but this has simultaneously caused the deaths of numerous invertebrates including predators. In Uganda this resulted in an increase not a decrease in *Simulium neavei*. Temephos (Abate) is a non-persistent organophosphate insecticide with very low toxicity to fish and other aquatic fauna, and has replaced DDT as the most important chemical for control of simuliid blackflies. It is sprayed weekly on streams and rivers over an area of 764 000 km^2 in the Volta River basin of West Africa to kill larvae of the vectors of river blindness (onchocerciasis). Spraying must continue longer than the life-span of the parasite in man, (about 15–17 years) before reduced transmission can be maintained. Unfortunately this has generated considerable controversy among environmentalists who fear this will destroy the aquatic ecosystem. Although, according to hydrobiologists, there has been a reduction in river invertebrates there has been no noticeable reduction in fish populations, and there is so far no evidence of gross ecological damage. Nevertheless repetitive spraying must be producing some ecological changes, but onchocerciasis transmission can at present be reduced only by insecticides. Therefore the benefits gained in freeing people from the threat of blindness, thus allowing improvements in their living conditions and health, must be carefully balanced against the side effects of environmental contamination.

The application of 2 g of DDT/m^2 twice a year to the interior surfaces of houses during malaria-control campaigns to kill indoor-resting anopheline vectors, results in depositing only about 10 g/ha of insecticide on the soil twice a year. This is negligible contamination compared to the 30–50 sprayings on cotton in Central America over a 90 day season, which may result in applications of 80 kg/ha of insecticide. Sometimes as much as 50–60 per cent of the insecticide sprayed against cotton pests falls outside the target fields. In El Salvador repeated sprayings of cotton with organochlorine, organophosphate and carbamate insecticides has resulted in insecticidal contamination of the larval habitats of *Anopheles albimanus*. As a consequence this important malaria vector has developed resistance to DDT, HCH (cross-resistance to dieldrin), malathion and propoxur. This serves as a dire warning of how control of agricultural pests with total disregard for other problems can create serious repercussions, and emphasizes the urgent need for greater liaison between agriculturalists and medical entomologists. Similarly, insecticide resistance arising from spraying crop pests has been promoted in other important malaria vectors such as *An. sundaicus* in Java, *An. sacharovi* in Greece and possibly in the *An. gambiae* complex in the Sudan and Upper Volta.

Whereas insecticidal house-spraying against adult mosquitoes poses few environmental hazards, repetitive spraying of mosquito larval habitats can result in indiscriminate mortality of large numbers and species of invertebrates, and sometimes fish and amphibia. Moreover, if organochlorine insecticides are used they will persist in the environment for many years and there is often ecological concentration of organochlorine residues in food chains. For example, the amount of DDT in the predatory fish *Gambusia affinis* may be 80 000 times greater than that necessary in the water to kill mosquito larvae. It is incorrect, however, to assume that increasing concentrations are inevitably found in successive animals in the food chain, because each animal will accumulate insecticide until a plateau of concentration is reached, after which further intake is balanced by excretion. Certain birds may in fact have lower concentrations than the organisms upon which they prey.

Sublethal doses of some insecticides may cause *Gambusia* to abort their young or produce atypical behaviour. Guppies (*Poecilia*) which are also used to control mosquito larvae, can become hypersensitive to insecticides leading to increased activity, abnormal behaviour and repression of reproduction. Because of the extreme persistence of organochlorine insecticides in the environment and their much greater toxicity to fish and other aquatic fauna, the World Health Organization recommends that organophosphates should be substituted for the control of mosquito larvae. *Gambusia* and *Poecilia* can tolerate exposure to doses of temephos 1000 times greater than those required to kill mosquito larvae; it is therefore one of the better insecticides to use in integrated vector management of mosquitoes.

Cost of pesticides and drugs

The control of vector-borne diseases is restricted by serious economic difficulties. Manufacturers need to produce pesticides that have either a very large market, as exemplified by DDT, or exceptional biological activity, in order to recover the immense costs incurred in research and development during the production of a new pesticide. There have been various estimates of the costs involved in producing a new pesticide, from about $US15 million in 1975 (spread over about 10 years) to $US15–20 million in 1978. Based on the latter figures there must be a market of about $US20 million a year for a profit, but only about one quarter of all pesticides have sales in excess of $US10 million. The chance of discovering a successful new pesticide was calculated as 1 in about 1800 in 1952, but as low as 1 in 10 000 in 1972. The patent life of a new product is only 15 years, after which companies that have not invested in its expensive development can manufacture it. It is not surprising therefore that the production of new pesticides has rapidly declined from a peak of 28 in 1966, to only 6 in 1974 (Fig. 3.1).

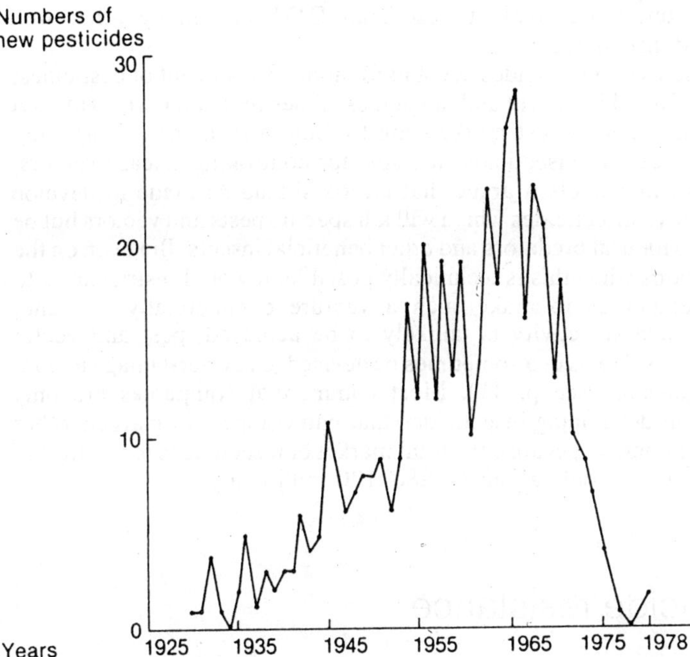

Fig. 3.1 Numbers of pesticides introduced each year from 1930–78. (Based on Watson and Brown, 1977; and Worthing, 1979, *The Pesticide Manual*, British Crop Protection Council, Croydon)

Similarly the development of new drugs for tropical diseases is making little progress. The increasing costs of research and development, the imposition of increased safety levels and the poor financial return of selling drugs to poorer parts of the world (where of course they are most urgently needed) means that few pharmaceutical companies are prepared to invest in such an expensive and hazardous gamble. They can get greater profits with less risk by marketing proprietary panaceas for coughs, headaches and other minor ailments, which inflict people worldwide, not just in the developing world.

Costs of buying insecticides are also increasing exponentially. The average price of a common insecticide has increased by about 20 per cent a year since 1970. DDT still remains the cheapest, and one of the more safe and effective insecticides for use against many vectors and pests, but developing countries are finding it increasingly difficult to afford to buy even this insecticide due to price increases. The cost of DDT for malaria control campaigns doubled from 1972–77, and if vectors became resistant and alternative insecticides such fenitrothion or propoxur had to be used, the costs were 23 and 39 times higher than the cost of DDT in 1972. The very effective synthetic pyrethroids (e.g. bioresmethrin, permethrin, deltamethrin) are about 100 times more expensive than DDT, and although dosage rates may be only a tenth of that required with DDT, they still work out more costly to use than DDT and many other well established insecticides.

World sales of insecticides account for about 36 per cent of pesticides; herbicides form 43 per cent and fungicides 19 per cent of the world market of pesticides. The largest markets are for cotton spraying (37%); only about 3 per cent of insecticides are sold for controlling disease vectors.

Conservationists often argue that efforts should be made to develop more selective insecticides which will kill specific pests and vectors but be harmless to natural predators and other beneficial insects. But even on the rare occasions when this is technically possible, it would restrict sales to such an extent as to make such a venture commercially unsound. Although such selectivity is unlikely to be achieved, pest and vector management systems can sometimes be devised to restrict damage to non-target organisms (see p. 41). Most commercial companies are only interested in developing insecticides that can compete in markets other than vector control, because the total market of insecticides for control of disease vectors is only about $US80–120 million a year.

Insecticide resistance

Much has been written about the problems created by insecticide resistance. It is important to remember that resistant individuals do not arise by subjecting insects to sublethal doses of insecticides, nor by mutations, but that a very few resistant individuals exist in populations *before* any insecticides are used. For some reason these insects already possess inherited mechanisms which will counteract insecticides, but they are usually at a disadvantage to non-resistant (susceptible) individuals and therefore form only a very small part of the population. When insecticides are applied their inherent ability to withstand them confers an advantage, so while their susceptible neighbours are killed, resistant individuals increase in numbers. There is therefore a gradual genetic selection favouring the increase of resistant individuals, and the population is finally said to have developed resistance.

The speed of the development of resistance depends on several factors including: (i) the proportion of resistant individuals already present in the population before spraying; (ii) the type of genetic system conferring resistance; (iii) the intensity of insecticidal applications; (iv) the degree of isolation of the population from neighbouring unsprayed populations; and (v) most importantly, the insect's reproductive rate and generation time. Houseflies and mosquitoes produce large numbers of eggs, and houseflies, and in the tropics mosquitoes, generally have a life-cycle of 2 weeks or less. In marked contrast, tsetse flies produce a single offspring at a time and the generation time is 5–7 weeks. It is not surprising therefore that DDT resistance appeared as early as 1946 in houseflies and about 4 years later in mosquitoes, whereas in tsetse it is as yet unknown although, we predict, equally inevitable. Some insects and mites such as certain populations of houseflies have developed resistance to DDT, HCH, dieldrin, other organochlorines, the organophosphates, carbamates, pyrethroids and even to the juvenile hormone, Altosid.

Using the same type of insecticide against larvae and adults will tend to speed up the selection of resistance. Similarly, if insecticides are applied to the entire population this prevents the dilution of resistance from nearby unsprayed areas by the immigration of susceptible individuals. Although cessation of insecticidal spraying may allow the population to revert

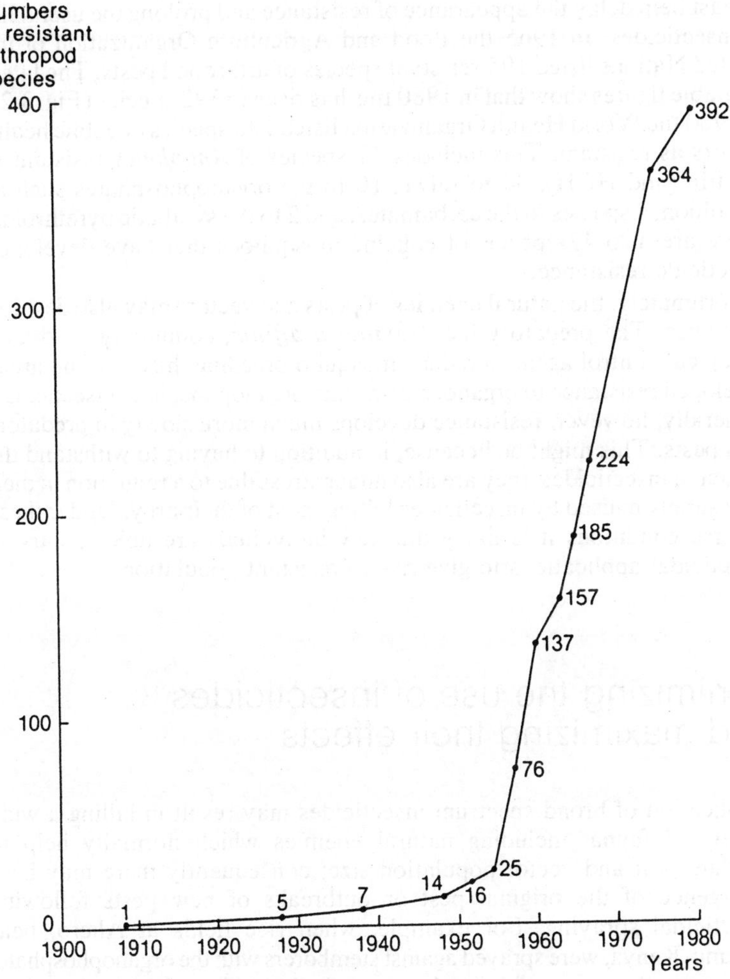

Fig. 3.2 Numbers of species of insects, mites and ticks resistant to one or more chemicals from 1908–80.

gradually to a predominantly susceptible one, the *genetic composition* remains altered, so that if the same insecticide is reused resistance reappears quickly. For example in Denmark, after more than 20 year's absence of DDT insecticidal spraying, houseflies developed resistance again within two months of the resumption of spraying. This was due to the persistence in the population of a changed genetic background that allowed resistance to express itself rapidly.

It seems likely that most major pests and vectors will eventually develop resistance to most insecticides if they continue to be used against them. This promotes the formation of a 'resistance race' which can only be slowed down, not stopped, by curtailing insecticidal usage, but this should at least help delay the appearance of resistance and prolong the useful life of insecticides. In 1965 the Food and Agriculture Organization of the United Nations listed 195 resistant species of arthropod pests. The latest available figures show that in 1980 this has risen to 392 species (Fig. 3.2). In 1980 the World Health Organization listed 136 species of public health vectors as resistant. This includes 47 species of *Anopheles* resistant to dieldrin (and HCH), 34 to DDT, 10 to the organophosphates such as malathion, 4 species to the carbamates, and 2 to the synthetic pyrethroids. There are also 42 species of culicine mosquitoes that have developed insecticide resistance.

Fortunately, the natural enemies of pests and vectors may also develop resistance. The predatory fish *Gambusia affinis*, commonly used as a biological control agent to reduce mosquito breeding, has in some areas developed resistance to organochlorine and organophosphate insecticides. Generally, however, resistance develops much more slowly in predators than pests. This might be because, in addition to having to withstand the impact of insecticides, they are also under stress due to a reduction in their food supply caused by insecticides killing most of their prey. Under these adverse conditions it is likely that few individuals are able to survive insecticidal applications to give rise to resistant populations.

Minimizing the use of insecticides and maximizing their effects

Application of broad-spectrum insecticides may result in killing a wide variety of fauna, including natural enemies which normally help to regulate pest and vector population size; consequently there may be a resurgence of the original pest or outbreaks of new pests following insecticidal spraying. For example, when rice fields at Ahero, near Kisumu, Kenya, were sprayed against stemborers with the organophosphate, Dimecron, there was initially almost total elimination of the larvae of the *Anopheles gambiae* complex from the sprayed fields. At the same time spraying killed numerous aquatic mosquito predators, and in their absence when mosquitoes once again oviposited in the rice fields the resulting larval population greatly exceeded the numbers prior to spraying.

Another example of insecticides causing unwanted ecological changes has taken place in West Africa. Here, the pest status of some insect species in the cocoa ecosystem changed following large-scale applications of organochlorine compounds to control cocoa mirids (*Sahlbergella singularis* and *Distantiella theobroma*). The most serious cases in Nigeria, Ghana and Cameroun involved the shield bug, *Bathycoelia thalassina*, which was originally a minor pest of cocoa. By 1965, however, populations of this insect had become high and they severely damaged the cocoa crop, because its natural enemies had been eliminated by insecticides. There are abundant field data to show that *B. thalassina* is an important cocoa pest only in areas where cocoa has been sprayed regularly with organochlorine insecticides. In Ghana and Nigeria populations of *Marmara* spp., *Characoma stictigrapta*, *Eulophonotus myrmeleon* and *Tragocephala castina nobilis* changed from being minor to major cocoa pests following regular 'blanket'-type treatment of cocoa with organochlorine insecticides.

Very few insecticides are selective, but mortality of non-target organisms may sometimes be minimized by using appropriate formulations and application techniques. Generally only 10–20 per cent of insecticidal dusts and 25–50 per cent of liquid formulations are deposited on the plants when crops are sprayed, and less than 1 per cent of the quantity applied actually reaches the pest. Improvements can be made: for instance granular and micro-encapsulated insecticides are more selective than sprays; and ultra-low-volume spraying is usually less damaging than high volume spraying. Restricting insecticidal treatment to particular parts of the habitat, i.e. 'spot treatment' or discriminate spraying, will be less disturbing to the ecosystem than indiscriminate applications. If seeds can be treated with chemicals, especially the systemic insecticides which are taken up by plants as they germinate and grow, this greatly reduces dosage rates and environmental contaminations. A typical useful systemic insecticide is schradan. This is almost non-toxic on contact with most insects but when taken up by plants it is highly toxic to aphids, spider mites and other crop pests which feed by sucking up the plant juices.

Pesticide dosages are frequently too high because they aim at killing all crop pests, which is usually an impossible goal. The famous insect physiologist Sir Vincent Wigglesworth suggested as long ago as 1945 that an insecticide that killed 50 per cent of a pest population might well be better than one that killed 95 per cent of the pests but at the same time most of their natural enemies. There is little doubt that more long-lasting control would in many instances result from reducing application rates, but the desired application rates need to be based on the local conditions and interactions between pests and their natural enemies.

Insecticides can be applied in the evenings to minimize damage to bees and other beneficial insects. Similarly, seasonal timing of spraying may be important in minimizing killing of non-target organisms. Whenever feasible, the frequency of applications should be reduced. Various strains of the spore-producing bacterium *Bacillus thuringiensis* have for many years been formulated as a biological (microbial) insecticide and sprayed on agricultural crops to kill certain pests, mainly lepidopterous larvae. More recently certain strains have been isolated that kill medically important insects such as larvae of mosquitoes and simuliid blackflies. An advantage of this microbial insecticide is that it is highly specific and does not kill harmless insects or natural enemies, and consequently is potentially suited for integrating into pest management programmes.

Pesticides in integrated pest and vector management

The account of pesticide usage for insect pest and vector control, together with the problems of pesticide use highlighted in this chapter, provide a sound basis for bringing into focus the role of pesticides in the integrated management of insect pests and vectors in the tropics.

Integrated pest and vector management does not advocate a total ban on the use of pesticides. Neither does it place pesticide use as the central feature of pest management, but rather permits the intelligent use of pesticides in a way which is compatible with other methods such as cultural, physical and biological control. Pesticides must be integrated with other pest and vector control methods in such a way that there is minimal disturbance of the ecological balance in the ecosystem. They must be used only when pest and vector populations cannot be reduced effectively by a combination of all other methods. Experience in several tropical developing countries shows a variety of major difficulties in developing sound, well co-ordinated, integrated management schemes for most insect pests and vectors.

Firstly, there is considerable ignorance of the details of the biology, ecology and behaviour of most tropical pests and vectors. Secondly, there do not yet exist efficient methods for monitoring and forecasting populations, nor for accurate assessment of the damage or losses attributable to various important pests and vectors. Thirdly, the relative importance of the variety of natural enemies regulating pest and vector populations in the field is poorly understood. Fourthly, there is an acute shortage of adequately trained local personnel to provide the research and extension services needed for the effective management of insect pests and vectors in developing countries. Fifthly, most tropical developing countries lack well co-ordinated institutional and organizational arrangements for dealing with specific major pests and vectors.

Because of these and other inadequacies, there tends to be a high dependence on pesticide use in the management of tropical insect pests and vectors. Nevertheless, with the development of relevant research and the rapid accumulation of tropical ecological information on insect pests and vectors, there will be less reliance on pesticides alone and more utilization of integrated control. Robert Metcalf has suggested the following ways in which pesticides can be fitted into modern integrated pest and vector control:

(a) Carefully timed and gauged suppressive applications aimed at a weak point in the insect's life-cycle;
(b) Emergency application to be used when economic threshold levels are exceeded because other measures are inadequate;
(c) Highly selective preventive treatments made with a minimal quantity of insecticide.

In effect, pesticides must never be used as routine prophylactic treatment in an agro-ecosystem or for regular application to the environment in which insect vectors live. Pesticides must be used only after careful consideration of the prevailing pest and vector circumstances together with an assessment of the relative merits of the variety of pest management methods available for use locally. The challenge therefore involves the intelligent adoption of selective pesticides in integrated pest or vector management strategies.

Examples of the integration of pesticide use together with other control methods for the management of selected tropical crop pests and disease vectors are outlined in Chapters 6 and 7 of this book.

Recommended reading

Brown, A.W.A. (1978) *Ecology of Pesticides*, John Wiley, New York, 525 pp.

Carson, R. (1962) *Silent Spring*, Houghton-Mifflin, Boston, 360 pp.

Davis, D.W., Hoyt, S.C., McMurtry, J.A. and **AliNiazee, M.T.** (1979) *Biological Control and Insect Pest Management*, University of California, 102 pp.

FAO (1974) Proceedings of the FAO Conference on Ecology in Relation to Plant Pest Control, FAO, Rome, 1974, 326 pp.

FAO (1975) *The Development and Application of Integrated Pest Control in Agriculture*. Report of an *ad hoc* session of the FAO Panel of Experts on Integrated Pest Control, Rome 1974, 15 pp.

Flint, M.L. and **van den Bosch, R.** (1981) *Introduction to Integrated Pest Management*, Plenum, New York, 240 pp.

Green, M.B., Hartley, G.S. and **West T.F.** (1977) *Chemicals for Crop Protection and Pest Control*, Pergamon Press, Oxford, 291 pp.

Gunn, D.L. and Stevens, J.G.R. (eds) (1976) *Pesticides and Human Welfare*, Oxford University Press, 278 pp.

Holdgate, M.W. (1980) *A Perspective of Environmental Pollution*, 2nd ed., Cambridge University Press, 278 pp.

McEwen, F.L. and Stephenson, G.R. (1979) *The Use and Significance of Pesticides in the Environment*, John Wiley, New York, 538 pp.

Mellanby, K. (1967) *Pesticides and Pollution*, New Naturalist Series, No. **50**, Collins, London, 221 pp.

Mellanby, K. (1980) *The Biology of Pollution*, 2nd ed., Studies in Biology No. **38**, Edward Arnold, London, 68 pp.

Metcalf, R.L. (1980) 'Changing role of insecticides in crop protection'. *Annual Review of Entomology*, **25**, 219–56.

Metcalf, R.L. and Luckmann, W.H. (eds.) (1975) *Introduction to Insect Pest Management*, John Wiley, New York, 585 pp.

Moriarty, F. (1975) *Pollutants and Animals, A Factual Perspective*, George Allen & Unwin, London, 140 pp.

Perring, F.H. and Mellanby, K. (eds.) (1977) *Ecological Effects of Pesticides*, Linnaen Society Symposium Series No. **5**, Academic Press, London, 193 pp.

Pimental, D. and Goodman, N. (1978) 'Ecological basis for the management of insect populations'. *Oikos*, **30**. 422–37.

van den Bosch, R. (1980) *The Pesticide Conspiracy*, Anchor Press/Doubleday, New York, 212 pp.

van den Bosch, R., Messenger, P.S. and Gutierrez, A.P. (1981) *An Introduction to Biological Control,* Plenum, New York, 247 pp.

Watson, D.L. and Brown, A.W.A. (eds.) (1977) *Pesticide Management and Insecticide Resistance*, Academic Press, London, 638 pp.

World Health Organization (1980) 'Resistance of vectors of disease to pesticides'. WHO Expert Committee, Vector Biology and Control. *Technical Report Series*, No. **655**, 82 pp.

4 Sampling methods for pest and vector management

Theoretical aspects of sampling

D.J. Thompson

The sampling of populations has been used for a variety of purposes ranging from extensive surveys designed for predicting plant damage, to intensive studies of the dynamics of animal populations. The same basic data on frequency distribution, major sources of variance and optimum size of sample unit are required for either extensive or intensive sampling.

Extensive surveys are carried out over large areas and are concerned primarily with distribution of arthropod species and population levels in relation to the epidemiology of disease transmission by vectors, or predicting crop damage. The chosen area is sampled only once or twice and attention is focused on a particular important part of the insect's, tick's or mite's life-history. Intensive studies involve more frequent sampling in smaller areas, the purpose of which is first to determine the factors that cause the major fluctuations in population size, and second those that regulate it. The object of the study will determine the exact type of sampling method to be used, so it is important from the outset to define clearly the objectives.

In fact defining the objective is perhaps the first criterion by which a sampling programme should be appraised. Other criteria are: Was the sampling design selected the most appropriate one for the objective? Was the optimal allocation of sampling resources determined? (i.e. Did we take too few or too many samples in relation to what we got out of the project?). Did we choose the correct method of analysing the data prior to deciding the sampling design?

In this first section of Chapter 4, we aim to provide some background to the kind of choices available when beginning a sampling programme.

Density: absolute and relative estimates

Ecologists frequently use the term 'density of a population' when they refer to the number of animals in a particular area. There are essentially two ways of measuring the density of a population. If it is possible to count or estimate all the animals in the area the result is the *absolute* density of the population. However, there are occasions when it is sufficient to know the ratio between two populations without knowing the actual size of either population. This is especially true when studying fluctuations over time in a population living in the same place. Sometimes technical difficulties may make it impossible to measure the absolute density of the population no matter how desirable this may be, and the best we can do may be to estimate the *relative* densities. It is almost impossible to construct an energy budget or a life-table without the conversion of population estimates to absolute figures in terms of numbers per unit area of ground.

It is important to draw the distinction between population density and what we might call population intensity, a measure of the number of animals in a unit of habitat. Population intensity would be measured in terms of number of insects per leaf or shoot or host. Such measures are frequently the first estimates obtained from a sampling programme. They are also more meaningful than absolute densities when we are trying to relate the level of an insect population to damage to a crop or a host animal. There are many pitfalls to avoid when considering intensity rather than absolute density. For instance, it is important to measure the number of habitat units per unit area since differences in crop density, for example, can lead to the population with the highest intensity having the lowest density when expressed in absolute terms. The additional labour involved in obtaining data on the density of the crop, as well as the population intensity of the crop will almost invariably be worthwhile. The population estimates obtained then become more meaningful and useful for understanding the population dynamics of the pest.

The characteristic feature of all methods for measuring relative density is that they depend on the collection of samples that represent some relatively constant but unknown relationship to total population size. Thus, by themselves, they provide no estimate of density but rather an index of abundance of more or less accuracy. Some of the techniques which yield relative estimates of population density are described later in this chapter. These include light traps for night-flying insects, pitfall traps in the ground for beetles and suction traps for aerial insects. The number of organisms trapped will depend not only on the population density but also on their activity and range of movement, as well as human skill at placing the traps. Further relative estimates of population size can sometimes be made by counting not the animals themselves, but their artifacts, for example frass, webs, exuviae, nests. A final category of techniques depends on measuring the effects of an insect, for example the levels of damage to a plant that it causes.

The main advantages of relative methods are that they frequently require simple equipment, often concentrate animals, and permit the collection of a large amount of data. It is the fact that many animals may be caught that makes relative methods particularly attractive. In many cases, when population densities are low, absolute methods with unit area sampling, bring poor returns. In addition, many of the traps that give relative density estimates are cheap to run, in that they will collect samples continuously without wasting man-hours.

As well as the obvious disadvantage of not being able to produce absolute population estimates, relative estimates by virtue of the techniques used to get them, run into other serious problems. What is really being estimated, at least with many of the insect traps used, is the proportion of those members of the population that were in the correct condition (age, physiological or behavioural condition, etc.) to respond to the trap. The animals caught are those that respond under the prevailing climatic conditions and current level of efficiency of the trap. It is sometimes possible to obtain estimates of absolute population density from relative methods; essentially a calibration procedure is used. When a series of indices of population density has been made at the same time as absolute estimates by another method, a regression of the index on the absolute population estimate can be made. The regression line can then be used to provide absolute population estimates direct from the indices. The danger of such a method is that conditions must not vary from those existing during the period when the initial estimates were made.

A second, similar method allows for a correction to be made due to variations in trap efficiency. If trap catches and absolute estimates are made simultaneously under a variety of conditions which are known to influence trap efficiency, then a table of correction terms can be drawn up which will give greater precision than the regression method described above.

The sampling programme

It is rarely possible to count all the insects in a population and therefore samples must be taken to estimate population size. The aim of a sampling programme must be to provide a population estimate with the highest accuracy in relation to the amount of effort (i.e. work or cost) expended. There is no perfect sampling scheme that will work every time; each problem and each arthropod population will require its own sampling programme, appropriate to the distribution and life-cycle of the arthropod in question. However, for each sampling programme it is necessary to decide upon three things: (i) the size of the sampling unit to be used; (ii) the number of sampling units to be taken; and (iii) the location of these sampling units in the sampling universe. It is these three decisions that form the main subject matter of the first section of this chapter.

Even assuming the life-cycle of the insect is known, some preliminary information is required before a sensible sampling programme can be formulated. This may take the form of some preliminary sampling or information from previous experience, or the information may come from the scientific literature. Exactly what information is needed is discussed in the appropriate sections which follow.

Before going on to discuss the sampling programme in more detail we need to consider the meaning of some terms whose statistical meaning differs from their everyday meaning.

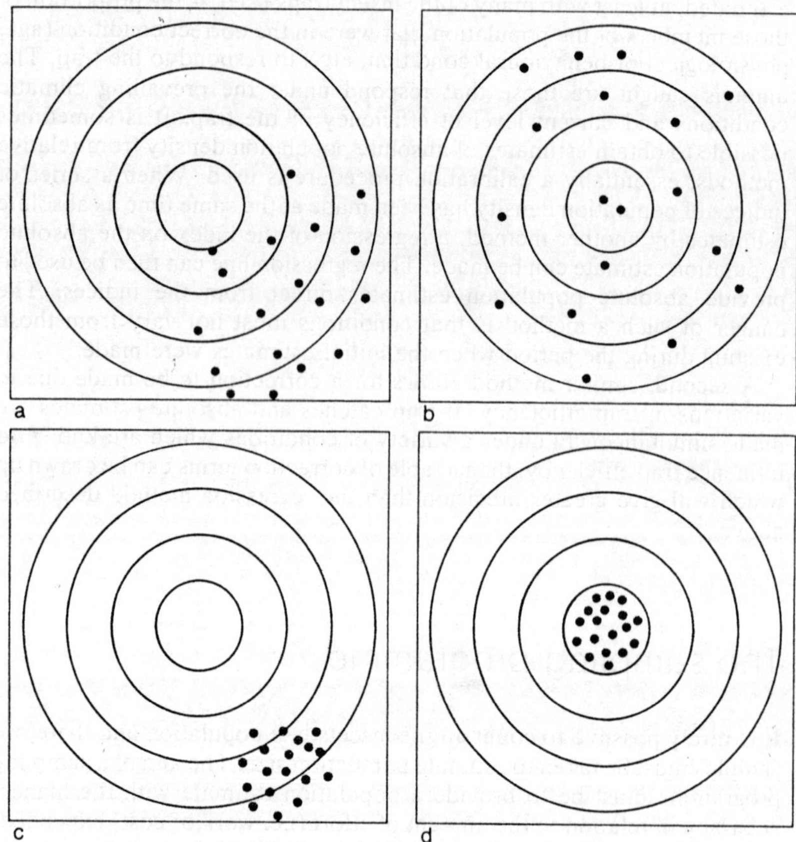

Fig. 4.1 Various patterns of shots at a target illustrating the meaning of some basic terms used in sampling programmes: (a) low precision, high bias, low accuracy; (b) low precision, low bias, low accuracy; (c) high precision, high bias, low accuracy; (d) high precision, low bias, high accuracy. (Based on R.J. Jessen (1978) *Statistical Survey Techniques*, Wiley, New York.)

An estimate This is a value constructed by some rules from sample results that stands for some true but unknown value; an estimator is a formula for obtaining an estimate.

Precision This is a measure of the closeness of each observation or sample to its own average over repeated sampling.

Accuracy This refers to a measure of closeness of the sample mean to the true value. For example, an estimator may have high precision but low accuracy (Fig. 4.1).

Bias If, on repeated sampling, the average value of estimates differs from the true value of the quantity being estimated, the estimator is regarded as biased. As a working rule, the effect of bias on the accuracy of an estimate is negligible if the bias is less than one tenth of the standard deviation of the estimate. Since the standard deviation of an estimate, as obtained from a sample, does not include the contribution of the bias, it is preferable to speak of this standard deviation as measuring the precision of the estimate, rather than its accuracy.

The relation between precision, accuracy and bias is illustrated in Fig. 4.1. which shows the imaginary shooting performance of four guns, A, B, C and D. Each gun is fired at a target a number of times, and the record of hits depicted in the figure.

Type of sampling unit

The type (size and shape) of a sampling unit used can affect both the cost of taking the sample and the precision obtained from it. The determination of the optimum type of unit may therefore be important in the economics of sampling.

The optimum unit is that which gives the desired precision for the sample estimates at the smallest cost, or more usually, the greatest precision for a fixed cost.

Before we consider the statistical desirability of sampling units of various shapes and sizes, it is worth noting a number of criteria which should be met by sample units. These criteria were first recognized by entomologists working on defoliating insects in Canadian forests.

(a) Each sample unit of the sampling universe must have an equal chance of selection.
(b) If sampling is to proceed through time, the sample unit must be stable. This is particularly important when the sample unit is an individual plant or parts of a plant.
(c) It is desirable that the proportion of the insect population using the sample unit as a habitat must remain constant, otherwise an additional sampling bias is introduced.

(d) The sampling unit must be easily recognizable as such in the field and following from this must lend itself to conversion to unit areas (see discussion on density p. 46).

Several workers have investigated the effects of the size of the sampling unit on the efficiency of sampling. The general conclusion is that a small unit is more efficient than a larger one when the dispersion of the population is contagious or aggregated, that is when there are clumps of individuals resulting in the variance being significantly greater than the mean. (See Ch. 5 for a further discussion of contagious distributions). The advantages of a small sampling unit over a larger unit are: (i) that more small units can be taken for the same amount of labour in dealing with the catch; (ii) as a sample of many small units has more degrees of freedom than a sample of a few large units, the statistical error is reduced; and (iii) since many small units cover a wider range of the habitat than a few large units, the catch of the small units is more representative. In general then, the smaller the sampling units employed, the more accurate and representative will be the results.

Many practical factors will determine the lower limit to the dimensions of the sampling unit. With a small sampling unit, the sampling error at the edge of the unit is proportionally greater. The size of the sampling unit must also depend on the size of the animal to be sampled. If the sampling unit is too small in relation to the animal's size, this will increase edge-effect errors. The bias introduced by edge-effect errors is of greater significance when the shape of the sampling unit is of the quadrant type rather than a biological unit. Edge-effect errors are minimal with circles, maximal with squares and rectangles and intermediate with other polygons, because the errors are proportional to the ratio of sampling unit boundary length to sampling unit area. For the purpose of selecting random numbers, the sampling universe is divided up into numbered sampling units. Circular units are impractical because of the gaps between them and hexagons are difficult to lay out and do not justify the reduction in error from their use. We are left then with rectangles or squares. Edge effect can, however, be minimized by conventions such as only counting animals that cross certain boundaries of the rectangles.

The number of samples

In planning a sampling programme, one of the most important decisions that has to be made is how many samples should be taken. Too large a sample implies a waste of resources, and too small a sample diminishes the utility of the results.

There are a number of steps involved in the choice of sample size. First, we must remind ourselves what is expected of the sample. This statement may be in terms of desired limits of error, or in terms of some decision that is to be made or action taken when the sample results are known. Second, some equation must be found which connects n, the sample size, with the desired precision of the sample. Precision may be expressed either in terms of achieving a standard error of a predetermined size, or in getting confidence limits of a predetermined half-width which is a percentage of the mean. We shall return to the statistical methods of determining sample size. First, it is worthwhile to describe a simple non-statistical method which gives an indication of what we are trying to achieve with our more refined methods.

Suppose we take five sampling units at random and calculate the arithmetic mean number of insects caught. Then we take five more units at random and calculate the mean of all ten units. Continue to increase the sample size by five units at a time and plot the mean number of insects caught per sampling unit against sample size. When the mean value ceases to fluctuate beyond an acceptable level, this sample size can be taken to be the appropriate one for this situation. This method is of limited application in reality because it is often impossible to calculate means at the time of sampling, though pocket calculators can be an asset.

Sample size can be calculated for a specific degree of precision. For many purposes a standard error of 5 per cent of the mean is an acceptable level. In a homogeneous habitat, the number of samples required to give this level of precision is given by:

$$n = \left(\frac{s}{D\bar{x}} \right)^2 \qquad [4.1]$$

Where s is the standard deviation, $\bar{x}$ is the arithmetic mean and D is the index of precision. D is given by:

$$D = \frac{\text{standard error}}{\text{arithmetic mean}} = \sqrt{\left(\frac{s^2}{n} \times \frac{1}{\bar{x}} \right)}$$

If 5 per cent is an acceptable level, then $D = 0.05$. What this equation does is compare the standard deviation of the sampling data (s) with the acceptable standard error ($D\bar{x}$) chosen on the basis of the particular needs of the sampler. It is important to note that the value of the standard error will change with the square root of the number of samples which means that to improve (reduce) the value of the standard deviation, s, a large increase is required in the sample size, n.

If the mathematical Poisson series (see Ch. 5) is known to be a suitable model for the samples, D is given by:

$$D = \frac{1}{\bar{x}} \sqrt{\frac{\bar{x}}{n}} = \frac{1}{\sqrt{n\bar{x}}} \qquad [4.2]$$

Since $n\bar{x} = \sum x$, (i.e. the total count for the sample), then the precision of the estimated population mean depends upon the total number ($n\bar{x}$) of animals in the sample, rather than sample size.

If the negative binomial series is known to be a suitable mathematical model for the samples, then D is given by:

$$D = \frac{1}{\bar{x}}\sqrt{\left(\frac{\bar{x}}{n} + \frac{\bar{x}^2}{nk}\right)} = \sqrt{\left(\frac{1}{n\bar{x}} + \frac{1}{nk}\right)} \qquad [4.3]$$

$$n = \frac{1}{D^2}\left(\frac{1}{\bar{x}} + \frac{1}{k}\right)$$

where k is the dispersion parameter of the negative binomial (see Ch. 5). These formulae require either $\bar{x}$ *or* s^2 (for general formula), $\bar{x}$ (for Poisson), or $\bar{x}$ and k (for negative binomial). Values of these statistics are either guessed from the results of previous sampling, or calculated from a preliminary sample. As n probably varies from one sampling occasion to the next, a rough estimate of sample size will have to suffice for most purposes. It is usual then to risk approximations assuming a normal distribution and thus the general formula [4.1] is used for most purposes. The special formulae for a Poisson series and negative binomial are easier to apply when either of these distributions is known to be a suitable model for the samples.

Where sampling is necessary at two different levels, e.g. number of leaves per tree and number of trees in a forest, the number of units (n_t) that need to be sampled at the higher level, e.g. the number of trees is given by:

$$n_t = \frac{s_s^2/n_s + s_p^2}{D\bar{x}^2}$$

where n_s is the number of samples *within* the habitat unit (calculated from [4.1] and s_p^2 is the variance *between* the habitat unit (inter-plant variance). When D is the relative error in terms of percentage confidence limits of the mean, the formulae given above, [4.1] to [4.3], must be multiplied by t^2, where t is found in Student's t-distribution ($t \simeq 2$ for more than 10 samples at the 5 per cent level). We then obtain the following equations. For general purposes we have:

$$n = \frac{t^2\, s^2}{D^2\, \bar{x}^2} \simeq \frac{4s^2}{D^2\, \bar{x}^2}$$

If the Poisson series is a suitable model for the samples, then:

$$n = \frac{t^2}{D^2 \bar{x}} \simeq \frac{4}{D^2 \bar{x}}$$

If a negative binomial is a suitable model for the samples, then:

$$n = \frac{t^2}{D^2}\left(\frac{1}{\bar{x}} + \frac{1}{k}\right) \simeq \frac{4}{D^2}\left(\frac{1}{\bar{x}} + \frac{1}{k}\right)$$

With $D = 0.1$, these equations would give the sample size required to obtain an estimate of the population mean within $\pm$ 10 per cent of the true value.

On some occasions we are not concerned with estimates of population density, but rather in the frequency of occurrence of a particular organism or event, for example the frequency of a particular genotype in the population. Before an estimate can be made of the total number of samples required, an approximate value of the probability of occurrence must be obtained (from a preliminary survey or past experience). The number of samples, n, is then given by:

$$n = \frac{t^2 p\, q}{D^2}$$

where p is the probability of occurrence (for example, of the genotype in question), $q = 1\text{-}p$ and t and D are as above.

The location of sampling units in the sampling area

Unrestricted random sample

To obtain an unbiased estimate of the population on the sampling data should be collected at random, that is so that every sampling unit in the universe has an equal chance of selection. Since many statistical methods require a random sample it is important to understand the principles of random sampling.

The universe or defined area for the investigation is the total available sampling area, and the number of available sampling units in this area is given by $N = A/a$ where A is the total sampling area and a is the area of the sampling unit (e.g. quadrat size). Therefore N sampling units form the population from which n sampling units are selected. As n is usually much smaller than N, the next decision concerns how to select the small sample of n units from the large population of N units. The sample must be representative of the whole population, and therefore the sampling units must be selected without bias. These conditions are satisfied when the sampling units are selected at random from the population.

True random selection is often difficult to achieve and the most reliable method is to use a table of random numbers. Suitable tables are to be found in many statistical text books. Ecologists often use methods for random sampling that are less precise than the use of random numbers, such as throwing a stick or quadrat or even worse, the selection of sites by eye. Such methods are not random in the statistical sense. The most serious problems are that they allow the intrusion of personal bias and even if that can be reduced to a minimum, still tend to under-sample marginal areas.

To use random number tables, select the first n numbers with values less than N. If all the available units in the population are listed from 1 to N, the n units in each sample are easily located. This method can be laborious in a large population. Therefore a two-dimensional grid is often used with each square equivalent in area to the area of a sampling unit. The squares of two adjacent sides of the grid are numbered, and each sampling unit is thus located by a pair of co-ordinates. Random numbers are then drawn in pairs and these are used as co-ordinates to locate each unit in the sample. The point is taken as either the centre or a specific corner of the sample.

Stratified random sampling

Unrestricted random sampling is not the most efficient way of minimizing the variance, since a majority of samples may turn out by pure chance to come from one area of the sampling universe. This is particularly true when n is much smaller than N. Therefore the more efficient method of stratified random sampling is to be preferred for most ecological work. The purpose of stratified random sampling is to increase sampling efficiency by dividing the population into several subpopulations or strata. These strata should be more homogeneous than the whole population, and should be well-defined areas of known size. Stratification also increases the accuracy of population estimates and ensures that subdivisions of the population are adequately represented.

In the simplest form of stratification, the sampling universe is divided into areas of equal size (strata). All the units in the sample are divided equally between strata and these units are located at random in each stratum. If the strata are unequal in area, the units in the sample are divided unequally between the strata, and the number of units allocated to each stratum is proportional to the area of the stratum. With proportional allocation of the units in a sample, the sampling fraction in each stratum is the same:

$$\frac{n_1}{N_1} = \frac{n_2}{N_2} = \cdots \frac{n_k}{N_k} = \frac{n}{N}$$

where we select random subsamples of n_1, n_2, ... n_k units from k strata containing N_1, N_2, ... N_k units.

The total numbers of sampling units in the sample (n) and in the population (N) are given by:

$$n = n_1 + n_2 + \cdots + n_k$$

$$N = N_1 + N_2 + \cdots + N_k$$

N_1/N, N_2/N, N_k/N are the relative weights attached to each stratum. With the proportional stratified sampling, the sample is self-weighting and the arithmetic mean of the whole sample is the best estimate of the population mean. Although proportional allocation is most frequently used in stratified sampling, the theoretical optimal allocation of the units in the sample is the one that minimizes the standard error of the estimated mean for a given total cost of taking the sample. This is achieved when the sampling fraction for each stratum is proportional to the standard deviation for the stratum. Of course, the standard deviation for each stratum is seldom known prior to sampling. When optimal allocation of sampling with strata is used, it is no longer self-weighting, and therefore a weighted mean and standard error must be calculated. When the habitat is stratified, biological knowledge can often be used to eliminate strata in which few insects would be found. The new restricted universe will give a greater level of precision for the calculation of a mean than an unrestricted and completely random sample.

Systematic sampling

Methods for selecting a sample from a population which do not involve random sampling have a more restricted use; systematic sampling is one such method. First the sampling units must be numbered. Then the first unit in the sample is drawn at random from the population and the next units are selected at fixed intervals. Thus, if a population contains 48 units to be sampled at a 10 per cent, or 1 in 10, rate we select a random number between 1 and 10, say 2, and then the sample would consist of the units numbered 2, 12, 22, 32, and 42.

The advantages of this type of sampling method are: (i) the sample is easier to select than a random sample, particularly if the selection of a random number has to be done in the field, so the cost will be less; (ii) the units in the sample are distributed evenly throughout the population; (iii) like the random sample, each unit has the same probability of inclusion at the outset.

Its disadvantages are: (i) the sample may be very biased when the interval between units in the sample coincides with a periodic variation in the population; and (ii) there is no reliable method for estimating the standard error of the sample mean, so that sampling data obtained in this way cannot be analysed statistically with any degree of reliability.

However, a variation of systematic sampling is the centric systematic area-sample which it is claimed can be treated as a random sample. The centric systematic sample is the one drawn from the exact centre of each area of stratum.

Two-stage sampling

It is usually possible to count all the insects in a sampling unit, but sometimes this task is laborious when the catch per sampling unit is very large as in some light-trap collections. The problem can be resolved by reducing the size of the sampling unit, but this is often impossible. A common feature is to select and measure a sample of the elements in any chosen sampling unit. The technique is called subsampling or two-stage sampling. The first stage is to select a sample of units, often called the primary units, and the second is to select a sample (subsample) of elements from each chosen primary unit.

Subsampling may be combined with any type of sampling of the primary units. The subsampling itself may employ stratification or systematic sampling.

The decision whether to subsample or not revolves around the now familiar issue of the balance between statistical precision and cost. One area in which subsampling is frequently carried out is in soil sampling where the extraction processes are often complex and expensive, but it is relatively inexpensive to collect samples.

Sequential sampling

Sequential sampling, a technique originally developed for quality control of manufactured goods, is characterized by its flexible sample size. Essentially, it is a procedure whereby samples are drawn in sequence with decisions being made on the information obtained from each observation. If no decision can be made on the basis of information from a given sample, the data are added to those obtained for the next, and the process repeated until a decision can be made. With insect sampling, the efficiency of the method lies in the absence of a fixed sample size; least work is required when populations are very low or very high, and extensive sampling is needed only at intermediate densitites. Sequential sampling thus lends itself to the rapid classification of populations into broad levels of infestation. It has been used extensively for surveys of forest insect defoliators.

A further application of the sequential method lies in the classification of infestation based on the need for control measures. Again, considerable time and expense can be saved by using the procedure to designate areas or fields as 'needing control' or 'not needing control'. Extensive preliminary work is necessary to establish the type of distribution (for the relation of the mean and variance see Ch. 5) and the density levels that are acceptable and those that we associate with extensive damage. From such data it

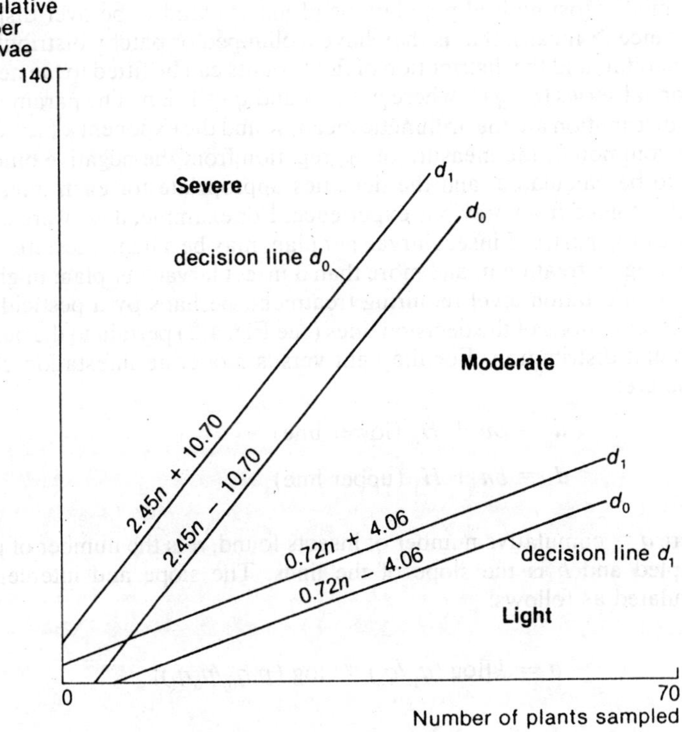

Fig. 4.2 Sequential graph showing decision lines as used in Canada for sampling late-instar (3rd-5th) of the butterfly *Pieris rapae*, a pest of brassica crops. (Based on D.G. Harcourt (1966) *Canadian Entomologist*, **98**, 741–6.)

would be possible to lay down a fixed number of samples that would enable one, in all instances, to be certain of the population level; however, this would frequently lead to unnecessary sampling and a sequential plan usually allows one to stop sampling as soon as enough data have been gathered.

Two types of error are involved in sequential sampling: the probability, α, of calling a light infestation moderate or a moderate infestation severe, and the probability, β, of calling a moderate infestation light or a severe infestation moderate. Let us say that the same level is acceptable for both, 1 false assessment in 20, i.e. a probability level of 0.05 for α and β. The levels of probability need not always be the same because of the costs of failing to apply control measures when an outbreak occurs may be much higher than those arising from the application of the treatment that is proved not to have been required.

The following account is appropriate for a sequential sampling programme when the distribution of insect species concerned follows a negative binomial. Most natural populations of insects tend to be over-dispersed (variance > mean), that is they have a clumped or patchy distribution in the habitat, and the distribution of field counts can be fitted to the negative binomial series $(q - q)^{-k}$, where $p = \mu/k$ and $q = 1 + p$. The parameters of this distribution are the arithmetic mean, μ and the exponent k (see Ch. 5). The common k, the measure of aggregation from the negative binomial, has to be calculated, and the densities appropriate for each infestation level assigned from previous experience. For example, if we were dealing with a crop pest, 0.5 insect larvae per plant may be a light infestation level requiring no treatment, and more than 3 insect larvae per plant might be a severe infestation level requiring treatment, perhaps by a pesticide.

The equations of the decision lines (see Fig. 4.2) pertain to the negative binomial distribution. For the light versus moderate infestation classes these are:

$$d_0 = bn + H_0 \text{ (lower line)}$$

$$d_1 = bn + H_1 \text{ (upper line)}$$

where d = cumulative number of insects found, n is the number of plants sampled and b is the slope of the lines. The slope and intercept are calculated as follows:

$$b = k[\log (q_1/q_0) \;/\; \log (p_1q_0/p_0q_1)]$$

$$H_0 = \log [\beta/(1 - \alpha)] \;/\; [\log (p_1q_0/p_0q_1)]$$

$$H_1 = \log [(1 - \beta)/\alpha] \;/\; [\log (p_1q_0/p_0q_1)]$$

The decision lines are plotted in Fig. 4.2 along with corresponding lines for the moderate versus severe infestation classes. These lines constitute the criteria for rating the infestation. Plants are examined in rapid succession and cumulative numbers of insects are plotted on the graph as a series of points. Sampling is discontinued when a point falls into one of the three zones, light, moderate or severe.

Mark–release–recapture

A variety of mathematical models exist under the generic name of mark–release–recapture models. There are many textbooks devoted to their use, some of which are listed in the recommended reading for this chapter. The purpose here is to explain the workings of the simplest of these models and state the assumptions implicit in most of them, which must be valid if the models are to give useful estimates of absolute population size.

Suppose we wish to estimate the size of a population into which there is neither birth nor immigration. On the first day we catch a random sample of r individuals, mark them so that we can recognize them in the future and return them to the population. Subsequently there is both death and emigration, to which marked and unmarked insects are equally prone. Thus the marked proportion remains the same as when the r marked individuals were initially captured. On another occasion a second random sample is caught of size n, containing m marked individuals.

It should be true that $\frac{m}{n} = \frac{r}{N}$ where N is the size of the whole population. An estimate of N which we write as $\hat{N}$ (pronounced 'N hat' becomes:

$$\hat{N} = \frac{rn}{m}$$

This is the Petersen estimate first advocated by him in 1896 and on which other models are based. It has a restricted usage for two reasons. First it involves only one release and one recapture, when even the simplest alternatives use more data. Second there are more assumptions implicit in the Petersen estimate than in many of the other models.

The basic assumption of mark–release–recapture models are these:

(a) marks are permanent and are noted correctly on recapture;
(b) being caught, handled and marked on one or more occasions has no effect on an individual's subsequent chance of recapture;
(c) being caught, handled and marked has no effect on an individual's chance of dying or emigrating;
(d) all individuals, whether marked or not, have inherently an equal chance of being caught. This effectively means that the population

must be sampled at random with respect to age, sex, reproductive condition etc;

(e) individuals have an equal chance of dying or emigrating whether marked or not.

In addition the Petersen estimate assumes either there are no births or immigrations, or that there are no deaths or emigrations, or that there is none of these. Finally, in all models that do not assume that there is birth, death, immigration or emigration, the period between successive samples must be short.

The interpretation of mark–release–recapture results is just the same as any absolute population estimate, a process of several stages. First the accuracy required of the results should be determined. Then the results can be examined using a purely statistical approach to see if the desired accuracy has been achieved.

Sampling crop pests

P.T. Walker

In the management of pests, a quantitative assessment of pest density is almost essential. Without it, there is no advance on such qualitative assessments as light, moderate or severe pest damage to crops. Personal judgements are subjective and inaccurate, and numerical treatment impossible. Quantitative assessments of pests and damage are a basic need in many aspects of pest management, such as studies of pest biology, pest distribution and forecasting pest outbreaks.

Pest management is the planned regulation of pest density by different methods, based on pest monitoring. Control is applied at an 'economic threshold', itself based on the relation between the pest infestation and the crop yield. How this information is obtained, the form it takes and how it is used are the subjects of this and the next section.

Pest management can be regarded as a decision-making system which answers the question: Do I apply a control measure or not, and if so how much and when? Control measures may be chemical, or otherwise, but any decision must be based on the economic ratio of benefit to cost, and this depends on the quantitative pest density.

The decision-making system can be considered in five phases (Fig. 4.3). Firstly, the infestation (i) is surveyed to assess the pest density. This survey phase (I) requires a standard and reliable method of assessment, a knowledge of sampling statistics and surveying, and information on the pest's biology and its distribution in place and time. Second is the experimental phase (II), when trials are needed to relate infestation rates to yield (Y), and to examine the effects and costs of different control measures at different times and at different 'economic

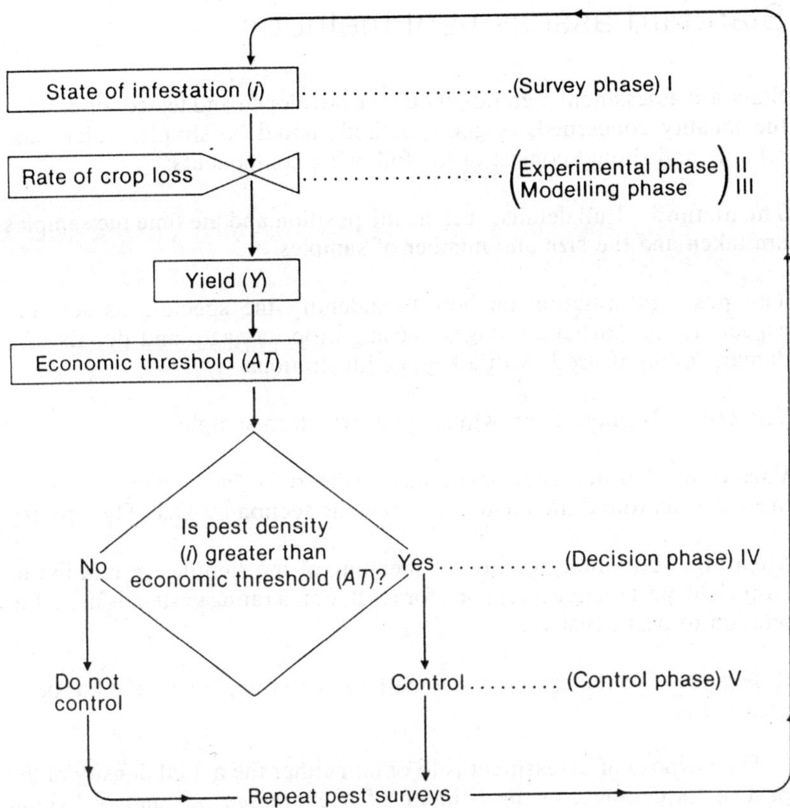

Fig. 4.3 An example of the structure and various phases (I–V) of a pest management decision-making system.

thresholds'. Thirdly in the modelling phase (III), infestation rates are related to yield (Y) in a model or equation, from which the expected yield (Y) can be obtained, the benefits of control can be compared with the cost, and an 'economic' or 'action threshold' (AT) can be derived. Fourthly in the decision phase (IV), survey results are compared with the action threshold and a decision made to control or not. In a final control phase (V), if control measures are applied they will affect the pest density (i), which must be surveyed again. If control is not applied, the pest density must still be surveyed continuously in order to detect when infestations reach the action threshold (AT).

The aim of pest control is to produce higher crop yields or improved quality at less cost, bearing in mind any adverse effects on the social or natural environment.

Standard assessment methods

Standard assessment methods should be developed and tested for pests in the locality concerned. A good method should be simple, quick, and reliable and should consist of the following components:

The method Full details, such as the position and the time the samples are taken and the size and number of samples.

The pest Information on how to indentify the species, its sex and especially the particular stages causing crop damage, and details of a damage rating if used, with a key, or illustrations.

The crop Numerical growth stage at which to sample.

Variation The accuracy or nearness to the true value, and the precision, the variation round the mean (see previous section by D.J. Thompson).

Meaning The interpretation or meaning of the samples in relation to crop yield, pest management or a forecast, or if a rating system is used, the relation to pest density.

Reference To the source of the method, for example where it has been used before.

The purpose of assessment is to obtain either the actual density of the pest or its damage, or an estimate of it expressed per unit of habitat (usually area) that represents the true situation as much as possible. If another unit, such as time, plant or plant part is used, it should be related to area. The terms incidence and intensity are used frequently in pest assessment. *Incidence* is the number of pests, lesions or attacked plants per unit, while *intensity* is a measure of the degree of attack, such as percentage of plants damaged in an area, or the percentage of an individual plant or stand damaged.

Various methods of pest assessment are now described.

Direct assessment

Pests can be counted directly, for example, the number of stemborer larvae per m^2, or indirectly, such as the number of damaged plants per m^2. Both may be either on plants or in the environment. In the environment the sampling unit may be time, volume of air or area. Reference to an absolute unit, such as area of ground, should be given where possible. If samples are based on a plant or plant part, secondary information on the number of grains, heads, tillers, stems, plants or hills per m^2 should be recorded. *The sampling efficiency is given as the ratio of the estimate of pest density to actual pest or damage density.*

Direct sampling on plants

Individual pests are counted often directly on leaves, stems or fruits, for example bollworm eggs on leaves of cotton and scale insects on sugarcane stems or citrus fruits. Plant parts may also be cut open to count stemborer larvae per stem, fly larvae per fruit, or weevil larvae per tuber. For small insects such as aphids or mites the plants can be crushed on glossy paper or filter paper to simplify counting. Ninhydrin paper can also be used to show the presence of live insects in grain. Pests inside plants, such as *Contarinia*, midge larvae in sorghum, lepidopterous larvae in stems, or beetles in stored grain, may be detected by X-rays.

The number of pests on whole or parts of plants can often be counted by beating the plant over a sheet of cloth or an inverted umbrella. An insecticidal knockdown spray of pyrethrum or dichlorvos can be used. For example, sorghum heads placed in a box and treated with such insecticides flush out hemipterous bugs or thrips which are then counted. Small pests may be brushed off by hand or machine, or scraped from the plant, as with white scale insects on sugarcane. Pests may also be washed off, perhaps with a detergent or egg cement solvent, and after filtering assessed by volume or weight.

Plants, or their parts, can be enclosed within walls or in a container, and a suction sampler used to collect all the pests on them. Alternatively some pests can be caught in a sweepnet and extracted with a hand pooter suction tube (aspirator).

Direct samples in the environment

Crop pest sampling may be based on simple counts, passive trapping, active suction of air into a trap, or attraction of pests to a trap sited in the habitat.

Sampling from the soil

Soil pests are sampled by digging in a quadrat, or drilling soil with a soil core-sampler to a standard depth to sample a constant soil volume. Pests are then separated from the soil by dry methods, such as the hot-water-jacketed Berlese funnel, or the Tullgren funnel heated by an electric light bulb (Fig. 4.4) or oil heater. Pests move to the collecting tube at the bottom of the funnel. In wet extraction methods, such as the Salt-Hollick system (Fig. 4.5), samples are washed, agitated, and sieved. Flotation agents such as magnesium sulphate and wetting agents are added, and then the

Fig. 4.4 Sampling from soil by dry extraction method: a longitudinal section of a Tullgren funnel; heat is supplied by an electric light bulb. (P.T. Walker.)

Fig. 4.5 Sampling from soil by wet extraction method. The Salt-Hollick soil washing apparatus. (P.T. Walker.)

pests are separated. Pests may be driven from the soil by repellents such as formalin, or attracted, in the case of termites, for example, by paper placed on the soil. When soil samples are taken along the length of a row of crops, pest densities are expressed on the basis of area. Sometimes sampling can be limited to clumps, hills or rows of plants where pests are ecologically confined. If pests are found only in these areas, this approach is acceptable, but otherwise area sampling is necessary. The essential point is whether the samples taken represent accurately the true pest population.

Sampling with traps

Trapping methods are used frequently for sampling pests and a wide variety of insect traps are available.

Fig. 4.6 A typical water trap which is painted yellow. (P.T. Walker.)

Water traps (Fig. 4.6) are shallow, water-filled plastic or metal trays, sometimes incorporating a particular colour, pheromone or some other attractant. Detergent in the water is useful in that it aids wetting. Traps must be protected from the weather and the catch removed and the traps refilled regularly. Shape and size should be standardized. Sticky traps (Fig. 4.7) made by spreading proprietary sticky material on plastic cards or sheets can also combine the attraction of colour or pheromones, etc. The shape may be flat, cylindrical, triangular or spheroid, depending on the habits and size of the pest and the form of the attractant. Cylinders and spheres are efficient for passive collection, while triangular or flat structures are used for pheromone trapping. Depending on the adhesive used the catch may be washed off with kerosene or other suitable solvents. Sometimes the sticky surface is sent to a laboratory for examination.

Fig. 4.7 A sticky trap made from yellow-coloured plastic cards smeared with an adhesive. (P.T. Walker.)

Pitfall traps are simple to construct, consisting of smooth-sided containers (e.g. glass jars) sunk into the ground. They must be protected from predators which will devour the catch, and rain which floods the traps. Emergence traps are placed on the ground to catch emerging pests, which collect in a transparent vial placed at the top.

Sweepnetting of crops can be used for general population assessment. If the diameter of the net, number of sweeps and the technique can be standardized, consistent results can be obtained from sample to sample. The catch is expressed as the average number of pests per sweep.

Many traps are coloured yellow (e.g. the Moericke trap) because many field pests are attracted to this colour, but traps used to sample fruit pests are often red as this seems to be more attractive to such pests. Standardization of the trap colour to an international standard is recommended. Traps may be flat, cylindrical or incorporate a complicated roof design: they can be combined with pheromones, or used with water or sticky substances. Baffles usually increase the catch.

Attractant traps may be based on the crop itself, on natural or synthetic chemicals which simulate the attraction of the crop, or on an attractant (e.g. pheromone) derived from the pest itself. The banana weevil, *Cosmopolites sordidus*, is assessed by using cut pieces of banana stem, Rhinoceros beetles, *Oryctes* spp., by cut pieces of coconut stem, tomato pests by pieces of tomato, etc. Sugar or molasses are used to attract moths, fermenting protein or commercially available protein hydrolysate are incorporated in fruit fly traps, and other chemicals are used for certain aphids, brassica pests and various beetles. Pheromones or sex attractants are now available commercially for a wide range of lepidopterous pests, and also for certain beetles and scale insects. Chemical attraction to the opposite sex must be demonstrated before the chemical is isolated, identified and synthesized. The main advantage of pheromones is their selectivity for a particular pest. Trap design and height above the crop affect attraction, as well as the condition of the pheromone, form of container or dispenser, the rate of release, and prevailing weather conditions (Fig. 4.8).

Fig. 4.8 A simple experimental pheromone trap sited in a maize field. This trap is baited with a female moth, *Busseola fusca*. (P.T. Walker.)

Light-traps have long been used to sample agricultural insects. Basically they consist of a light above a funnel and a container below to collect the

Fig. 4.9 A Robinson light-trap which is operated from mains electricity. The light source is a mercury vapour discharge lamp. (P.T. Walker.)

catch. Fig. 4.9 shows a round metal drum in the easily made and rugged Robinson trap. This trap employs a mercury discharge ultraviolet or mixed radiation electric bulb needing a ballast (= choke), unless a special blended bulb is used. In other traps such as the 'Entech' (Fig. 4.10) the container is a metal box which can be collapsed. This trap has a fluorescent tube and a daylight-controlled switch which allows it to be switched on and off automatically when light intensity changes. It can be operated from mains electricity or from batteries. The 'Heistands' trap and its modification have a fluorescent discharge tube, often emitting only ultraviolet or black light. Light-traps may have vanes, a suction fan, a roof, and an electric grid to kill insects attracted to them. An insecticide, such as dichlorvos, is sometimes placed in the container of light-traps to kill the insects and thus prevent them from damaging themselves. Some traps, such as the Monks Wood light-trap (Fig. 4.24) are designed specially to run on a 12-volt car battery.

The catch from a light-trap will vary with the height of the trap above the ground, the light intensity, the angle of the light and such natural factors as temperature, wind and the amount of moonlight. The catch from one trap may not be the same as from another nearby one, or from the same trap if moved. Light-traps are used mainly to assess the number of pest species, to monitor relative changes in their population size, to time these changes, and to relate them to other methods of assessment as an aid to pest management.

Fig. 4.10 The 'Entech' light-trap. This is a portable trap using a small fluorescent tube and working from a 12-volt car battery. As shown it can be dismantled for transportation. (P.T. Walker.)

Suction traps may be portable or fixed. For small insects and restricted habitats such as the base of rice plants, a hand-held portable battery-driven car vacuum cleaner type of sampler, with a 2.5 cm diameter nozzle is useful. For larger areas, a variety of bigger electric machines are available. The 'Johnson-Southwood' sampler having a 3 cm diameter nozzle, the petrol-driven 'Univac' sampler with a 6 cm diameter nozzle, and the larger 'D-vac' machine with a 13.5 in diameter nozzle which equals a square foot in area, are all examples of larger portable machines.

Insects in flight can be sampled with fixed suction traps. They usually consist of a metal gauze cone into which air is sucked by an electric fan. In the 'Johnson-Taylor' trap, with unenclosed cone, there is either a 23 cm or 30 cm diameter opening, and a segregating device of metal discs coated with pyrethrum, which drop and separate the catch hourly, or as timed (Fig. 4.11). The more rugged enclosed 'Johnson-Taylor' trap has a 45 cm diameter opening, can also have a segregating device, and is often used to sample at different heights above the ground.

The catch in these suction traps is affected by wind speed and size of the insect, and correction factors are available for these variables as a logarithm which is added to the log catch/hour. Corrections are also made for the trap's height above ground level and volume of air it samples.

Fig. 4.11 A 'Johnson-Taylor' 23 cm diameter suction trap with unenclosed cone. The small metal box contains the time clock which operates the disc mechanism by which the catch can be segregated into hourly or other intervals. (P.T. Walker.)

General considerations

In general, any type of trap catch may be affected by the following five factors:

Pest population This concerns the variations in the number of pests available for trapping.

Pest stage or condition The stage, age, sex or condition of the pests affects the number available and their response to the trap.

Pest activity The activity of the pest is related to temperature, reproductive or feeding phase, and the other factors above.

Trap factors The type and position of the trap itself is very important, as is the age and condition of any attractant (e.g. pheromone) that may be used.

Weather Weather conditions can affect both the activity of the pest and its response to the trap, and also the efficiency of the trap itself.

In all cases, regular servicing of the trap, attention to the attractant, and regular removal of the catch are essential for good results.

For mobile pests, the true population can be estimated by the mark–release–recapture technique. This method is explained briefly by D.J. Thompson in the previous section.

Indirect assessment

Pest density can sometimes be assessed more easily or quickly from the effect of pests on crop plants, or by some other indirect measure. There are many examples, a few of which are described below.

Whole plants The number or percentage of plants attacked, missing or showing signs of damage is recorded as an indication of the population of the pest, for example cutworms or beetle larvae in the soil or stemborers of cereals.

Stems Borer damage to cereals or sugarcane is assessed as the number or percentage of stems attacked, nodes or internodes bored, while breakage due to borers or rodents is used in cereal or banana damage assessment. Borer assessment by 'dead-heart' counts is also used.

Fig. 4.12 Typical pest damage to rice inflorescence causing the condition known as whiteheads. (P.T. Walker.)

Leaves Eating or scarification of leaves by adult or larval beetles, and damage by mites or bugs can be assessed by the incidence or intensity of damage. The actual leaf area damaged or removed can be measured by a variety of techniques such as copying onto paper and weighing, by photography or photocopying, by a planimeter or electronic scanning apparatus, by counting visible points on a grid beneath the leaf, or simply by comparison of the dry weight of a damaged leaf with that of an undamaged leaf.

Fruits and seeds Damage to maize cobs by borers; to sorghum or millet seeds by midges; to legume pods, coconuts, bananas, rice inflorescences (whiteheads) (Fig. 4.12), to citrus fruits by scale insects or mites; and to cotton bolls by bollworms, can be directly counted and recorded. As with leaves, the area of damage can also be measured. There are many other examples of this type of assessment.

Other methods Arthropod pest populations can also be assessed from the amount of their larval or adult frass, or honeydew from bugs, etc. Assessment from a distance, by satellite picture or aerial photography, is being investigated. Such techniques are based on the fact that damage by pests or diseases affects the plant's water balance and changes the reflectance of a crop which may be seen by normal or infrared (false colour) colour photography. Careful interpretation and background data are necessary to relate pest attack to recorded pictures.

Scale or damage rating Infestations, damage of various percentages are often rated on a scale for speed and simplicity. Such scales or scores can be added and analysed like original counts, but in all cases they give less information than actual counts or percentages, and can even be misleading. If a scale is used, the divisions should not be too small or too numerous. Classes can represent arithmetic increases, for example up to 5, 10, 15, 20 pests, or geometric increases, such as up to 1, 10, 100, 1000 pests, or by a root or other progression, depending on the relation between pest numbers and the aim of the assessment. A standard diagram showing known areas of damage or numbers of lesions can be prepared easily and used for a more accurate and objective assessment.

If an indirect method of assessment is used, its relation to a direct method or to pest density per unit area of ground should be examined, so that quick and simple indirect assessments can be made to measure the actual population. For example, if the relation between the percentage of rice tillers attacked by borers and the borer population per m^2 is known, percentage attack can be used to estimate the borer population.

Pest surveillance

In surveys of pests or their damage, for general monitoring, for estimating crop losses due to pests, or for forecasting pest outbreaks, attention should be given to a number of points:

(a) A handbook of assessment methods, with standard crop growth charts and standard area diagrams should be obtained, for example that issued by the International Rice Research Institute for rice pests.

(b) Pest identification notes are needed for the use of field staff. The notes prepared for the African armyworm surveys are a good example.

(c) Standard forms, a communication network and a central data handling office and facilities must be available.

Assessment of crop losses

P.T. Walker

The previous section describes methods of assessing pest density, which provide essential information for rational decisions in crop pest management. Decisions to control pests are based on the concept of the threshold pest density at which to initiate control to prevent economic losses. Consequently the economic effects of different pest densities or degrees of damage are needed for decision-making. Similarly, the economic effects of vectors of human or animal diseases on human or animal health and productivity are required, and the reduction in economic crop yield due to diseases and weeds.

Action and damage economic thresholds

The pest density at which economic loss occurs is called the 'economic injury level' (*EIL*), and the pest density at which control must be applied to prevent the infestation rising to the *EIL* is called the 'economic threshold' (*ET*). The terms 'damage threshold' (*DT*) and 'action threshold' (*AT*) seem less ambiguous terms for *EIL* and *ET* respectively, and are used here.

Economic loss must be defined, and is usually taken as being when the benefits of crop production are less than costs, or in the case of loss due to pests, when the loss caused by the pest, or the benefits of controlling it, are less than the costs of control. The degree of control is not proportional to the cost – increased control for a given cost tending to decrease. Moreover, the relationship between yield, its value and increasing infestation is also seldom proportional – the increased benefit from increasing control tending to decrease. The action threshold (*AT*) has also been defined as the infestation at which the marginal, or extra unit of, cost of the control curve crosses the marginal value of the yield curve, with increasing infestations. An *AT* must therefore be based on quantitative assessments of crop loss, and on the rate of yield reduction with changes in infestation levels,or the relation between yield (Y) and infestation (i), or if negative, the loss function (see Fig. 4.3). This can be expressed very simply:

$$Y = f(i)$$

where f denotes 'a function of ' in this case, infestation (i). There are accepted ways of obtaining this information, which will be described below.

Crop losses

Crop loss assessment is important in economic planning and in the management of food and fibre resources. A knowledge of crop losses may also give valuable information on how they are caused and how reduction in yield due to pests may be prevented. Consequently, research efforts can be directed to those pests which cause the greater loss.

Crop loss due to pests may be defined as the reduction caused in the maximum potential yield in the absence of pests, and is usually expressed as a percentage of the maximum. Several causes of yield reduction can be drawn in a diagram as a crop loss profile, which, if quantified, will show the relative importance of different causes of loss. Losses may be divided into those caused before harvest and those occurring during storage or processing, although some post-harvest pests originate in the field, such as the potato tuber moth, *Phthorimaea operculella*, the maize weevil, *Sitophilus zeamais*, and the bean weevil, *Callosobruchus maculatus*. The yield should be defined clearly and is normally the product harvested, although it may be a derived or extracted by-product such as vegetable oil or protein. Loss may also be reflected in lowering of the grade or the quality of the crop. Individual losses are not necessarily additive because of the complicated and dynamic growth system in plants, in which the functions of one part or process may be replaced by another in producing the final yield. This interaction may, for example, cause greater losses from leaf and root damage together, than from the sum of either alone.

Pests cause loss in yield in many ways, for example: (i) by destruction of the photosynthetic area of leaves (Fig. 4.13); (ii) by interruption of water and nutrient uptake through roots and transport systems in stems; (iii) by the destruction of storage contents in organs such as yams or sweet potato

Fig. 4.13 Damage to castor oil plants by moths of the genus *Achaea* in Thailand resulting in a great loss of leaves. (P.T. Walker.)

tubers or in sugarcane stems; and (iv) by damage to reproductive parts such as flowers, fruit or seeds. Secondary attacks by pests or diseases can occur, and diseases which cause loss can be introduced by vectors. Some pests, such as whitefly on cotton, increase the difficulty and reduce the efficiency of harvesting; others cause loss due to spoilage with excreta or insect parts, for example the cotton stainer on cotton lint. The effects of different pests or diseases cannot always be separated, in which case the attack will be regarded as a pest or disease complex.

Crop yields are often measured and expressed on the basis of area of ground, or per plant, plot or tree. Loss is then expressed as the percentage loss per unit of pest attack.

Surveys of crop losses may be made directly in the field by selecting pest outbreaks at random, but stratifying if necessary for different areas or types of crop. The natural infestation levels may be used, or pests on part of the plots or on other plots can be controlled with insecticides or by the other methods described below, for comparative studies to assess the crop loss due to different numbers of pests. Yields are taken and averaged, often weighted by area of crops. The difference between yields of pest-free and pest-attacked plots represents the loss due to pests. If, however, an accurate yield-infestation relationship has been found, ($Y = \mathrm{f}(i)$), then only pest assessment may be needed, because expected yields and hence losses can be derived from this.

Methods of loss assessment

Yield must be measured under natural farming conditions, so if assessment is based on an experimental research station, a correction factor will need to be introduced in the calculations. The relation between yield and infestation in as wide a range of pest densities as possible is required, perhaps in logarithmically increasing numbers, depending on the mode of yield reduction by the pests. Some examples are given below.

Natural infestation levels

Natural infestation and yields may be studied by using (i) randomly selected single or paired plants (ii) a series of field plots or (iii) by sampling infestations and yields over a wide area. The yields obtained are then related to infestation. Plants or fields as similar as possible, except for infestation, should be chosen. The main source of error will be natural yield differences due to weather and growing conditions. Yields may be related to infestation the year before, as is often the case in cocoa or coconut plantations. If the cause of natural variation can be identified, stratification may be possible to remove the effects of different soils or different treatments by farmers.

There are two ways of examining yield, or loss, from natural or other infestations: (i) on a standard area base, for example measuring kg sugar/stemborer larva/ha; (ii) on a plant base, for example kg sugar/larva/stem. If the latter is used, the number of stems per hectare is needed to relate the two methods. If the yields of damaged and undamaged stems are recorded, the result may be inaccurate due to the wide variation in the category 'damaged' and because of compensation by the undamaged stems for the damaged ones (see p. 82). The second method is quick and easy, but needs to be interpreted with caution. Losses of grain weight in store can similarly be assessed on a standard volume base, or on a single grain base multiplied by the number of grains.

Chemicals for controlling pest densities

Chemicals are often used to obtain varying pest densities. Different insecticides may be applied at different times or in different concentrations and formulations. Different pests may be selectively controlled, for example with a systemic insecticide for aphids, with *Bacillus thuringiensis* for Lepidoptera and a soil insecticide for soil pests. Acaricides, nematicides, fungicides, rodenticides and avicides are available for the selective control of different pests. Chemicals have also been used to attract or repel pest attacks. The obvious difficulties are: first, an effect of the chemical on plant yield, which if the chemical is toxic will be reduced, but will be increased if it is a stimulant; and second, the effect on possibly unknown pests or diseases in addition to the pest being examined, for example the effect of the carbamate insecticide, carbofuran, which also kills nematodes.

Other methods for controlling pest densities

Pests may be removed by hand, or alternatively cages, nets or barriers can be placed round plants, plots or even fields to prevent attack (Fig. 4.14). Cages have been used to control the attack of cotton by whitefly, *Bemisia tabaci*, and metal walls have been erected round plants to enclose different numbers of cutworm larvae around them. Infestation levels inside cages can be controlled with a knock-down insecticide, and by introducing a known number of pests, often in a logarithmic series above 0, for example: 3, 6, 12, 24, pests/m^2 caged area.

Fig. 4.14 Rice plants in Bangladesh confined in floating cages to assess losses caused by rice stemborers, *Scirpophaga* spp. (H.D. Cathing)

It should be remembered that cages affect plant growth and pest development by reducing light, transpiration and perhaps rainfall. Cages should be compared either with control cages without pests, or with cages that allow pest movement in and out. If pests do not move readily cages may not be needed, for example with beetle grubs in the soil around plant roots or with stemborers in cereals or sugarcane plants. Traps employing light, pheromones or some other attractant, or trap crops have sometimes been used to reduce attack on nearby plots. Another approach involves infested or susceptible crops being placed or grown with the test crop to increase attack, for example with sorghum shootfly, *Atherigona soccata*. There are difficulties, however, in re-infesting crops. For accurate assessment of the reduction in natural yield caused by the pests, the pests must either be collected in quantity or bred in culture, and artificial infestations must be at the same time and with the same distribution in space as natural infestations.

Varieties of crop which are resistant or susceptible to a pest can be grown and their yields compared. Varieties with similar yields must be used or else a correction must be applied to the natural differences in their yields.

When pests are able to migrate easily between plants or plots, errors due to inter-plot interference may occur. As populations are decreased in a treated plot, this may in itself result in smaller populations in the next plot because of insects migrating between plots; or the converse may arise so that an increase in migration occurs from the treated to untreated plot. Another problem is that because of repeated attacks on the treated plots before pests are killed, damage on them may even be greater. Screening and separation are the usual methods of avoiding these difficulties. In all cases it is better to monitor the actual pest densities on all plants or plots to check the actual number of pests, or the degree of damage causing the loss being measured.

Simulated damage

Leaves can be cut, stems damaged, roots destroyed or even complete plants removed to imitate pest attack. The amount of injury should be recorded by a suitable measure such as leaf area, leaf area index – that is the ratio of leaf area to area of ground – or dry weight. The distribution in space and time of artificial damage should, as with artificial infestations, be as similar to natural attacks as possible, for example in a grouped pattern of plant attack or at the same stage in relation to the growing point. The growing conditions before and after damage will also influence the final yield.

The relation between yield and infestation

By these methods the regression of yield on infestation density can be determined and the effect of changes in density on yield and the economic effect of control measures can be examined.

The reduction in yield with increasing infestation is seldom proportional.

This happens only when one unit of infestation reduces yield by one unit, without any threshold level or compensation by unattacked units, for example when late stemborer attack destroys a stem and the percentage yield reduction is proportional to the percentage of plants attacked. When a single attack on seeds or fruits, such as millet or coffee, reduces yield by one unit, it is usually too late for compensation to take place. Yield reduction in this case is proportional to infestation rate.

Usually the relation between yield and infestation is sigmoid (Fig. 4.15), the infestation increasing to a threshold pest density below which pest attack is compensated for and no overall yield is lost. Thereafter the following part of the curve shows yield loss almost proportional to infestation increase, and finally there is a part of the curve where increase in infestation causes little further loss of yield. If attack is late on reproductive parts such as millet seeds or coffee berries, when compensation cannot occur, then the curve will be concave upwards.

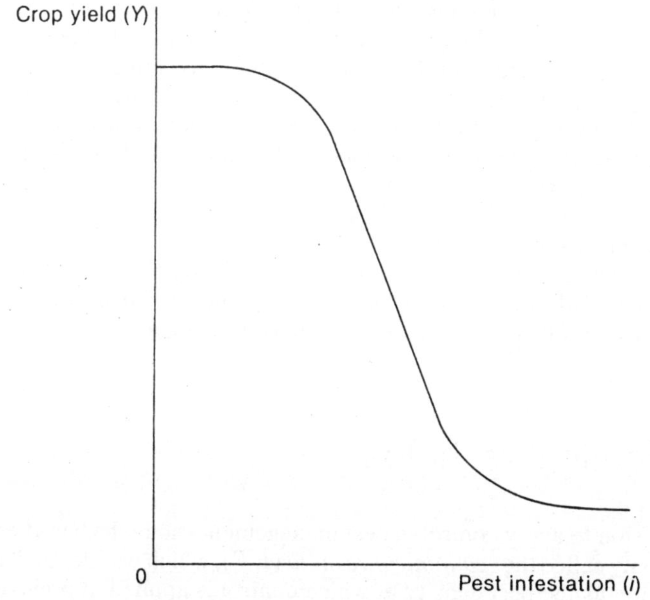

Fig. 4.15 The relation between yield (Y) and infestation (i) usually follows a sigmoid curve.

If yield reduction is related to logarithmic and not arithmetic pest population increase, the logarithms of yield and infestation may be related. Yields can increase with small increases in pest population, due to attacks affecting apical dominance, when more tillers or fruiting points are produced. This occurs in sorghum when attacked by *Atherigona soccata*, the shootfly, or after 'Lygus' attack on cotton. Damage can also increase plant maturity, as in *Eldana saccharina* stemborer attack on sugarcane which affects the stalk sugar content. Plants sometimes receive more solar radiation because of reduction in shading caused by pest destruction of foliage, and in other cases plant growth can be stimulated by attack. If a few vectors transmit a disease which seriously reduces yields, or a few pests cause loss in yield or quality out of proportion to their numbers, then a great loss in yield may result from very low pest infestations. This form of the relationship may considerably affect economic decisions on reducing pest numbers to obtain maximum yield. Single regressions of yield on infestation are commonly undertaken, but if several factors are present, a multivariate regression can be used.

Compensation

The distribution of pest attack on single plants or groups of plants affects the plant yield or the total plot yield. Unattacked plants or parts of plants can increase their contribution to the plant or plot yield up to the potential maximum. The plant can be regarded as a dynamic system with interchangeable sources of nutrients and energy which can compensate for damage to one source from another, up to the potential maximum of internal or external nutrients or energy available. Similarly an area of ground will, for any variety of crop, yield a potential maximum, limited by nutrient and energy resources. Yield loss due to damage to some plants will be compensated by increased yield from others up to the limits of resources available. Only when plants are sufficiently close to interact and the nutrient and energy supply is adequate, that is when climate and growing conditions are good, can compensation occur.

Economic thresholds

The decision to apply control in pest management should be based on the pest density called the *economic threshold* (*ET*), which was defined in the last section as the pest density at which control is applied to prevent the infestation rising to the economically damaging economic injury level (*EIL*). However, as already stated the terms action threshold (*AT*) and damage threshold (*DT*) are less ambiguous, and are consequently used below in place of the two older terms.

The action threshold (*AT*) is usually at a lower pest density and appears earlier than the damage threshold (*DT*), but depending on the development of the infestation the level may be the same. Action thresholds may be formulated in three ways: (i) by adapting an *AT* used for the same pest elsewhere; (ii) by a consensus of the opinions of experts on the pest and the crop; or (iii) by empirically balancing costs against benefits to work out the pest density at which it is economic to control. Once an *AT* has been proposed, it must be tested to see whether controls applied at that level do in fact: (i) produce increased yields; (ii) give more economic or social benefits; (iii) use less pesticide; and (iv) have less harmful environmental impact. If they do not, the *AT* must be altered and tested again.

The *AT* is not a fixed pest density, but is a dynamic concept which varies with the type of pest and crops, as well as with environmental, economic and social factors. Any factor which increases costs or lowers control efficiency, such as ineffective control methods, non-critical times of applying control, or more concern for the environment than for high yields, will raise the *AT*, and this means that control will be economic only at high pest densities. Any factor which increases benefits, such as higher yields, increased value of yield, increased control efficiency – perhaps by pest monitoring, better application methods, a favourable yield–infestation relationship, increased expectation of profit, or even a farmer's willingness to accept risks – will lower the *AT*, resulting in control becoming economic at low pest densities. The *AT* will therefore probably change during the life of the crop because of critical periods for pest control during the life of the crop and the pest, and perhaps because of changing conditions.

Another approach is to apply controls according to a *yield development threshold*, which uses the number and time of appearance of fruiting parts needed to produce an economic yield. Guidelines for these, for example squares and bolls of cotton, are drawn up taking into consideration the particular variety, yield and growing conditions. If the numbers of squares or bolls fall below the guidelines, pests are controlled on a basis of action thresholds to restore crop production to the required levels. In this way account is taken of the growth and development of the crop as well as the effect of pests on it.

The use of an economic *AT* as a base for pest management by small African farms has been criticized because pests, crops and yields are very variable and communication may be poor. It is also difficult to establish a suitable *AT* for a wide area. Consequently, either the farmer must assess his own pest density and his risk of making a loss, or an efficient *AT* must be provided from local or national resources, qualified and corrected for different farming conditions.

Management decisions may be analysed by the methods of game theory and systems analysis; and the effect of different decisions in different situations, with different probabilities of occurrence, can be compared in a matrix. Simulations using changing inputs and conditions, such as management decisions, climate and pest densities, can be generated but no matter how advanced the technique, adequate data on pest density and its relation to yield loss must be obtained in the field.

Sampling disease vectors

There are many reasons for sampling vector populations, for example to undertake surveys and compile lists of species present in an area; to identify those species which bite man or his livestock and to determine which of these are important or potential vectors of disease. Samples are also required to monitor seasonal build-up and decrease in vector numbers, their yearly changes in relative population size, and to study their distribution within habitats, their behaviour, dispersal, longevity, host preferences and other biological aspects, especially those relevant to the epidemiology of disease transmission. Samples are also needed to compare and evaluate different vector control measures, and occasionally to try to attempt to get estimates of absolute population size. In planning a sampling programme, knowledge of the periodicity of the vectors is important. It is, for example, little use employing attractant traps at night if the vectors are diurnal. Location of collecting sites is of paramount importance. A change of only a few metres can greatly affect both the numbers and species composition of a catch.

It will become clear from the following sections on sampling procedures for disease vectors that in many cases the principal method consists of the human-bait catch. Samples from such collections are biased in that they will consist of only those insects that are orientated to blood-feeding, but this may not be a serious limitation because frequently these are the insects which are required. There are, however, ethical considerations. Is it right for instance, to expose collectors to the risk of contracting certain diseases by getting them to perform human-bait catches? While most workers appreciate this difficulty, they are faced with the problem that in many instances there are at present no satisfactory alternative sampling procedures. For example, as long ago as 1966 the World Health Organization recognized the need for alternative sampling procedures for simuliid blackflies (vectors of onchocerciasis), but some 16 years later the major sampling procedure still consists of catching them at human-bait (see section by J.F. Walsh below). Sometimes, however, other methods do exist. The Challier and Laveissiere biconical trap is, for example, exceptionally efficient at catching certain tsetse flies (vectors of trypanosomiasis) (see section by M.J. Roberts below), while light-traps are very useful for routine collection of certain mosquito vectors of arboviruses, for example Japanese encephalitis, see following section. Nevertheless, it is true to say that there remains an urgent need to devise alternative methods for trapping host-seeking vectors, as well as to improve existing techniques for sampling other components of vector populations.

Sampling mosquitoes

M.W. Service

Sampling adult populations

Human-bait collections

The most direct and sometimes simplest method of catching mosquitoes biting people is to undertake human-bait collections. There are many variations of this method but one of the better is for two people to sit together at a selected site and allow mosquitoes attracted to them to settle, so that they can be caught in test tubes. A small handnet may be useful for catching mosquitoes flying around the collectors. When mosquitoes are biting at night, such as *Anopheles* malaria vectors, it is usual to perform bait catches from about 18.00–06.00 hours, whereas with species biting during the day collections may be made from 06.00–18.00 hours. In some species biting is restricted to much shorter periods, such as around dusk, and bait collections of only 3–4 hours may be sufficient to adequately sample populations. If, however, a survey is being made of all man-biting species, then 24-hour catches are required. Torches may be used intermittently at night to assist in catching mosquitoes (Fig. 4.16). One pair of collectors should work for only 3–4 hours before being replaced by another pair, and the same collectors should not catch mosquitoes during the same time on successive sampling occasions, as this is likely to introduce sampling bias. This is because some collectors may be more attractive than others, or be better at catching mosquitoes. Depending on local circumstances it may be important to ensure that those collecting mosquitoes take prophylactic antimalarial drugs, are immunized against yellow fever, and are periodically checked for the presence of microfilarial parasites in their blood.

A series of catches is usually needed to sample the population of anthropophagic mosquitoes in any area. The numbers of different species caught each hour should be segregated so that a graph can be made of the numbers biting at different times of the day or night (Fig. 4.17).

The location of the catches is important. When collections are made inside houses mosquitoes biting man indoors are sampled and these may differ from those biting out of doors in village compounds. If information is required on the species biting people on farms or in forests then collections must be made in these habitats. There may also be differences in both the species and numbers of mosquitoes biting at different seasons of the year, and it may therefore be necessary to undertake human-bait catches once a week for a year or more to measure seasonal fluctuations of mosquito populations.

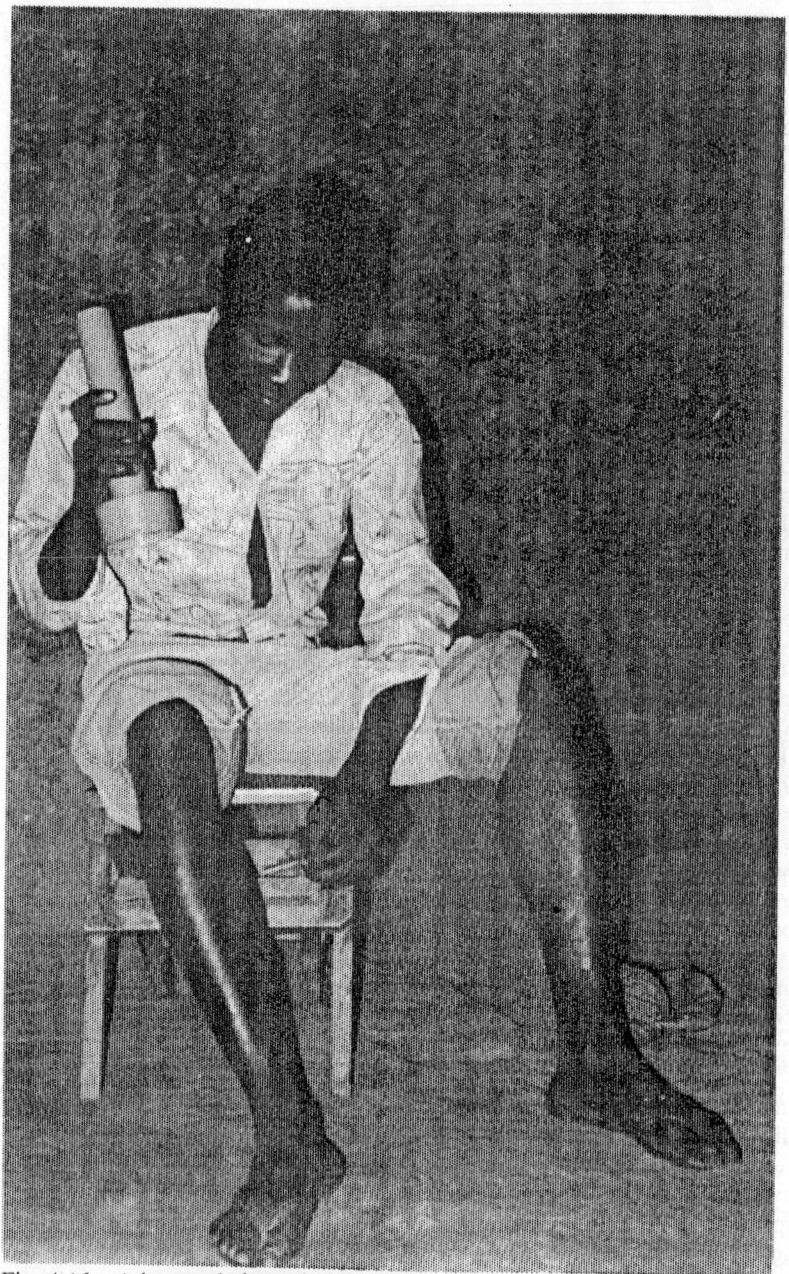

Fig. 4.16 A human-bait catch for indoor-biting mosquitoes. (M.W. Service.)

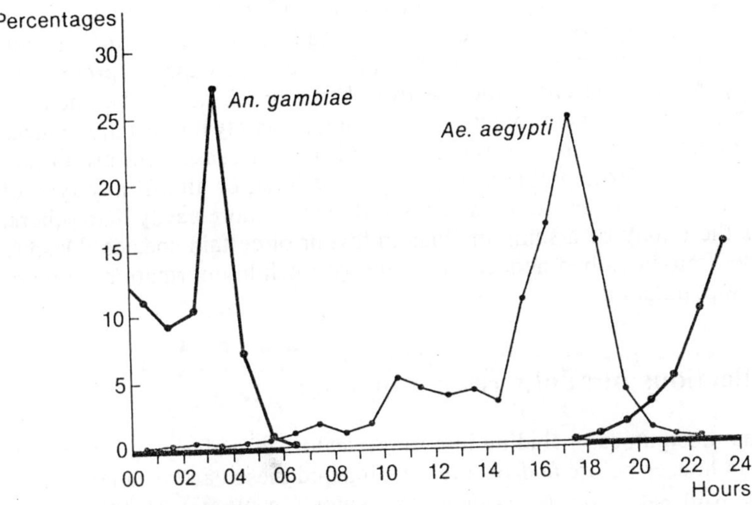

Fig. 4.17 The biting cycles of two important African mosquito vectors. *Aedes aegypti* which bites mainly in the late afternoons and is a vector of yellow fever and *Anopheles gambiae* a nocturnal vector of malaria in Africa.

After identification, mosquitoes caught in bait collections can be dissected to see whether they are infected with malarial or filarial parasites. To determine whether they are vectors of arboviruses, that is arthropod-borne viruses such as yellow fever, dengue, Japanese encephalitis and o'nyong-nyong, mosquitoes should be kept at −196 °C until they can be sent to specially equipped laboratories for virus isolation studies.

Human-bait collections can provide valuable information on the mosquito species biting man, their times of biting, seasonal occurrence, proportions of different species biting indoors and outdoors, and their *relative* abundance. But the numbers caught should not be interpreted as the absolute numbers biting an 'average man' during the day or night, because bait collections are artificial and cannot reflect a natural situation.

Bed-net collections

A mosquito net can be made into a trap for collecting mosquitoes attracted to man. A common procedure is for someone to sleep in a house or out of doors under a mosquito net that is raised about 15 cm from the ground (Fig. 4.19a). Many of the hungry mosquitoes attracted to host odours will eventually find their way into the net to feed on the occupant. After engorging the blood-fed mosquitoes will be relatively inactive and rest inside the net, from which they can be collected in the morning or at intervals throughout the night. This method overcomes the necessity of employing a number of people (about 8) to catch mosquitoes throughout the night. However, the numbers caught will usually be less than in more direct bait collections because not all mosquitoes attracted to the bait will enter the net, and moreover, after feeding some will escape. To prevent the occupant sleeping under the net from being bitten he can be enclosed within a smaller inner protective net but unfortunately this usually reduces the numbers entering the outer net and also increases the proportion escaping. Another problem of using the bed-net, or almost any type of trap, is that some mosquito species will enter it more easily than others, thus there may be a sampling bias in favour of certain species. Despite these limitations bed-nets can provide worthwhile information on man-biting populations.

Collections of adults resting indoors

Some mosquitoes, including important vectors of malaria and filariasis, rest in houses before and/or after feeding and these can be caught either alive with aspirators or as dead mosquitoes in pyrethrum spray-sheet collections.

In the former method, searches are made during the day, with the aid of torches, for mosquitoes resting underneath beds, on walls, ceilings and roofs of houses. They are collected with an aspirator and gently blown into a small container. Sometimes attempts are made to collect all mosquitoes resting in houses, but this will rarely if ever be possible. At other times collections are continued for a standard time, such as 15 minutes per house. Although this latter method does not attempt to estimate the total indoor resting population, it can nevertheless provide a useful comparative sampling index.

Before a pyrethrum spray-sheet collection is undertaken, food and easily removable objects are taken from the house or room after which strong white calico sheets are placed over the floor, bed and other items of furniture. Then, after windows and doors have been shut, one or two people using small hand pumps (e.g. 'flit guns') space-spray the room with 0.1–0.2 per cent pyrethrum in kerosene (paraffin) synergized with piperonyl butoxide. If there are large gaps between the tops of the walls and roofs, these can be either filled temporarily with spare spray sheets or one or two people can walk round outside the house spraying the eaves to prevent mosquitoes escaping. After 10–15 minutes dead mosquitoes are removed from the spray-sheets for identification and dissection. It will be impossible to catch all mosquitoes resting in houses because despite precautions some will escape through various cracks and openings, while others will fail to fall onto the spray-sheets. The percentage caught will depend much on house constructions, but is unlikely to exceed 80 per cent of the total indoor resting population. Some houses have too many gaps in their structure for pyrethrum catches to be of much value, and mosquitoes have to be sampled by aspirator collections.

Blood in the stomachs of engorged mosquitoes collected by aspirators or in pyrethrum collections can be squashed onto filter paper and later serologically examined, such as by the precipitin test, to determine whether they have fed on man or some other host. Blood-meal analyses provide useful information on the natural host preferences of mosquitoes as well as for other blood-sucking insects.

Exit traps

Exit traps fitted to windows, doors or the eaves of houses and animal shelters catch a proportion of the mosquitoes flying out of the houses or

Fig. 4.18 A simple mosquito-netting trap fitted to a door to sample mosquitoes entering houses. When turned round it acts as an exit trap and catches mosquitoes leaving houses. (M.W. Service.)

shelters. The catch can consist of male mosquitoes, unfed and blood-fed mosquitoes and gravid females ready to lay eggs. Such traps provide valuable comparative information on both the numbers and physiological condition of the female mosquitoes leaving insecticide-sprayed and unsprayed houses. As such they aid in assessing the effective duration of insecticides sprayed on the interior surfaces of houses. A useful exit trap consists of a 30 cm cube cage covered with mosquito netting and having one side inverted to form a funnel through which mosquitoes enter the trap (Fig. 4.18).

Animal baits

If mosquitoes which bite man also feed on cattle, donkeys or goats, etc. these animals can be used as attractant baits to monitor seasonal fluctuations of mosquitoes and to assess relative changes in population size due to control measures. The bait animal can be tethered and mosquitoes attracted to it caught in either aspirators or small handnets. Collections can be made continuously, or for 15 minute periods in each hour. Alternatively the animal can be placed under a bed-net or in some other trap (Fig. 4.19b) and mosquitoes collected periodically from the trap. It should be remembered, however, that any type of baited trap can introduce sampling bias (see p. 88).

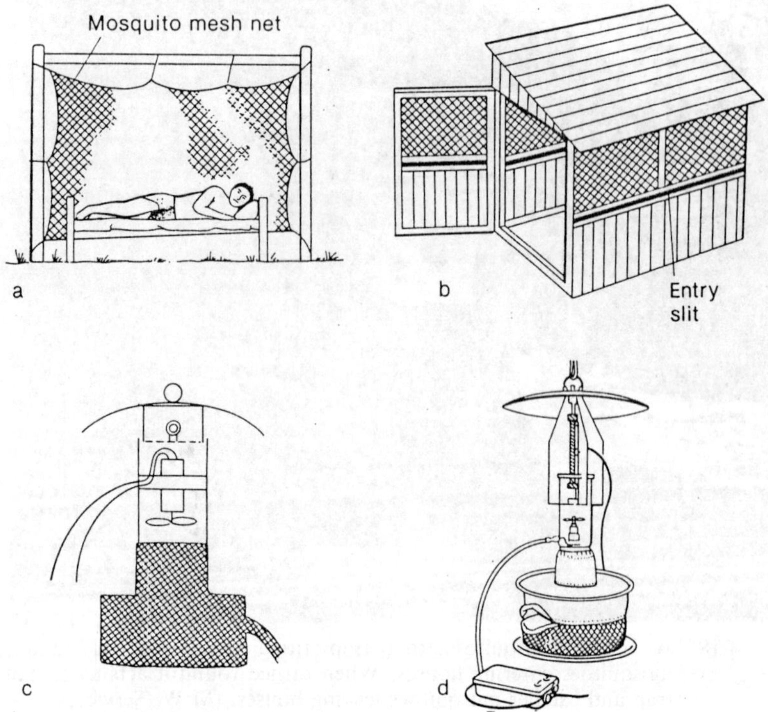

Fig. 4.19 Examples of mosquito traps: (a) mosquito bed-net raised from the ground; (b) a wooden stable (Magoon) trap which can be baited with a medium-sized animal such as a donkey, calf or goat; (c) a CDC-type battery-operated light-trap; (d) a Monks Wood battery-operated light-trap with a fluorescent tube. ((a), (b) and (c) redrawn from M.W. Service (1976); (d) redrawn from Marshall Laird (1981) *Blackflies: The Future for Biological Methods in Integrated Control*, Academic Press (London.)

Light-traps

Mosquitoes such as *Aedes taeniorhynchus*, *Ae. tritaeniorhynchus* and *Culex tarsalis* are attracted from some distance to light-traps, whereas others, such as those of the *Anopheles gambiae* complex and *Culex quinquefasciatus* (= *C. fatigans*) are not usually caught unless the trap is placed in or very near their flight paths, such as inside huts or under the eaves, in animal shelters or cattle corals, or near other aggregations of hosts. Light-traps are very selective in the species they catch, and do not of course catch day flying mosquitoes. Trap location is very critical, a change of only a few metres may greatly alter the numbers caught. Although light-traps may not give representative samples of all mosquitoes in an area, they can nevertheless be very useful in monitoring population changes of certain vectors, in addition to providing large collections of mosquitoes for arbovirus isolation studies.

There are many different designs of light-traps but the most useful two are the CDC (Fig. 4.19c) and the Monks Wood (Fig. 4.19d) traps. The former employs a small torch bulb which operates from 6-volt dry cell batteries or a motor-cycle battery. Since it does not produce a bright light, the CDC trap can be placed at night in houses to catch mosquitoes without disturbing the occupants, as well as in a variety of outdoor situations. The Monks Wood light trap has a 23 cm, 6-watt fluorescent light tube (ultraviolet or white light tubes) and operates through a transitor ballast (inverter) from a 12-volt car battery. Because of its bright light it is mainly used out of doors, but if used indoors light intensity can be reduced without affecting spectral emission by masking most of the tube with metal foil. Both traps incorporate a miniature fan and motor to suck the mosquitoes attracted to their lights into a collecting bag. The numbers of mosquitoes caught and the species composition, not only depends on trap location but on the intensity and spectral emission of the light source.

If light-traps are proved to sample adequately local man-biting mosquitoes it may be possible to use them instead of human-bait collections to sample indoor and outdoor-biting vector populations. One of the advantages of light-traps is that a single person can set up several traps in the evening which can then be left unattended until the mosquitoes are removed in the morning.

Sampling larval populations

The great diversity of different types of larval habitats and the highly aggregated distributions (see p. 133) of the immature stages makes it very difficult to standardize sampling techniques and obtain reliable sampling data. Larvae of mosquitoes which oviposit in ground collections of water, such as pools, rice fields, marshes and ponds and in water-storage pots, can be collected with a ladle (Fig. 4.20) or net, whereas larvae living in smaller habitats such as water-filled tree-holes, leaf axils and bamboo stumps have to be pipetted or siphoned out. Specialized collecting techniques are required for the larvae of *Mansonia* species which insert their siphons into submerged roots or stems of aquatic plants. They rarely detach themselves and swim up to the water surface, and consequently are not caught by normal sampling methods.

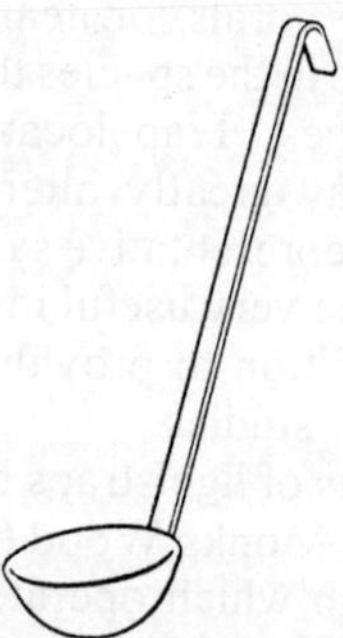

Fig. 4.20 A typical ladle or 'dipper' used for sampling mosquito larvae and pupae.

One of the main purposes of larval collections is to enable the principal larval habitats of man-biting mosquitoes to be identified so that control measures can be applied.

Sampling egg populations

Mosquito eggs are not usually sampled in mosquito surveys, but oviposition traps may sometimes be used to detect breeding of *Aedes aegypti*. A useful ovitrap consists of a small discarded food tin or glass jar painted black and partially filled with water and containing a small strip (2 X 12 cm) of hardboard termed a paddle (Fig. 4.21). This soaks up some of the water in the container and its rough side provides an attractive oviposition surface for *Ae. aegypti* and a few other *Aedes* mosquitoes. Ovitraps can be placed in various situations in villages, towns, forests, etc. and the paddles collected from them at regular intervals. The number of traps having eggs on the paddles provides a useful comparative index of seasonal variations of abundance of *Ae. aegypti* in different situations and localities.

Fig. 4.21 An *Aedes* ovitrap consisting of a tin painted black on the outside and partially filled with water. A narrow strip of hardboard is inserted as a suitable surface on which *Aedes* eggs are deposited. (M.W. Service.)

Sampling simuliid blackflies

J.F. Walsh

Sampling the immature stages

Satisfactory standardized quantitative methods for sampling the fauna of running waters are lacking and this is particularly so for blackfly larvae. *Simulium* larvae and pupae have a very patchy distribution in streams and rivers, so there may be numerous larvae on some submerged rocks, stones

Fig. 4.22 Polythene strips suspended from string tied to a small branch overhanging a stream. The plastic strips in the turbulent water act as artificial attachment sites for *Simulium* larvae and pupae. (Courtesy of R.H.L. Disney.)

and weeds while other similar substrates have few or no larvae. This can result in considerable sampling bias. Recently attempts have been made, by employing random sampling techniques, to obtain reliable quantitative samples by trying to standardize methods of searching for larvae and pupae occurring on natural substrates.

Using another approach to try to overcome the difficulties of sampling a variety of different natural attachment sites, many workers have exploited a variety of artificial substrates for the larvae to attach themselves to. The use of standardized artificial substrate units (SASUs) has unfortunately not been uniformly applied. For example, some ecologists consider that a bottom sampler, such as a ceramic tile, is adequate while others think that floating substrates such as vegetation or polythene strips are better (Fig. 4.22).

Early work with artificial substrates was concerned mainly with comparative estimates of larval density or with the reductions in infestation levels caused by larviciding. More recent studies have centred on measuring the influence of hydrological conditions, light and the presence of other animals on the distribution and movements of simuliid larvae. The depth at which larvae of the *Simulium damnosum* complex occur has also been studied. Few of the methods tried are practicable in large rivers where so many of the major pests and vector species occur (for example, *S. arcticum* in North America; *S. metallicum* and the *S. amazonicum* complex in South America; and the *S. damnosum* complex in Africa). The problems of sampling phoretic species, such as the East African *S. neavei* whose larvae and pupae are found attached to crabs, particularly those of the genus *Potamonautes*, have not been solved.

Because of the difficulties of getting reliable information on even relative population sizes of the immature stages, attention has been concentrated on sampling adults to monitor population changes, brought about by, for instance, larvicides.

Sampling adult blackflies

Blackflies are strong fliers and essentially diurnal, though many are markedly crepuscular and there may be some nocturnal activity. Normally adults do not enter buildings to rest or feed on the occupants. Although many different methods of sampling adult blackfly populations have been evaluated, few, except for bait catches, have met with much success. In most control programmes the collection of adults at human-bait has been the preferred method of assessing efficiency of control.

Human-bait collections

Adult blackflies are commonly caught by one person acting as both bait and collector. In the simplest situation a person with part of his or her body, usually the legs and feet, exposed sits on a convenient rock or tree stump and catches each blackfly in a small glass or plastic vial as it lands to feed. This system works especially well with vectors such as those of the *Simulium damnosum* complex which often are not particularly numerous and which prefer to bite close to the ground. In contrast, with species such as the Central American *S. ochraceum* which prefers to bite the head and which can occur in very large numbers, it may be necessary to use two people, one acting as bait while the other catches those settling on the head and shoulders with an aspirator or small net.

There have been many variations of the human-bait catch technique. In early work in Africa a 'fly-round' system borrowed from tsetse workers was favoured (see p. 101). It is now generally accepted that a stationary all-day catch (10 or 12 hours duration) is preferable. For not only do biting densities vary with time of the day and from season to season but flies of different physiological ages do not necessarily bite at the same time of day. Thus, to obtain valid comparative data or make assessments of transmission, standardized catches of several hours duration are usually necessary.

In the large Onchocerciasis Control Programme (OCP) in the Volta River basin of West Africa (see p. 284) bait collections are carried out by two people working alternate hours from 07.00–18.00 hours. They remain seated at a particular site and each hour's catch is labelled separately and stored in an ice-box for subsequent identification and dissection. There is much to be said for expressing the results as the mean number of flies biting a person per day, i.e. a fly/man/day density index.

Animal-bait catches

Many species of blackflies are of veterinary importance and others, including some vectors of human onchocerciasis, feed on animals as well as man. Animals, ranging from poultry to cattle have consequently often been used as baits. Frequently catches are made directly from the bait animal by a man using an aspirator or small net but this is not recommended because the presence of the collector may influence both the numbers and species of blackflies attracted, and thus caught. It is preferable to use some sort of cage or bed-net with animal baits, in other words an animal-baited trap (Fig. 4.23).

Chemical attractants

Attempts have been made, especially by Canadian workers, to find chemical attractants which could replace baits. Such studies are intrinsically difficult and so far have met with only limited success. Thus carbon dioxide has been shown to be of great value in attracting many species of blackflies, for example *Simulium venustum* and *S. vittatum* in North

Fig. 4.23 A mosquito bed-net strung between two parked Landrovers. A suitable bait animal such as a cow (shown here outside the net) is placed within the net to attract adult *Simulium*. The host-seeking females enter the net through the gap left between its lower edge and the ground. (Courtesy of M.S. Omar, and M.S. Omar *et al.* (1979) *Tropenmedizin und Parasitologie*, **30**, 157–62.)

America and *S. adersi* in Africa, but unfortunately carbon dioxide has not proved very attractive for the *S. damnosum* complex. Most striking of the results on chemical attraction was the discovery that *S. euryadminiculum* is specifically attracted to extracts of the uropygial glands of its avian host, the common loon, *Gavia immer*.

Light-traps

Many attempts have been made to sample blackflies other than hungry unfed individuals in search of a host. Rothamsted light-traps in Scotland were surprisingly successful in catching males, gravid and blood-engorged females of several *Simulium* species in addition to unfed females. These results stimulated interest in light-traps, but most subsequent studies have not been very rewarding. In Ghana, however, Monks Wood light-traps using fluorescent tubes have caught thousands of ovipositing *S. squamosum* when placed near a forest stream (Fig. 4.24), and also many *S. sirbanum* when the trap was sited near savannah breeding places. The exact location of the light-traps seems to be of great importance and this makes it difficult to use them for quantitative sampling.

Fig. 4.24 A Monks Wood light-trap placed near a river in Ghana where caught thousands of *Simulium squamosum* (a species of the *S. damnosum* complex) in a single night. Most adults were caught in the early evening when they were attracted to the river to lay their eggs. (Courtesy of M.W. Service.)

Sticky traps

Sticky traps have also been used in Africa in studies on the *S. damnosum* complex. Traps consisting of plywood boards and lengths of plastic tubing coated with sticky compounds have been hung at various heights in riverine vegetation near *S. damnosum* breeding sites. Although useful information has been obtained concerning the probable location of resting flies, the results have only rarely been commensurate with the effort involved. In Canada, tins painted blue and white, smeared with sticky material and baited with dry-ice (i.e. solid carbon dioxide) were used successfully to trap adults such as *S. venustum*, in mark–release–recapture studies which involved marking thousands of adults with radionuclides to study their dispersal.

Sheets of aluminium coated with sticky material and placed beside breeding sites in West Africa have resulted in trapping ovipositing adults of the *S. damnosum* complex. Providing such traps are adjacent to oviposition sites they may trap much larger numbers of gravid flies than unfed females caught nearby in human-bait collections. However, with minor fluctuations in water level the catch may drop to zero.

Other traps

Non-attractant traps such as suction traps also have interesting possibilities though they have been used without much success in Scotland. Recent results from West Africa suggest that even with such sampling the catch remains highly selective and the small numbers caught usually make the method unattractive.

Sampling tsetse flies

M.J. Roberts

Tsetse flies (*Glossina* spp.) have an unusual life-cycle in which the egg and larva are retained within the uterus of the maternal fly throughout growth and development. The larvae are nourished on 'milk' produced by specialized maternal milk glands and the third and final stage larva is only finally released, or borne, when fully grown. It then burrows into the soil and hardens up to form a puparium within approximately one hour of deposition. Practically speaking, therefore, it is only the puparium and the adult fly which may be sampled directly. The eggs and the larvae may only be sampled through the collection of pregnant female flies.

Puparial sampling

Tsetse puparia are found to be widely scattered in their natural habitat but they are not randomly distributed. The female fly actively seeks out suitable sites with a loose, friable soil protected from extremes of temperature and humidity, such as under leaf litter, fallen logs, overhanging rocks, and in cavities.

Fig. 4.25 Searching for tsetse fly *(Glossina)* puparia in northern Nigeria. (M.J. Roberts.)

The most widely-used technique (Fig. 4.25) is the searching of known puparial sites by hand-sorting of soil. This method is tedious, and time-consuming but skilled workers are able to recognize good sites and fairly high numbers of puparia may be found. This, is however, a highly subjective technique which is not susceptible to quantification. In an attempt to introduce a more objective element into the sampling, artificial pupariation sites have been introduced into likely habitats and some, especially logs with shading, have proved quite successful.

Another quantitative sampling technique that has been used on few occasions consists of dividing an area into 1 m squares and the soil from a random sample of squares is then examined to a depth of 5 cm. The results of some of these studies indicate that a high proportion of pupariation may take place outside traditionally recognized sites, and they give an estimate for converting the results of puparial counts to an estimate of absolute puparial population size for a given area.

Adult sampling

Fly rounds

A man walking through tsetse fly-infested countryside attracts a swarm of flies, some of which settle on his back and others on surrounding vegetation when he stops. This observation is the basis of the most commonly-used tsetse sampling technique, the 'fly round'. The fly round involves the regular collection of flies on a definite route which is usually of a length which can be comfortably covered at slow walking pace in one day, for example 8 km. The route taken depends on the type of information sought but it is often a linear transect chosen to pass through a variety of representative vegetational communities and tsetse habitats (Fig. 4.26). Other routes which have been used include straight paths, circular, polygonal, and concentric or spiral paths.

Two or three men with handnets walk along the route and stop at regular intervals of, for example, 100 m to collect all flies seen at that point whether resting on the catchers themselves or on adjacent vegetation. In East Africa, a black hessian screen draped over a pole may be carried by two men to increase the size of the target offered to the tsetse. A fly round may also be carried out by one man on a bicycle with a screen mounted behind the back wheel or by two men in a motor vehicle travelling at about 15 km/h. Flies collected on the fly round are identified, and details of sex, age, hunger-stage, and pregnancy are recorded. A characteristic of the 'following swarm' on a fly round is the very high proportion of males, many of which are not hungry but which join the swarm to mate with young, teneral females attracted to their first blood-meal. If a fly round is used for the estimation of population size, the small number of females of various ages and also the teneral males (together often no more than 20 per cent of the catch), are often discounted from computations as a variable quantity whose contribution is difficult to assess. The estimate obtained is one of relative population size or rather of 'density-activity' since it depends on attraction of a small proportion of active flies. An absolute population estimate can be obtained through mark–release–recapture studies.

A fly round which is sampled at regular intervals of, for example, once a week throughout the year, shows the importance of different vegetational communities to different species of tsetse and of any seasonal fluctuations in different vegetational zones. Fly rounds are also widely used for large-scale tsetse surveys over rough terrain where there are few or no paths, for the detection of most of the common species at low densities, and for the assessment of insecticide treatments in control campaigns.

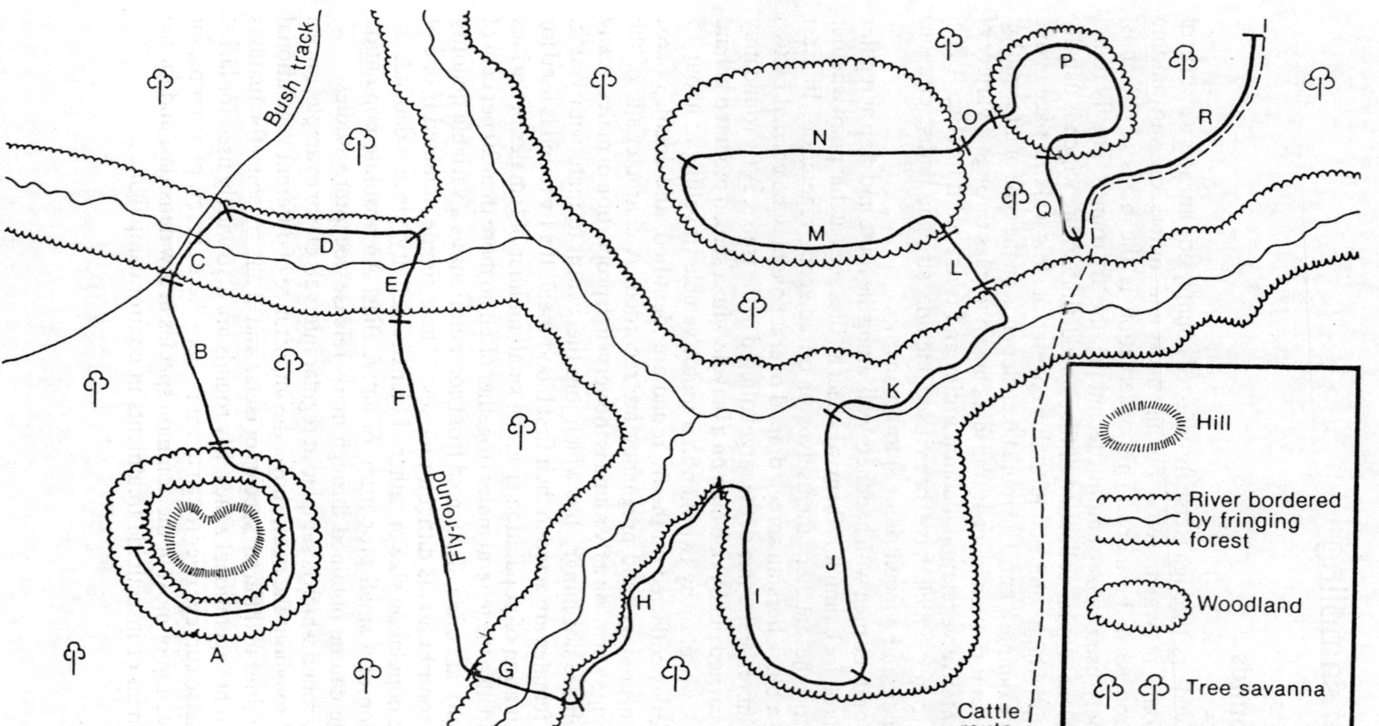

Fig. 4.26 A typical tsetse fly round consisting of 18 sections (A–R) starting amongst wooded vegetation at the bottom of a hill and continuing mainly along riverine vegetation and ending in open savanna. (From H. Davies (1977) *Tsetse Flies in Nigeria*, Oxford University Press.)

Stationary captures at bait

Fly rounds are widely used for sampling *Glossina palpalis* grp. flies and for some *G. morsitans* grp. flies such as *G. morsitans, G. swynnertoni* and *G. pallidipes*, but it has been shown that the presence of man reduces catches of *G. morsitans* and *G. pallidipes* at bait oxen, possibly due to the presence of lactic acid on human skin.

A number of large mammals are good attractants for some species of tsetse, cattle being most frequently used for this purpose. In a recent study in Zimbabwe, 2000 tsetse per hour were attracted to the odour of 11 500 kg of livestock kept in a covered pit. The fly round is not suitable for recording the diurnal rhythm of biting behaviour, and therefore stationary captures from a bait ox, using either handnets or a descending cage, are frequently used for this purpose. Bait oxen may also be the only practical method of detecting *G. fusca* grp. flies which are frequently not attracted to man.

Sentinel animals

A particular problem in tsetse sampling is proving the absence of flies or detecting very low density populations. *Glossina* populations may survive at densities of only 100 flies per km^2 and many species are not readily attracted to man or easily trapped at low density. The only practical technique in such circumstances may be to introduce a small herd of uninfected, susceptible animals, for example exotic cattle, into the area. These may require protection with wire-mesh cages during transit to ensure that they are uninfected when introduced. If they can be maintained for 12 months without developing trypanosomiasis, there is strong circumstantial evidence for the absence of *Glossina*, although there may of course be *G. palpalis* grp. flies present feeding largely on reptiles.

Traps

A skilled tsetse worker with a handnet is very efficient in the collection of flies at low and medium densities but as the density increases his efficiency declines. At medium and high densities, and for unmanned, continuous sampling, a trap is often the most useful method of sampling *G. palpalis* grp. and *G. morsitans* grp. populations. The earliest traps (e.g. Harris trap, Morris trap and Langridge trap) were designed to resemble common mammalian hosts in some aspects of size, shape and colour. Recent research on trap technology has adopted a more analytical approach with attention to individual aspects of trap design. Most traps in current use are attractant traps and the first problem is therefore to provide suitable cues to increase the number of flies visiting the trap. On arrival close-range attraction and other factors result in the flies alighting on the trap surface from which their escape must be prevented.

Fig. 4.27 A Challier-Laveissiere biconical tsetse trap. The lower blue section attracts the flies which enter the trap and fly up into white netting section and finally into the collecting cage positioned on top of the trap. (M.W. Service.)

The Challier-Laveissiere biconical trap (Fig. 4.27) has proved to be a very effective device for sampling *G. palpalis* grp. flies and for *G. pallidipes*. Its development has shown the attractiveness of dark colours, especially blue, and the importance of a dark–light interface for attracting these flies. Alighting and retention are controlled by instinctive upward movement to light and by a non-return system. Nevertheless a proportion of flies do not enter the trap or else leave it after entering.

A second line of investigation has been the development in Zimbabwe and Nigeria of electric traps in which a grid of wires at high voltage stuns insects impinging on the trap surface. These traps are not inherently attractive and therefore are combined usually with another technique. For example, they can be worn as back-packs on fly rounds or used as a catching surface with ox-odour bait. They provide very efficient retention of flies which alight and can, therefore, be used to test the efficiency of other sampling techniques, particularly in areas of high fly density.

Current progress in improving the attractiveness of traps includes the analysis of the active components of ox-odour and possible future synthesis, and the use of pheromones. The only adult pheromones detected so far, however, are essentially 'contact' pheromones having a very small range of action which probably limits their usefulness to the retention of flies which have alighted on a trap, in much the same way as the incorporation of adhesives in some tsetse traps.

Resting sites

Tsetse flies spend only about 30 minutes each day in flight and the identification of the sites where they spend the remainder of the day is, therefore, very important for control, particularly for insecticide applications and for an appreciation of the epidemiology of trypanosomiasis. Examination of resting sites is also important for the detection of engorged flies which do not come to fly rounds, baits, or traps, and for *G. fusca* grp. flies.

Two major groups of resting sites can be distinguished for *G. morsitans* and *G. palpalis* grp. flies: day-time resting sites, often low down on trunks and branches but varying considerably with species, location and season; and night-time resting sites usually on leaves. Initial location of flies at their day-time resting sites requires careful examination of the vegetation. Artificial resting sites such as refuges and pits are sometimes employed and have been shown to provide samples with a wide range of nutritional states and a reasonable sex ratio and species composition.

Location of flies at night-time resting sites has been facilitated by the release of flies sprayed with reflective paint containing minute glass beads, and more recently by the introduction of small portable ultraviolet light sources to detect flies which have been dusted with fluorescent powders and released.

Sample composition

Tsetse flies are collected for one of several conflicting, and often incompatible, objectives. Large numbers of tsetses may be required for control or eradication campaigns, based on depleting the local population by capturing the flies. Such an approach led to the eradication of *G. palpalis* on the Island of Principe (Gulf of Guinea) in 1911–14, achieved largely by men wearing black sticky screens on their backs to attract and catch the flies. The same principle applies to present research in Zimbabwe on the combination of an automatic chemo-sterilizing apparatus with an attractant trap baited with natural or synthetic ox-odour.

Sampling may also be undertaken to detect the presence of small residual populations of tsetse flies after insecticide spraying. For example, a combination of fly rounds and the age-grading of all specimens in northern Botswana in 1976 showed that the very small (less than 0.2%) residual populations of *G. morsitans* remaining after spraying endosulfan, were old pregnant females which were considerably less susceptible to the insecticide than the majority of the population. Where there are apparently no flies, confirmation can only be sought at present from the introduction of herds of uninfected, susceptible cattle, although the development of a synthetic odour attractant could replace this approach.

In the past most attention has been applied to the development of attractant traps and to increasing sample size. For ecological studies and to determine the composition and size of a wild population both prior to and after insecticide applications, however, a random unbiased sample is required which reflects the true species composition, sexual balance, and represents all age-classes and nutritional groups. In practice, such a technique is not available but recent research has attempted to rectify some of the more obvious deficiencies especially in the unequal recorded sex-ratio. A feature of most sampling techniques is an under-recording of the proportion of females due in part to their reduced activity during pregnancy.

Sampling triatomine vectors

O.P. Forattini

Triatominae bugs are sampled for several epidemiological purposes. For example, to carry out surveillance of the species and numbers in an area, to determine the vectors of Chagas' disease of which the causative organism is *Trypanosoma cruzi*, and to evaluate the effectiveness of control campaigns. In addition, efficient sampling is needed to study the behaviour and ecology of triatomine bugs. It is often necessary to sample both those species and individuals that rest indoors (domiciliary species) and also those resting in a variety of outdoor situations (including sylvatic species). Most methods used to sample triatomine vectors are simple, inexpensive and require little apparatus, but because the bugs hide away in a variety of cracks and crevices in houses and other structures they are difficult to find, and hence many sampling methods are imprecise.

Sampling indoor populations

Collections will be necessary not only in houses but in other buildings used as shelters for animals, or for the storage of food, tools and equipment, and also among piles of rubble (Fig. 4.28a). In all cases infestations may be detected by direct manual search (Fig. 4.28c) and capturing the adults and nymphs, and by noticing their faecal deposits, exuviae and empty egg

Fig. 4.28a

Fig. 4.28b

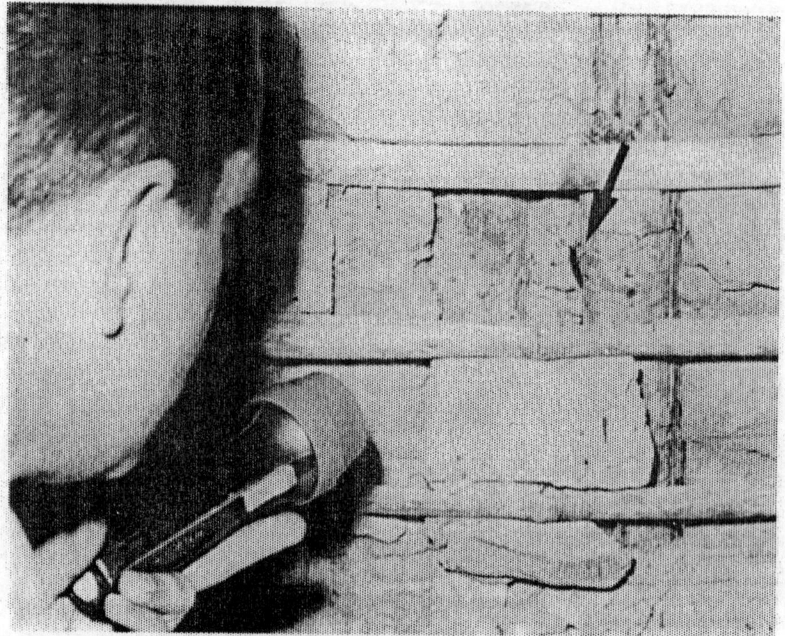

Fig. 4.28c Searching for triatomine bugs in South America. (a) searching out of doors potential resting places; (b) an experimental fowl-house to attract natural colonies of triatominae, small structures which can be dismantled easily to enable all bugs sheltering in them to be collected; (c) direct searching amongst cracks and crevices inside a house for nymphs and adults. (O.P. Forattini.)

shells. Only simple equipment such as tweezers and torches are needed for such surveys but for a thorough examination personnel must be properly trained so as to know exactly where to look. To increase the efficiency and speed of these searches potential hiding places can be sprayed with insecticides, such as pyrethrum or synthetic pyrethroids like bioresmethrin, to flush the insects out from cracks and crevices. These chemicals are applied usually as water-dispersable powders or as dusting powders.

Results of such surveys are used to establish an index of local infestation. A common procedure is to calculate the average number of dwellings (houses, animal shelters, etc.) infested with bugs out of the total number searched. The density of infestation may be calculated by dividing the number of bugs collected by the total number of dwellings searched. Another estimate, the agglomeration rate, is obtained by dividing the number of triatominae collected by the number of infested dwellings. Yet another index that is used sometimes is the 'apparent density index' which is expressed as the average number of bugs collected by one hour's search by one man, i.e. triatomine bugs/man/hour.

When sampling is being undertaken to evaluate advanced stages of control operations, it is more common to take a representative sample of dwellings in a village or area and search these rather than all dwellings. Moreover, in the houses or shelters chosen for examination it is frequently satisfactory to examine only a few favoured resting sites like bedrooms, and not all rooms.

To obtain more reliable quantitative results, the numbers of triatominae resting on a wall that has previously been measured and its surface delineated by grids can be counted. This assists in standardizing the sampling procedure and thus reduces variability in sampling. The technique of using grids is used both in general surveillance and in assessing control strategies. The presence of faecal droppings, exuviae and eggs, even hatched ones represented by their empty shells, can only be counted on specific areas of walls. These survey methods are particularly suited to infestations with highly domestic species such as *Triatoma infestans* and *Rhodnius prolixus*; the latter species also commonly occurs in the jungle.

The so-called 'Gómez-Nuñez box' (Fig. 4.29) is an artificial resting site that has proved particularly useful for sampling highly domiciliated populations. These boxes are made of cardboard and have a number of holes (usually 48) cut out of one side, and contain a sheet of folded paper inside them (Fig. 4.29 b). The boxes are hung on a nail with the perforated side touching the wall. At regular intervals during routine surveillance these attractive resting sites are examined and the presence of bugs, their faeces or exuviae counted and recorded. Gómez-Nuñez boxes provide a quick, easy, cheap and very effective method of detecting the presence of bugs in houses and other shelters. They also help greatly to standardize sampling methods and thus obtain more reliable comparative population indices. Larger collections could probably be obtained if the boxes incorporated a suitable attractant, such as pheromones, ultraviolet light or even the low intensity light emitted by radioactive betalights. Such possibilities deserve attention but as yet have been little explored.

Because most methods of collecting bugs in houses fail to find the complete indoor resting population, a few attempts have been made to collect all bugs resting inside a house by carefully dismantling it. In Brazil as many as 8000 bugs have been collected when a single house was taken apart. Clearly, however, this drastic sampling method cannot be routinely used! Attempts have also been made to estimate the total population of bugs in houses by marking a few bugs with paints, releasing them in houses and then recording the numbers of marked ones caught in samples. Such mark–release–recapture population studies have not proved very successful mainly because very few marked bugs are recaptured.

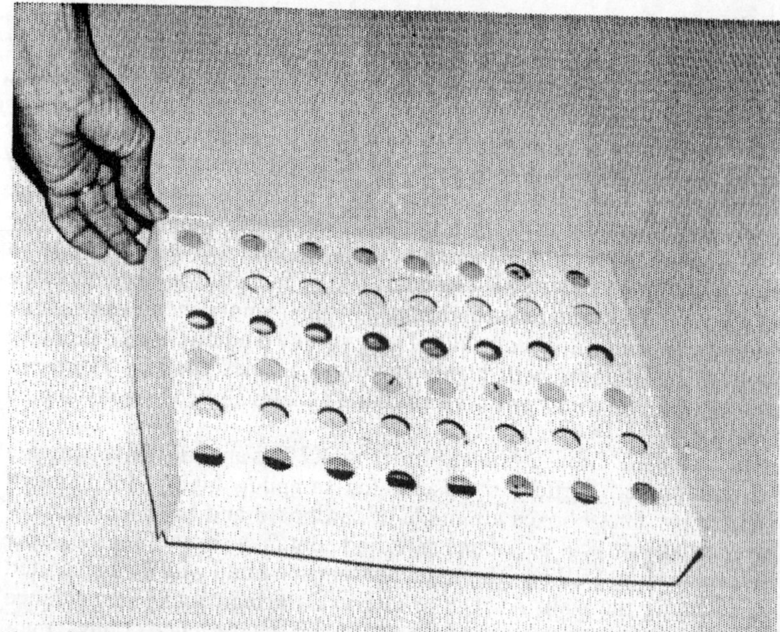

Fig. 4.29a

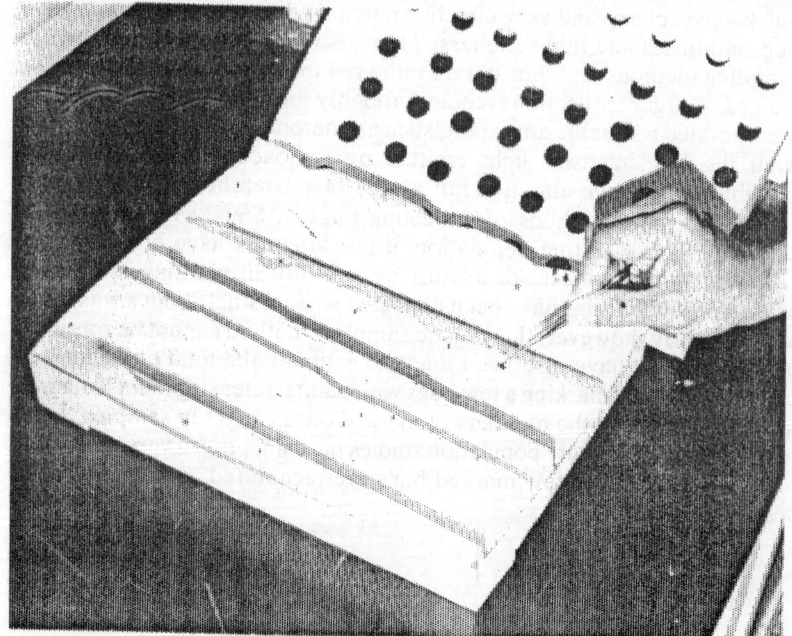

Fig. 4.29b

Fig. 4.29c The use of a Gómez-Nuñex box which acts as an artificial resting site for triatomine bugs: (a) the complete cardboard box; (b) box with lid removed to show folded paper; (c) box fixed to wall of a bedroom to act as a trap. (O.P. Forattini.)

Sampling outdoor populations

Sampling triatominae in their outdoor habitats involves examination of their natural resting places, such as birds' nests, rodent burrows, other animal shelters, tree-holes and palm trees. It is also possible to catch some species in attractant traps such as light-traps, especially those using ultraviolet light.

A profitable method used to study bug behaviour, including the tendency of wild triatominae populations to rest indoors, is based on specially built artificial shelters. One of the more practical and commonly employed shelters is a chicken house which can be easily erected and baited with poultry (Fig. 4.28b). Moreover, the simple design enables it to be readily dismantled and this allows a thorough search to be made for triatominae hiding in cracks and crevices and also enables the shelter to be transported to other sites. These shelters act as traps and several can be used at various sites, then examined at regular intervals. This method can also allow natural colonies of triatominae to become established in the shelters and facilitates studies on the development times, population dynamics and yearly cycles of important vectors of Chagas' disease.

Sampling ticks

R. Tatchell

Effective tick management has come to depend in recent years more on programmes based on ecological information than on the blind acceptance of rigid control schedules. In this way money and effort may be saved to improve the economics of the exercise and any harmful environmental impact is minimized. Significant improvements in control without risk to the livestock owner can be gained only after the collection of adequate data on which management decisions may be based. The acquisition of tick data is a largely straightforward operation and involves no advanced technology, but both parasitic and non-parasitic phases of the life-cycle need to be collected and analysed.

Life-cycles of ticks can be depicted diagrammatically as shown in Fig. 4.30. This diagram divides the life-cycle into three components, two of which are non-parasitic – the developmental stages and host-seeking

Fig. 4.30 (*opposite*) Diagrammatic representation of the different types of life-cycles and development phases of both one- and three- host ticks (Ixodidae) and multi-host soft ticks (Argasidae). In the diagram *n* is the number of nymphal instars and is variable. (Redrawn from Sutherst, Wharton and Utech (1978).)

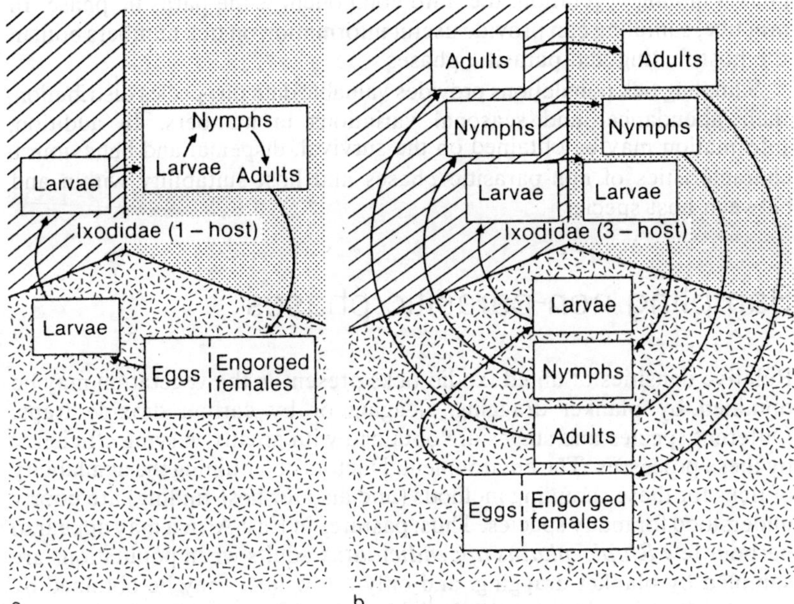

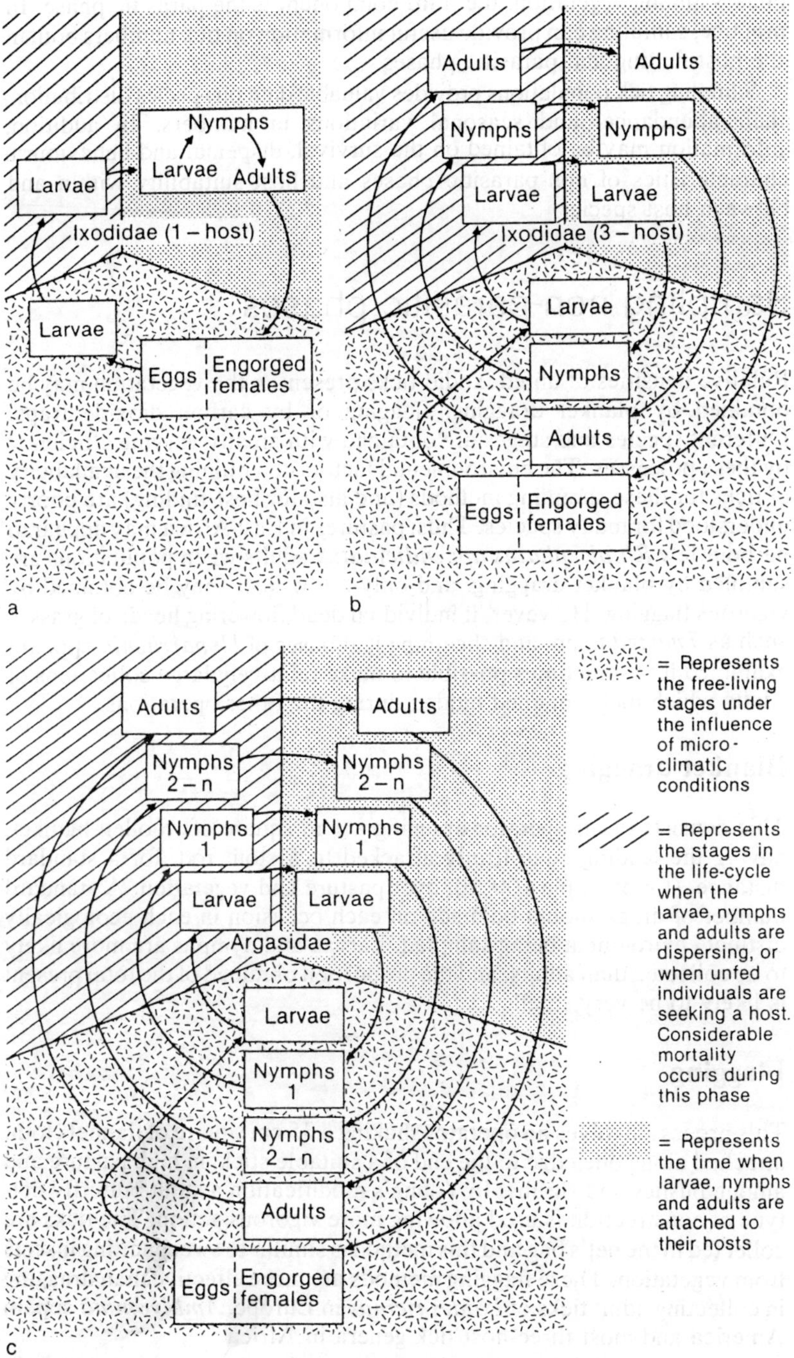

Adults
Adults
Nymphs 2 – n
Nymphs 2 – n
Nymphs 1
Nymphs 1
Larvae
Larvae
Argasidae
Larvae
Nymphs 1
Nymphs 2 – n
Adults
Eggs
Engorged females
c

= Represents the free-living stages under the influence of micro-climatic conditions

= Represents the stages in the life-cycle when the larvae, nymphs and adults are dispersing, or when unfed individuals are seeking a host. Considerable mortality occurs during this phase

= Represents the time when larvae, nymphs and adults are attached to their hosts phases among vegetation; the third component is the parasitic phase. In practice, sampling can provide useful information on tick numbers both in the host-finding and parasitic phases.

Sampling tick populations provides valuable data on species distribution, their abundance, and seasonal variations in numbers. In addition, information may be obtained on the survival, dispersal and behavioural characteristics of non-parasitic phases and host suitability within and between host species.

Sampling non-parasitic phases

Relative densities of unfed ixodid ticks present in the vegetation may be obtained by blanket dragging, flagging or by carbon dioxide traps. Different species and stages are obtained with greatly varying efficiency by these processes. The environmental factors determining the availability of ticks for these sampling methods (and also for host-finding) are poorly understood for many species. Thus, pastures in East Africa may appear to harbour few *Rhipicephalus appendiculatus* adults when populations are sampled by blanket dragging, and only a few ticks may be obtained by vigorous flagging. However, if individual dead flowering heads of grasses such as *Themeda* spp. and sheathing leaf bases of *Hyparrhenia* spp. are examined it is sometimes possible to make relatively large collections of adults which had remained hidden during the other operations.

Blanket dragging

This consists of an operator dragging a 1 m square of woollen blanket, having the leading edge firmly attacked to a stout rod, for a standard distance, (for example 100 m), over pasture and vegetation. A standard number of drags should be made on each occasion in each ecologically distinct environment being sampled. Larvae and nymphs are more likely to be collected than adults, and the proportion obtained of the total present is likely to be very small.

Flagging

This process involves an operator shaking a 35 cm square of flannel cloth, attached along one edge to the end of a suitable stick, in high pasture and among bushes and shrubs. In Young's modification a stout butterfly net-type of construction is used allowing more vigorous action. The ticks are collected in the net's bag and this method is similar to sweepnetting insects from vegetation. These flagging methods are more effective than dragging in collecting adult ticks of *Ixodes ricinus* in Europe, *Amblyomma* spp. in America and most three-host tick genera in Africa.

Carbon dioxide traps

These traps use carbon dioxide as an attractant, and basically consist of placing the gas in a container to which the ticks have easy access, for example a basin sunk flush into the soil. Ticks are stimulated to migrate to the carbon dioxide source. Adults of more active species might be expected to travel 10–20 m but the method has as a disadvantage the comparatively slow rate of progress of larvae and nymphs of many small species. The carbon dioxide source can be simply a block of solid dry-ice, or gas released through tubing from a suitable gas cylinder. The latter method allows control over the rate of release, although sometimes reduction in gas pressure causes the formation of ice and this can decrease the flow of gas making controlled regulation of carbon dioxide rather difficult. If ticks are required alive they must be prevented from contacting the solid carbon dioxide. Carbon dioxide trapping in the USA (Oklahoma) has been shown to give more consistent results and larger collections of *Amblyomma americanum* than other methods. Its relative lack of operator-induced bias suggests that this method should be used more frequently, but varying wind speeds and direction introduce other variables making standardization difficult.

Argasid ticks disperse over much smaller distances than ixodid ticks and sampling their populations usually demands knowledge of the restricted sites likely to be favoured for pheromone induced tick aggregation. Thus, in warm climates *Argas persicus* will be found in dense clusters in cracks and under the plaster of poorly constructed poultry houses, while man-biting ticks of the *Ornithodoros moubata* complex are usually found in and around houses.

Sampling parasitic phases

To evaluate the seriousness of an ixodid tick problem there are obvious advantages in making collections from hosts if these are easily sampled. Thus, for ticks of veterinary importance, cattle, sheep, goats and other hosts can be regularly examined wherever good stock handling facilties exist.

Different sites should be selected to cover the range of ecological zones in the country or study area. If possible broadly based ecological zones (such as the six proposed for East Africa by D.J. Pratt and P.J. Gwynne (1981) *Journal of Applied Ecology*, **3**, 369–82) should be used rather than those based on a multiplicity of vegetation and soil types. In East Africa, for example, ecological zones II and III almost exactly coincide with the distribution of *Rhipicephalus appendiculatus*. At each site collections should be made at least once a month throughout the year. Less frequent collecting can miss transient peaks of abundance or indeed even the presence of a tick species entirely. It is not necessary to collect from large numbers of hosts although they should come from more than one herd and, if possible tick control should be poor or lacking. Experience has shown that collecting ticks from only 3 individuals from each of 3 herds of hosts at each site can give statistically significant results. The selection of hosts should not be entirely random because very young or very old animals tend to provide much lower and much higher numbers respectively than the herd mean. Little training is required for the removal of most engorging females but skill and concentration are needed to find males of the smaller species. It is almost impossible to collect all larvae and nymphs from cattle; trained teams of collectors searching each animal for 2 hours produce counts that are probably still very low. It is better not to examine the entire host, but to clip all hair from sample areas, and then scrape the hair and skin with a sharp scalpel and transfer the debris to 70 per cent alcohol for subsequent recovery of the less easily found immature stages.

Preferably these collections should be carried out for at least five years in order to obtain reliable data on seasonal or other climatic effects on the incidence of all stages in the tick's life-cycle.

It is desirable that an assessment is also made of the favourability of hosts within and between host species (Fig. 4.31). In natural host–parasite relationships the frequency of ticks on individual hosts within a population is often uneven, with a few hosts having heavy infestations but the majority with only small numbers of ticks. Thus the distribution of ticks among hosts is skewed (see p. 306). Significant control benefits may result if it is possible to eliminate the heavily infested animals.

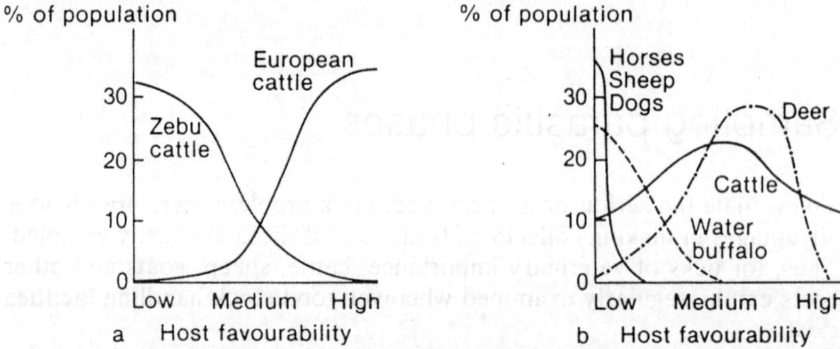

Fig. 4.31 A simplified representation of host favourability or selection: (a) within hosts of the same taxonomic groups and (b) between different species of hosts. For example, it is clear that most African Zebu cattle have few ticks and are thus more resistant to tick infestations than many European breeds. Similarly, horses, sheep and dogs are infested usually with fewer ticks than deer and some breeds of cattle. (Redrawn from Sutherst, Wharton and Utech (1978).)

The favourability of different hosts within an ixodid host species may be measured by counting: (i) the number of engorging ticks from 'standard' artificial infestations; or (ii) the number of 'standard' ticks present in infestations on naturally infested hosts. A so-called 'standard' tick may be defined as one which will engorge and detach within 24 hours. The proportion of standard ticks in an infestation on a highly favourable host should reflect exactly the number of days required for engorgement and hence the number of days on which additional ticks will have been picked up from vegetation. Any reduction in this proportion indicates a degree of unfavourability, that is host resistance. It must be emphasized that this procedure requires the use of hosts with sufficient previous tick infestations to have acquired their normal level of resistance, because all hosts are fully susceptible on their first exposure. An additional factor to be taken into account is stress, as this can cause loss of resistance in many hosts.

For *Boophilus microplus* in Australia the standard tick is a female measuring between 4.5–8.0 mm, but for other species some experimental measurements will be needed to determine the correct size range (oversized ticks represent individuals engorging the previous day but for some reason failing to detach). Once known, simple measuring gauges can be made to enable staff to make counts *in situ* or from collections that have been fixed in 68 per cent ethanol.

Collecting and counting ticks on domestic or wild animals requires: (i) adequate restraint of the host; and (ii) a systematic approach by the collectors, who should collect from a set sequence of body zones in order to avoid areas being missed.

More details of ancillary techniques and approaches can be found in Sutherst, Wharton and Utech (1978).

Sampling chigger mites

M. Nadchatram

Scrub-typhus, or chigger-borne typhus as it is sometimes known, is a febrile illness widely distributed in the Asiatic–Pacific region caused by the rickettsial organism, *Rickettsia tsutsugamushi*. The vectors are the larval stages of trombiculid mites of the genus *Leptotrombidium*, which are commonly known as chiggers.

There are approximately 2200 trombiculid species and the genus *Leptotrombidium* alone consists of over 240 species most of which occur in the Indian subcontinent and south-east Asia and adjacent islands. The subgenus *Leptotrombidium* contains more than 160 species, and all known vectors are in this subgenus. The most important vectors are *Leptotrombidium deliense, L. fletcheri, L. arenicola, L. akamushi, L. pallidum, L. pavlovskyi* and *L. scutellare*.

Scrub-typhus is closely associated with the ecology of the countryside. It is largely an occupational disease. The populations at risk are mostly rural people, such as plantation and field workers, but military personnel on field exercises are also exposed to the disease. Scrub-typhus gained military importance among both Allied and Japanese troops during the Second World War (1939–45). It is estimated that 30 000 casualties suffered from the disease with fatality rates ranging from 1–35 per cent in American troops.

Biology of Leptotrombidium mites

The life-history of the chigger is somewhat complicated (Fig. 4.32). Eggs are laid singly in the soil, debris or in nests of birds and small mammals,

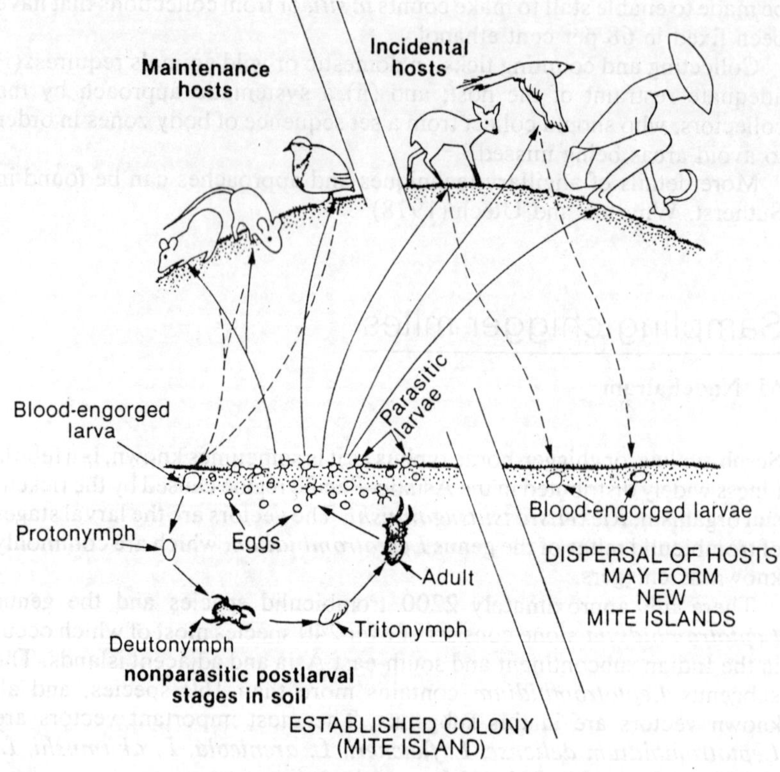

Fig. 4.32 Diagrammatic representation of the life-cycle of *Leptotrombidium* mites, the vectors of scrub-typhus in the Asiatic-Pacific region. (Redrawn and modified from J.R. Audy (1968).)

especially rodents. They go through a developmental stage called the deutovum or prelarva, after which a 6-legged larva emerges which is either pallid (i.e. almost colourless), yellow, orange or red, depending on the species. A few days after emergence the larvae attach to hosts (usually birds, rodents, or other small mammals) and feed continuously for a few days (Fig. 4.33). Blood-engorged larvae detach from their hosts and undergo a moulting stage (protonymph) and at the end of about 2 weeks active 8-legged deutonymphs emerge. The deutonymphs are usually shaped like a figure 8, are very hairy and are similar in appearance to adults, but are asexual and smaller. Within 10–14 days there is a second 'pupation' (tritonymph) and after usually 2–3 weeks the adult emerges. Both deutonymphs and adults are non-parasitic on vertebrates, feeding on tissue fluids of soft-bodied arthropods or their eggs. The duration of the life-cycle varies according to species and ecological conditions, but the widespread vector species, *Leptotrombidium deliense*, takes approximately 2 months.

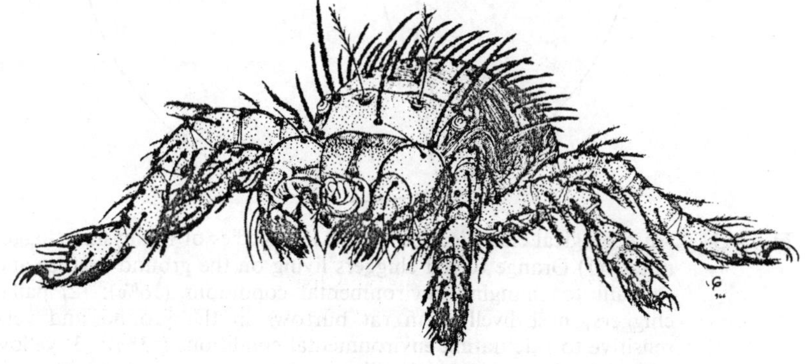

Fig. 4.33 Larva of *Leptotrombidium akamushi*, an important vector of scrub-typhus. (Courtesy of P.H. Vercammen-Grandjean.)

Vector species are found most commonly in forest clearings where secondary vegetation has become established, e.g. scrub, grassland and plantations overgrown with undergrowth. Here the unfed larvae are found in clusters on grass, leaves, and litter, or on the surface of the ground. A useful ecological classification has been developed which shows that vector species, and other species of chiggers that bite man, are ground-surface dwellers and always orange to red in colour. This classification is of great epidemiological importance because it provides useful knowledge of their natural distribution and their ability or inability to come in contact with man. In Malaysia, this ecological category of chiggers represents only 16 per cent of the 150 known *Leptotrombidium* species (Fig. 4.34). Chiggers are more habitat-specific than host-specific. For example, the three known vectors in Malaysia are *Leptotrombidium deliense, L. fletcheri* and *L. arenicola* and each occupies a habitat distinct from the others, although some overlap may occur in their distribution. *L. deliense* is common along forest fringes and in plantations such as oil palms, rubber and coconut. *L. fletcheri* occurs mainly in grassland (*Imperata cylindrica*) mixed with such bushes as *Lantana* and *Melastoma*, while *L. arenicola* is usually to be found in vegetation along sandy coastal areas.

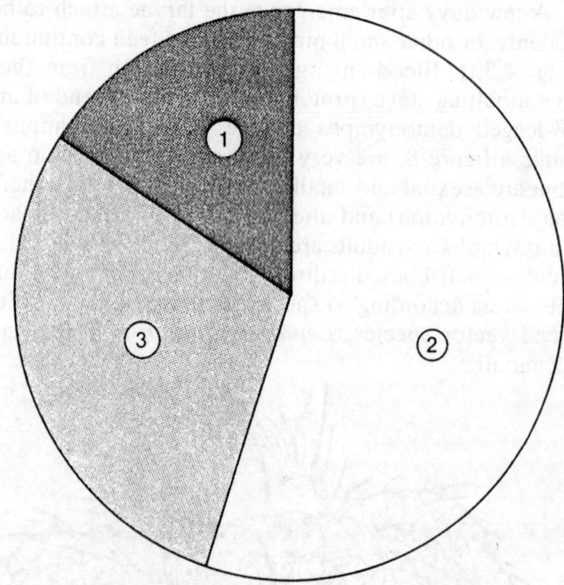

Fig. 4.34 An ecological classification of the 150 species of Malaysian chigger mites. (1) Orange to red chiggers living on the ground surface and tolerant to changing environmental conditions (16%); (2) pallid chiggers, nest-dwelling in rat burrows in the ground and very sensitive to fluctuating environmental conditions (53%); (3) yellow and pink chiggers, nest-dwelling and closely associated with their hosts (i.e. they exhibit ecological specificity) and sensitive to fluctuating environmental conditions (31%).

Sampling larval *Leptotrombidium* mites

There are various sampling techniques for collecting chiggers. The methods chosen depend upon the information required and the intended use of the chigger material. The main collecting method for systematic studies involves trapping or shooting animals and removing the parasitic larvae. (Trombiculid systematics is based almost entirely on parasitic larvae, because post-larval stages are free-living and difficult to find). The collection of animals for chiggers also provides information on host–parasite relationships and the habitats preferred. Live trapping also supplies engorged larvae for laboratory rearing of post-larval stages for systematic and biological studies.

Another simple sampling method consists of exposing small (15 X 15 cm) black formica plates in areas suspected to be infested with newly hatched larvae. Hungry unfed larvae are stimulated by, or attracted to, carbon dioxide and 1–2 minutes after putting the plates in position on the surface of the ground, larvae, if present, can be seen as tiny orange or red dots crawling over them. It is believed that the small amounts of carbon dioxide released by man when handling the plates is responsible for them attracting the chiggers. The larvae are collected easily with an artist's small paint brush and transferred to vials containing water or alcohol. *Leptotrombidium* larvae occur in pockets or little 'islands', that is they are not randomly scattered over an area but have a very patchy distribution. Black plates should therefore be placed on the ground in several different sites in a sampling area. The best times for exposing the plates are between 08.00–11.00 hours and 16.00–18.00 hours. Bowls of water placed overnight on the ground in selected sites may also trap chiggers which after climbing up and into the bowls are trapped in the water. Larvae are removed from the water the next morning.

Sampling free-living stages (nymphs and adults)

Tullgren-type equipment (Fig. 4.35) based on the Berlese funnel is effective for collecting post-larval stages especially adult trombiculids.

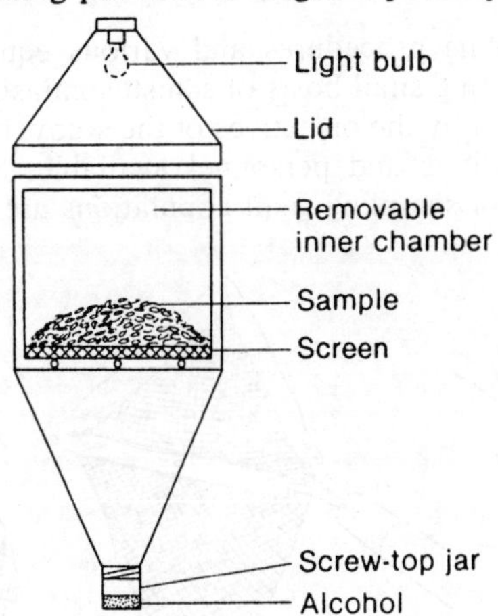

Fig. 4.35 Diagrammatic representation of a Tullgren extraction funnel. Leaf litter and debris is placed below the light bulb on the metal screen. Heat from the bulb drives the post-larval stages of *Leptotrombidium* to bury deeper in the debris and to fall eventually down the funnel into the jar containing alcohol or water.

Samples of top soil and litter collected from areas suspected to be infested with mites are placed on a screen within the funnel and a light bulb suspended above. Heat from the bulb induces the mites to burrow deeper into the substrate with the result that they eventually reach the screen, fall through, and are collected in a container placed underneath the funnel.

Alternatively if Tullgren funnels are not available, mites can be extracted from soil and litter samples by the flotation method. For this, samples are flooded with a solution having a specific gravity of 1.2 (for example, solutions containing magnesium sulphate, household salt or cane sugar). This causes soil and inorganic particles to sink while the mites together with other arthropods and botanical debris float on the surface, from which they can be removed. It may be necessary to centrifuge the immersed samples or place them under reduced pressure to aid separation. It can be a tedious process.

In summary, the two methods that have been used most widely for sampling *Leptotrombidium* mites have been the trapping of small mammals and the use of small formica black plates.

Sampling snail intermediate hosts of schistosomiasis

M.A. Amin

Several snail sampling procedures and various equipment have been developed for collecting snail hosts of schistosomiasis. Selection of any method is determined by the objectives of the study, by the nature of the habitat and by facilities and personnel available. Some of the more important methods of sampling snail populations are described here.

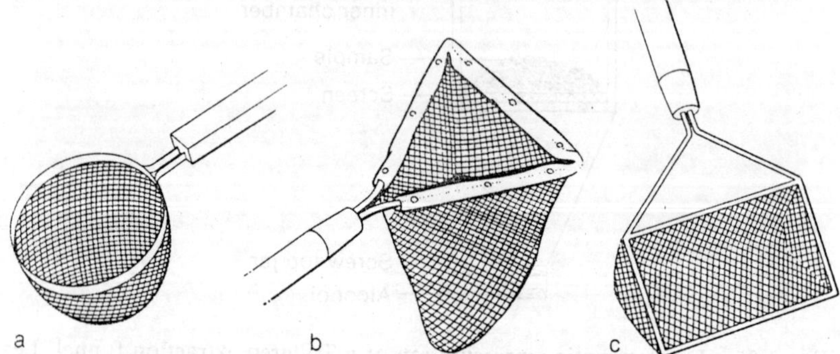

Fig. 4.36 Various simple scoops and nets for collecting aquatic snails: (a) a scoop for collecting snails at the edge of habitats and from floating vegetation; (b) an aquatic net with a strong and coarse mesh bag; (c) a scoop designed mainly for collecting snails from the bottom of ditches and ponds.

Fig. 4.37 A typical spade-like scoop being used to collect snails from ditches in an irrigated cotton field in the Sudan. (M.A. Amin.)

Sampling methods

Dip-net or scoop method

Various designs of nets and scoops have been described. Almost any convenient sieve or aquatic net with a coarse mesh as used by hydrobiologists can be used to collect aquatic snails (Fig. 4.36). Even a kitchen sieve or strainer can be useful in small accumulations of water. Essentially the dip-net, or scoop as it may also be called, consists of a wire-mesh or net scoop mounted on a strong metal frame having a long handle (Figs. 4.36a, c, 4.37). Such a scoop is useful both for collecting snails from the edges of aquatic habitats, such as irrigation canals, streams, ditches and small rain pools where snails are found in association with vegetation, and for collecting snails buried in mud at the bottom of the water (Fig. 4.36c).

Drag-scoop method

The drag-scoop consists of a deep wire-mesh net at the end of a long handle (Fig. 4.38). While the dip-net (scoop) is designed mainly to obtain samples from the edges of collections of water, the drag-scoop can be either drawn across the water surface or dragged face downwards over the bottom mud up to distances of 2–3 metres. However, sometimes a large amount of debris and other material is caught in the scoop which makes subsequent examinations for snails very laborious.

Snails collected by these two methods can be expressed as 'snails per dip'. This index has been widely used and it has the advantage of simplicity and usefulness in a variety of habitats. It is better if the habitat is divided into sampling stations and a standardized number of samples are taken from each station. After the snails have been collected, identified and counted the average number per dip is obtained. This method is commonly used to evaluate control measures. It has been shown that there is a reasonable statistical correlation between snail population density and the number of snails that can be collected from a defined area in a unit of time by a given number of persons. Thus another comparative population index is 'snails per unit of time', and results are usually given as the number of snails collected by one man in an hour.

Palm-leaf method

This method is designed mainly for habitats not amenable to sampling with the dip-net or drag-scoop, such as lakes and other large water-bodies.

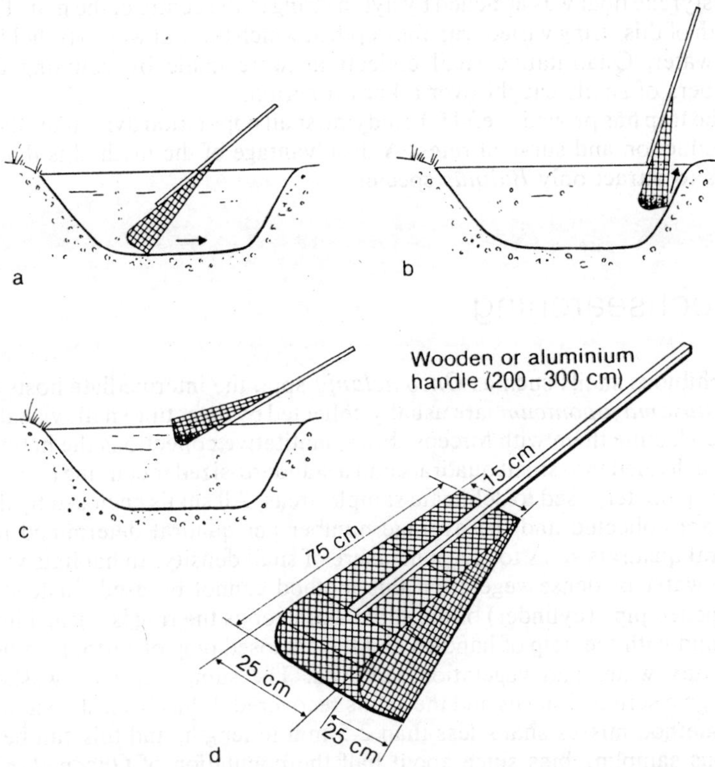

Fig. 4.38 Diagrammatic representation of a scoop specially designed to collect snails buried in the mud of aquatic habitats: (a) the scoop is turned upside down and dragged over the bottom mud; (b) it is then pulled up the sides and out of the water; and, finally, (c) the scoop is turned over and the collection washed in water; (d) details of the construction and measurements of the scoop. (Redrawn and modified from Ansari (1973).)

Basically the method consists of semi-submerging in the water a branch of a palm tree having a known number of leaves and tying it by string to a pole or a bush. Snails are attracted to the palm leaves which can be removed periodically and the number of snails counted.

A modification of this method, developed to standardize the technique, was used for estimating the density and horizontal distribution of *Bulinus truncatus rohlfsi* in the Volta Lake, Ghana. Ribs of fresh palm fronds were cut into 42.5 cm lengths and then dried. Holes were drilled in the fronds and 4 lengths joined together by wire to form a square frame, and nylon cord was then strung in 7 parallel lines between opposite sides of the frame. The day before these traps were used fresh palm leaves about 43 cm long were interwoven through the nylon cords. A stone was fixed on opposite sides to submerge the trap, but in addition an expanded polystyrene float was attached by nylon string to the centre of the mat. The length of this string varied with the depth at which the mat was to be held in the water. Quantitative snail collections were made by counting the numbers of snails caught over a known period.

The trap has proved useful for studying snail population dynamics, their reproduction and survival rates. A disadvantage of the method is that it tends to attract only *Bulinus* species.

Direct searching

Amphibious snails such as *Oncomelania* spp., the intermediate hosts of *Schistosoma japonicum*, are usually collected by detecting snails visually and collecting them with forceps. For quantitative collections the habitat can be divided into small quadrats and a standard-sized metal ring (10–15 cm in diameter) used to delineate sample areas. All snails enclosed by the ring are collected and the average number per quadrat determined for several quadrats so as to give a measure of snail density. In habitats with deep water or dense vegetation this method cannot be used. Instead a sharpened pipe (cylinder) of the same diameter as the ring is pushed into the mud with the help of handles and the enclosed plug of earth, together with any water and vegetation removed. The sample is then washed through a series of sieves and the snails recovered. It has been shown that this method misses snails less than 2.5 mm in length, and this can be a serious sampling bias since about half the population of *Oncomelania* may consist of small snails such as this.

Other methods

Usually one or more of the above techniques are used in snail surveys and to evaluate the effectiveness of schistosomiasis control operations aimed at reducing snail populations. Sometimes, however, larger and more complicated equipment is employed to collect snails from deep water, for example standard limnological apparatus having hinged jaws acting as grabs such as the Petersen grab or the Ekman bottom sampler.

Snail habitats vary greatly from place to place and also seasonally, due in part to changing water level, and this makes it very difficult to devise a method that is efficient in all habitats and for all snail species and then to standardize it. At present this ideal has not been achieved and therefore sampling methods tend to change according to local environmental conditions and this sometimes makes comparisons between different sampling programmes difficult.

Recommended reading

Theoretical aspects of sampling

Begon, M. (1979) *Investigating Animal Abundance. Mark–recapture for Biologists*, Edward Arnold, London, 97 pp.

Cochran, W.G. (1963) *Sampling Techniques*, 2nd ed., Wiley, New York, 413 pp.

Elliott, J.M. (1977) *Some Methods for the Statistical Analysis of Samples of Benthic Invertebrates*, Freshwater Biological Association Scientific Publication No. **25**, 160 pp.

Morris, R.F. (1960) 'Sampling insect populations'. *Annual Review of Entomology,* **5**, 243–64.

Service, M.W. (1976) *Mosquito Ecology. Field Sampling Methods*, Applied Science, London, 583 pp.

Southwood, T.R.E. (1978) *Ecological Methods, with Particular Reference to the Study of Insect Populations, 2nd ed., Chapman & Hall, London, 524 pp.*

Sampling crop pests

Bardner, R. and Fletcher, K.E. (1974) 'Insect infestations and their effects on growth and yield of field crops: a review'. *Bulletin of Entomological Research*, **64**, 141–60.

Chiarappa, L. (1971) *Crop Loss Assessment Methods: FAO Manual on the Evaluation and Prevention of Losses by Pests, Diseases and Weeds*, FAO and Commonwealth Agricultural Bureaux, 276 pp; Supplement I, 1973; Supplement II, 1977.

Gomez, K.A. and Gomez, A.A. (1976) *Statistical Procedures for Agricultural Research with Emphasis on Rice,* International Rice Research Institute, Manila, 294 pp.

IRRI, (undated) *Standard Evaluation System for Rice,* International Rice Research Institute, Manila, 64 pp.

Nashida, T. and Torii, T. (1970) *Handbook of Field Methods for Research on Rice Stem Borers,* International Biological Programme Handbook No. **14**, Blackwell, Oxford, 132 pp.

Southwood, T.R.E. (1978) *Ecological Methods, with Particular Reference to the Study of Insect Populations*, 2nd ed., Chapman & Hall, London, 524 pp.

Strickland, A.H. (1961) 'Sampling crop pests and their host' *Annual Review of Entomology*, **6**, 201–20.

Crop assessment

Bardner, R.and Fletcher, K.E. (1974) 'Insect infestations and their effects on growth and yield of field crops: a review'. *Bulletin of Entomological Research*, **64**, 141–60.

Chiang H.C. (1979) 'A general model of the economic threshold level of pest populations'. *FAO Plant Protection Bulletin*, **27**, 71–3.

Chiarappa, L. (1971) *Crop Loss Assessment Methods: FAO Manual on the Evaluation and Prevention of Losses by Pests, Diseases and Weeds,* FAO and Commonwealth Agricultural Bureaux, 276 pp; Supplement I, 1973; Supplement II, 1977.

Farrington, J. (1977) 'Economic thresholds of insect pest infestation in peasant agriculture: a question of applicability'. *PANS*, **23**, 143–8.

Judenko, E. (1973) 'Analytical methods for assessing field losses caused by pests on cereals with and without pesticides'. *Tropical Pest Bulletin,* **2**, 31 pp.

Khosla, R.K. (1977) 'Techniques for assessment of losses due to pests and diseases of rice'. *Indian Journal of Agricultural Science*, **47**, 171–4.

Metcalf, R.L. and Luckmann, W.H. (eds.) (1975) *Introduction to Insect Pest Management*, Wiley, New York, 585 pp.

Norton, G.A. (1976) 'Analysis of decision making in crop protection'. *Agroecosystems*, **3**, 27–44.

Ordish, G. (1952) *Untaken Harvest: Man's Loss of Crops from Pest, Weed and Disease*, Constable, London, 171 pp.

Walker, P.T. (1977) 'Crop losses: some relations between infestation, cost of control and yield in pest management'. *Mededelingen van de Faculteit Landbouwwetenschappen Rijsuniversiteit Gent*, **42**, 919–26.

Sampling mosquitoes

Service, M.W. (1976) *Mosquito Ecology. Field Sampling Methods,* Applied Science, London, 583 pp.

Service, M.W. (1977) 'A critical review of procedures for sampling adult mosquitoes'. *Bulletin of Entomological Research*, **67**, 343–82.

Sampling blackflies

Disney, R.H.L. (1972) 'Observations on sampling pre-imaginal populations of blackflies (Dipt., Simuliidae) in West Cameroon'. *Bulletin of Entomological Research*, **61**, 485–503.

Service, M.W. (1977) 'Methods for sampling adult Simuliidae, with special reference to the *Simulium damnosum* complex'. *Tropical Pest Bulletin*, **5**, 48 pp.

Service, M.W. (1979) 'Light-trap collections of ovipositing *Simulium squamosum* in Ghana'. *Annals of Tropical Medicine and Parasitology*, **73**, 487–90.

Walsh, J.F. (1978) 'Light-trap studies on *Simulium damnosum* s.l. in northern Ghana'. *Tropenmedizin und Parasitologie*, **29**, 492–6.

Walsh, J.F. (1980) 'Sticky-trap studies on *Simulium damnosum* s.l. in northern Ghana'. *Tropenmedizin und Parasitologie*, **31**, 479–86.

Walsh, J.F. Davies, J.B., Le Berre, R. and Garms, R. (1978) 'Standardization of criteria for assessing the effect of *Simulium* control in onchocerciasis control programmes'. *Transactions of the Royal Society of Tropical Medicine and Hygiene*, **72**, 675–6.

Wotton, R.S. (1977) 'Sampling moorland stream blackfly larvae (Diptera: Simuliidae)'. *Archives für Hydrobiologie*, **79**, 404–12.

Sampling tsetse flies

Challier, A. (1977) 'Trapping technology'. In M. Laird. (ed.) *Tsetse: The Future for Biological Methods in Integrated Control*, International Development Research Centre-**077e**, Ottawa, 109–23.

Glasgow, J.P., and Phelps, R.J. (1970) 'Methods for the collecting and sampling of *Glossina*'. In H.W. Mulligan, (ed.) *The African Trypanosomiases*, Allen and Unwin, London, 395–415.

Hargrove, J.W. (1977) 'Some advances in the trapping of tsetse (*Glossina* spp.) and other flies'. *Ecological Entomology*, **2**, 123–37.

Jordan, A.M. (1974) 'Recent developments in the ecology and methods of control of tsetse flies (*Glossina* spp.) (Dipt., Glossinidae) - a review'. *Bulletin of Entomological Research*, **63**, 361–99.

Koch, K., and Spielberger, U. (1979) 'Comparison of hand-nets, biconical traps and electric trap for sampling *Glossina palpalis palpalis* (Robineau-Desvoidy) and *G. tachinoides* Westwood (Diptera:Glossinidae) in Nigeria'. *Bulletin of Entomological Research*, **69**, 243–53.

Vale, G.A., and Phelps, R.J. (1978) 'Sampling problems with tsetse flies (Diptera:Glossinidae)'. *Journal of Applied Ecology*, **15**, 715–26.

Sampling triatomine bugs

Forattini, O.P., Ferreira, O.A., Rocha e Silva, E.O. da and Rabello, E.X. (1973) 'Aspectos ecólogicos da tripanossomose americana. V - Observações sobre colonização espontânea de triatomińeos sil vestres em ecótopos artificiais, com especial refereñcia ao *Triatoma sordida'. Revista de Saude Publica*, Sao Paulo, **7**, 219–39.

Jimenez, J.S. (1974) 'Control de la enfermedad de Chagas'. Venezuela. Direccion de Malariologia y Saneamiento Ambiental. In *Bases ara un lano Interamericano de Saneamiento Ambiental*, Caracas, Pub. No. **70–03**, 98–106.

Minter, M.M., Minter-Goedbloed, E., Marsden, P.D., Miles, M.A. and Macedo, V. (1973) 'Domestic risk factor – an attempt to assess risk of infestion with *Trypanosoma cruzi* in houses in Brazil'. *Transactions of the Royal Society of Tropical Medicine and Hygiene*, **67**, 290.

Sampling ticks

Sutherst, R.W., Wharton, R.H. and Utech, K.B.W. (1978) *Guide to Studies on Tick Ecology*, Commonwealth Scientific and Industrial Research Organization, Australia, Division of Entomology, Technical Paper No. **14**, 59 pp.

Sampling scrub-typhus mites

Audy, J.R. (1968) *Red Mites and Typhus*, Athlone Press, London, 191 pp.

Nadchatram, M. (1968) 'A technique for rearing trombiculid mites (Acarina) developed in a tropical laboratory'. *Journal of Medical Entomology*, **5**, 465–9.

Nadchatram, M. (1970) 'Correlation of habitat, environment and color of chiggers, and their potential significance in the epidemiology of scrub typhus in Malaya (Prostigmata: Trombiculidae)'. *Journal of Medical Entomology*, 7, 131–44.

Suzuki, T. (1953) 'Studies on the bionomics of chemical control of tsutsugamushi (scrub typhus mites). Part 1. *Trombicula akamushi* (Brumpt) in Northwestern Honshu, Japan'. *Japanese Journal of Experimental Medicine*, **23**, 487–516.

Traub, R. and Wissemann, C.L. (1974) 'The ecology of chigger-borne rickettsiosis (scrub-typhus)'. *Journal of Medical Entomology*, **11**, 237–303.

Sampling aquatic snails

Ansari, N. (ed.) (1973) *Epidemiology and Control of Schistosomiasis (Bilharziasis)*, S. Karger, Basel, 752 pp.

Brown, D.S. (1980) *Freshwater Snails of Africa and their Medical Importance*, Taylor and Francis, London, 430 pp.

Chu, K.Y. and Vanderburg, J.A. (1976) 'Techniques for estimating densities of *Bulinus truncatus rohlfsi* and its horizontal distribution in Volta Lake, Ghana'. *Bulletin of the World Health Organization*, **54**, 411–6.

Hairston, N.G. Hubendick, B., Watson, J.M. and Olivier, L.J. (1958) 'An evaluation of techniques used in estimating snail populations'. *Bulletin of the World Health Organization*, **19**, 661–72.

Jordan, P. and Webbe, G. (1969) *Human Schistosomiasis*, William Heinemann Medical Books, London, 212 pp.

5 Quantitative procedures after sampling

This chapter is divided into two parts. The first considers some problems associated with the treatment of sampling data after they have been collected, particularly the problems of aggregated distributions, which are the main type that pest managers are likely to face. It is not intended to be a section on statistics, but rather a step in getting from the collection of sampling data to a position from which normal statistical methods can be applied. The second part of the chapter introduces the concepts of population modelling through an introduction to the basic principles of population dynamics. It goes far enough to show some of the pitfalls that await us in attempting to control populations without a good knowledge of their population dynamics.

Treatment of sample data

D.J. Thompson

Some basic terms

Since much of what follows is concerned with the relationship between the mean ($\bar{x}$) and variance (s^2) of a set of samples, we must first define what we understand by these basic terms.

The arithmetic mean is one of a number of measures of central tendency. It is the sum of the counts (Σx) in a sample divided by the number of sampling units (n).

$$\bar{x} = \frac{\sum x}{n}$$

On each sampling occasion there is never more than a limited number of measurements. We assume, however, that the sample is part of a population which exhibits normal variation. Each count (x) differs from the estimated arithmetic mean, $\bar{x}$, by a certain amount, called the deviation. The variance, s^2, of a sample is the mean of the squares of these deviations.

$$s^2 = \frac{\sum(x - \bar{x})^2}{n - 1}$$

The normal computational formula is:

$$s^2 = \frac{\sum x^2 - (\Sigma x)^2/n}{n - 1}$$

The standard deviation, s, is the square root of the variance.

The dispersion of a population

The dispersion of a population describes the pattern of distribution of individuals in space. The dispersion of a population is second in importance to population size in population dynamics studies. It is, for example, of vital importance in predator–prey and host–parasite interactions. One of the features of host–parasite systems which leads to their stability is the ability of some parasites to devote most of their search time to searching in areas of high density where hosts are aggregated. As we have already indicated in Chapter 4, the dispersion of a population influences the sampling programme required to produce population estimates of a given level of precision.

The individuals of a population follow one of three basic types of spatial distribution depending upon the relationship between the variance and the arithmetic means:

Random distribution There is an equal probability of an organism occupying any point in space so that the presence of one individual does not influence the presence or absence of another. The variance of the population (σ^2) is equal to the population mean, (μ) in a random distribution ($\sigma^2 = \mu$).

Regular distribution This is an under-dispersed, or uniform, distribution where the variance is less than the mean ($\sigma^2 < \mu$).

Contagious distribution This is an over-dispersed, clumped or aggregated distribution. The name was originally derived from epidemiological studies arising, no doubt, from the localized spread of an infection to contacts in the immediate vicinity of an infected individual. Most arthropod populations that have been studied conform to this type of distribution in which the variance is greater than the mean ($\sigma^2 > \mu$).

The three basic types of distribution are shown in Fig. 5.1. It is worth pointing out that the three types of distributions can overlap to some

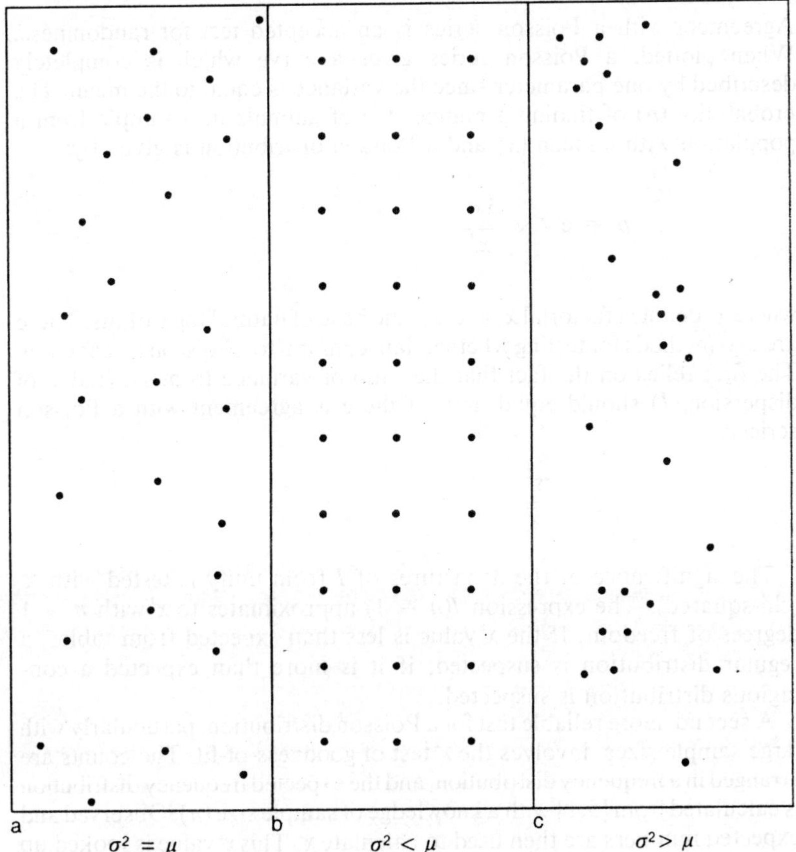

Fig. 5.1 Diagrammatic representation of three different types of distribution which may be found in insect populations: (a) here the insects are randomly distributed; (b) an unlikely situation where the insects are evenly spaced and thus have a regular distribution; (c) a contagious distribution where insects are aggregated into clumps.

extent. For example, the detection of a contagious distribution depends to some extent on the degree of clumping in relation to the size of the sampling unit.

Since there are well known mathematical descriptions of the three possible relationships between variance and mean, these mathematical models can be applied to the appropriate dispersion patterns. Before we come to the main problem with which a pest or vector manager is likely to be confronted, that of contagious distributions, we must first consider random and regular distributions.

Random distributions

Agreement with a Poisson series is an accepted test for randomness. When plotted, a Poisson series gives a curve which is completely described by one parameter since the variance is equal to the mean. The probability (p) of finding a number (x) of animals in a sample from a population with a mean (x) and a Poisson distribution is given by:

$$p_x = e^{-\bar{x}} \times \frac{\bar{x}^x}{x!}$$

where $x!$ denotes factorial x, and e is the base of natural logarithms. There are two methods for testing whether data conform to a Poisson distribution. The first relies on the fact that the ratio of variance to mean (index of dispersion, I) should equal unity if there is agreement with a Poisson series.

$$I = \frac{s^2}{\bar{x}}$$

The significance of the departures of I from unity is tested with χ^2 (chi-squared). The expression $I(n - 1)$ approximates to χ^2 with $n - 1$ degrees of freedom. If the χ^2 value is less than expected from tables, a regular distribution is suspected; if it is more than expected a contagious distribution is suspected.

A second, more reliable test for a Poisson distribution, particularly with large sample sizes, involves the χ^2 test of goodness-of-fit. The counts are arranged in a frequency distribution, and the expected frequency distribution is calculated from [5.1] with a knowledge of sample size (n). Observed and expected numbers are then used to calculate χ^2. This χ^2 value is looked up in tables with $n-2$ degrees of freedom. The null hypothesis is that the data do conform to a Poisson distribution. If the χ^2 value obtained is greater than the appropriate one in the tables, we reject this null hypothesis.

Regular distributions

The characteristic feature of a regular distribution is the uniform spacing of individuals in the population. Such a distribution tends to occur when the individuals are relatively crowded and tend to move away from one another. Therefore unless population density is extremely high, a regular distribution will rarely describe the dispersion of a population over a large area, but may describe it over a small area.

The positive binomial distribution is a suitable model for a regular distribution. The expected frequency distribution is given by $n\,(q+p)^k$ where n = number of sampling units, p = probability of any point in the sampling unit being occupied by an individual, $q = 1-p$, and k = maximum possible number of individuals a sampling unit could contain.

The parameters, k, p and q must be estimated from the sample. The best estimates of k, p and q are given by:

$$k = \frac{\bar{x}^2}{\bar{x} - s^2} \;;\quad q = \frac{\bar{x}}{k} \;;\quad q = 1 - p$$

Expected frequencies can then be compared with observed frequencies using the χ^2 goodness-of-fit test (as on previous page).

Contagious distributions

Most commonly in studying pests and vectors the variance will be found to be larger than the mean, that is the distribution is contagious. There are always definite clumps or patches of individuals in a contagious distribution, but there is much variation about this basic pattern. It is not surprising that so many arthropod species whose distribution depends so closely on environmental factors, themselves unevenly distributed, should conform to contagious distributions. In addition many species form aggregations regardless of environmental conditions. Of the many mathematical models for contagious distributions, the negative binomial distribution seems to be the most flexible. There are at least four ways in which a negative binomial can arise.

(a) True contagion, where the presence of one individual in a sampling unit increases the chance that another will occur there also. For example insects who lay their eggs in masses as opposed to individually will produce a contagious distribution.

(b) Constant birth, death and immigration rates, when the first two are expressed per number of individuals and the third per unit of time, will lead to a population whose size will form a negative binomial series.

(c) Clumps may be distributed randomly. If a number of clumps are distributed as a Poisson, but the number of individuals per clump follows a logarithmic distribution, the resulting distribution (per unit area) will be a negative binomial.

(d) In heterogeneous Poisson sampling , when the mean of a Poisson varies randomly from occasion to occasion, under certain conditions a negative binomial results. One example in which a negative binomial was shown to arise in this way was when a series of counts of gall-wasps on chestnut trees were distributed as a Poisson for each single tree, but when counts from all trees were combined they were described by a negative binomial. The flexibility of the negative binomial distribution is a very attractive feature, but it also serves as a warning that it is unwise to attempt to analyse the details of the biological processes involved in generating a distribution from the mathematical model it can be shown to fit.

The probability series of the negative binomial distribution is given by the expansion of $(q-p)^{k}$ where $p=\mu/k$ and $q=1+p$. The two parameters of the distribution are the arithmetic mean, μ and the exponent k. Unlike in the positive binomial distribution, k is no longer the maximum possible number of individuals a sampling unit could contain. This time k is a measure of the amount of clumping and is often referred to as the dispersion parameter. For the most part, values of k are in the region of 2. As k increases the distribution approaches a Poisson distribution. Fractional values of k (as clumping increases) indicate a distribution converging on the logarithmic series.

The value of k has been computed by a number of methods. Two methods are described below. The first is generally used to provide an approximate value of k which can be used in the second method to obtain a more efficient estimate of k.

$$k \doteq \frac{\bar{x}^2}{s^2 - \bar{x}}$$

This method is not very efficient for values of k below 4, unless the mean is also less than 4. This virtually limits the use of this method to low density populations.

$$n \log_e\left(1 + \frac{\bar{x}}{k}\right) = \sum\left(\frac{A_x}{k + x}\right)$$

where A_x is the sum of all frequencies of sampling units containing more than x individuals, for example $A_6 = x\ (f_7 + f_8 + f_9 \ldots)$, and n is the total number of sampling units. Different values of k must be tried in the equation until the two sides balance. If the left-hand side is too large, then the approximate estimate of k is too large.

Once x and k have been estimated, they can be substituted into the following equation for different values of x in order to obtain the expected frequency distribution:

$$P_{(x)} = \left(1+\frac{\bar{x}}{k}\right)^{-k} \frac{(k+x-1)!}{x!(k-1)!}\left(\frac{\bar{x}}{\bar{x}+k}\right)^{x}$$

where P_x is the probability of x individuals in a sampling unit. The expected numbers are then given by nP_x. Once again observed and expected numbers can be compared with the χ^2 goodness-of-fit test (this time with $n-3$ degrees of freedom). The null hypothesis is that the data do conform to a negative binomial distribution.

The normal distribution and transformations

The normal or Gaussian distribution is the distribution associated with continuous variables, which when frequency is plotted against amount gives a symmetrical bell-shaped curve. The normal distribution is often suitable for measurements such as the heights of men and women, but is seldom a suitable model for counts. It is certainly of no use in describing dispersion. The importance of the normal distribution is that a large number of statistical methods are associated with it. For many statistical techniques such as correlation, analysis of variance, t-tests, etc. the frequency distribution must have the following properties: (i) the variance is independent of the mean; (ii) the components of the variance are additive; and (iii) the data must follow a normal distribution.

In order to be able to use statistical methods based on the normal distribution, it is necessary therefore, to transform the data to another scale. To transform data is to replace actual numbers with a suitable mathematical function so that the transformed data are normalized. The choice of the correct transformation to use depends upon the original frequency distribution of the data. It is essential to transform data, however complicated the transformation may be, if the assumptions listed above are not met and if it is intended to use the statistics of the normal distribution. Before going on to consider some transformations, in detail, we must first look further into the relationship between the mean and variance. Taylor's power law states that the variance of a population is proportional to a fractional power of the arithmetic mean.

$$s^2 = a\,\bar{x}^{\,b}$$

where a and b are constants, a is largely a sampling factor depending on the size of the sampling unit, while b is an index of dispersion. The parameters a and b can be estimated by linear regression once the mean and variance of a set of samples have been converted to logarithms.

$$\log s^2 = \log a + b \log \bar{x}$$

In other words log variance is plotted against log mean. The parameter b varies from 0 for a regular distribution, through 1 for a random distribution (when a also is equal to 1), to infinity for a highly contagious distribution.

Once b has been estimated, it is possible to apply a transformation to the original data. The appropriate transformation is to replace each x by x^p where $p = 1 - b/2$. Taylor's power law seems to hold over a wide range of means and variances, so once b has been calculated, it is a simple matter to calculate p and thus transform the values of x. Some common transformations are, for a Poisson, $b = 1, p = 0.5$, therefore use the square roots of the counts, and for a log-normal distribution, $b = 2, p = 0$, use a logarithmic transformation. Tables exist of transformations for a range of values of p and for negative powers. It should be remembered that although based on data, these are precise transformations which validate the assumptions of the normal distribution. There are, however, some practical difficulties to consider. For example, in order to permit zero counts in logarithmic transformations, a constant, usually 1 is added to the original count. So that, if a count was x, the transformed count would be $\log (x + 1)$. Although Taylor's power law gives the most appropriate transformation, biologists rarely go to the extent of calculating such precise transformations for their data. Instead they rely on the general rule that a suitable transformation for samples collected from a regular distribution is the square of the original counts, likewise data from random or slightly contagious distributions can be adequately transformed with square roots, while a logarithmic transformation is usually suitable for a highly contagious distribution.

One additional transformation not associated with the dispersion, but against which a pest manager may be faced is the arcsin or angular transformation which is appropriate for proportions. This transformation is necessary since in a binomial distribution, the variance would be a function of the mean ($\sigma^2 = pq/k$, where $p = \mu$, $q = 1 - p$ and k is a constant for any one problem). In the angular scale, proportions near 0 or 1 (percentages near 0% or 100%) are spread out so as to increase their variance. Proportions are replaced by the angle whose sine is the square root of the proportion. Tables for the angular transformation can be found in many statistics textbooks. It is not considered worthwhile to make this transformation if all the proportions which make up the data lie between 0.2 and 0.8.

There are inevitably some disadvantages associated with the transformation of data. Once the arithmetic mean of the transformed data is back-transformed to the original scale, it may differ from the original mean. For example, the means in a square root transformation are reconverted to the original scale by squaring. Then they give values lower than the original means because the mean of a set of square roots is less than the square root of the original mean. However, small adjustments can be made to the *derived* means to render them comparable. These are described in standard texts such as those of Elliott (1977) and Siegel (1956). Since the transformations discussed above are non-linear, confidence limits computed in the transformed scale and changed back to the original scale would be asymmetrical. Stating the standard error in the original scale would therefore be misleading. It is most sensible in citing results from transformed data to provide means on the original scale followed by their asymmetrical confidence limits.

Interpretation of sample data

D.J. Rogers

All animal populations have a characteristic mean level of abundance, and changes about this mean can be related to seasonal or life-history phenomena. Ways of estimating levels of abundance were discussed in the previous chapter, and the present account outlines methods of interpreting changes in abundance that occur through time.

The methods of analysis to be described here were developed during essentially theoretical studies on animals which could rarely be called pests, but these methods now have immense practical relevance for the reasons discussed in Chapters 2 and 3. During the initial period of pest management, when only one control method was applied at any one time (insecticides, *or* parasites, *or* resistant varieties of plants), the empirical philosophy of 'the more, the better' was generally, though not always, the optimal one. Now that we are thinking in terms of integrating a variety of control methods (for example, pesticides and natural enemies) for any single species of pest, the question arises as to how much of each should we use, and in what order. The more sophisticated our integrated strategies become, the greater is the need for a basic understanding of the natural regulation of the populations under attack. This is where population analysis and models can help. It is no more realistic to expect to control a population with insecticides but with no knowledge of natural population regulation than it is to repair a motor car with a hammer but no idea of the workings of the internal combustion engine.

Setting the scene

Populations are dynamic units that increase through the processes of *birth* and *immigration* and decrease through the processes of *death* and *emigration*. Fig. 5.2 shows several relationships between the *rate* of population increase (i.e. the combined effects of birth and immigration) or decrease (death and emigration) and population size. All populations necessarily possess at least one of each sort of process, and these can be combined with the consequences listed in the lower part of Fig. 5.2. Clearly there can be regulation of a population at a characteristic equilibrium level in only a few of the possible combinations (cases A and B in the figure); other combinations lead to population explosion or extinction, or an equilibrium that is unstable.

In all cases where an equilibrium is both present and stable we find a

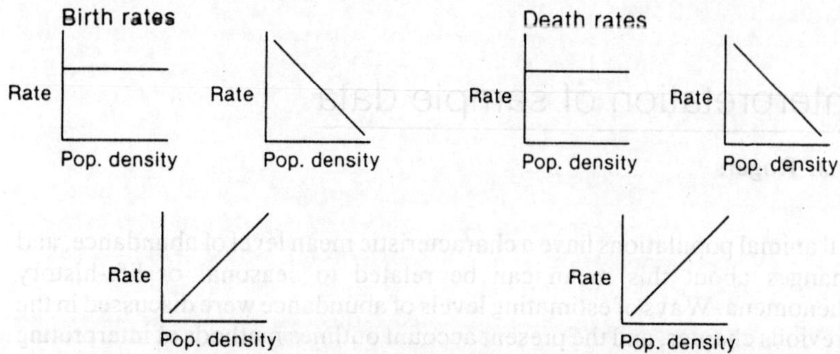

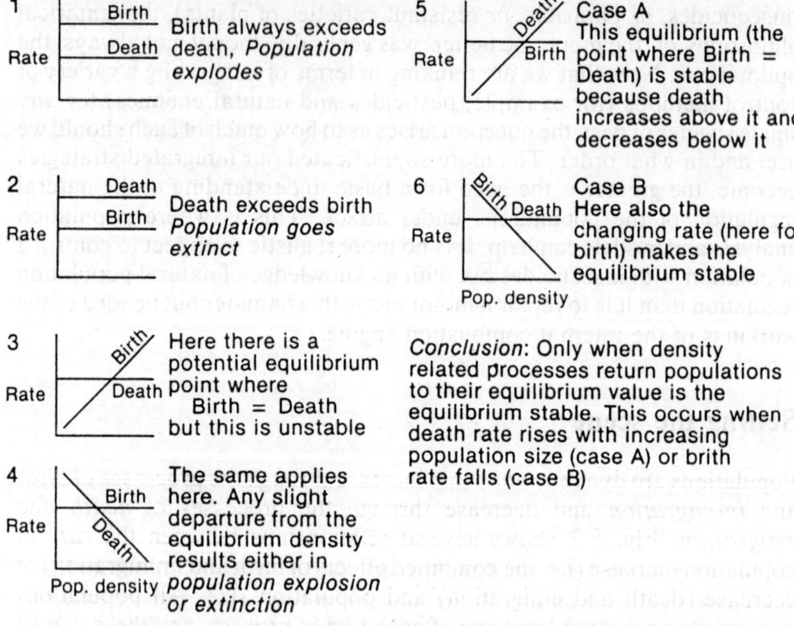

Fig. 5.2 Only a few of the possible combinations of natural birth and death rates (upper part of the figure) give rise to stable population equilibria (lower part of the figure).

direct density–dependent process operating. This has the effect of slowing down and halting population growth when density is high, and allowing it to speed up when density is low.

For the rest of this chapter we will be dealing mainly with density-dependent mortalities in which the *rate* of mortality (or emigration) increases as population density rises. This is typical of many insect life-histories (case A, Fig. 5.2). Exactly the same regulation is achieved by the action of a density-dependent birth rate, in which the birth rate decreases as population density rises (case B, Fig. 5.2), rather more typical of vertebrates.

Fig. 5.2 identifies density-dependent processes as vital for population regulation, so it becomes important to discover their natural causes. Much research now demonstrates that only competition between individuals for limited resources, and some forms of parasitism (including diseases caused by micro-organisms) and predation, can produce such density dependence. The agents of the mortality are thus always other living creatures (of the same or different species), and for this reason such mortalities are often described as 'biotic'. How to quantify them is described in the next section.

Population growth curves and density dependence

Fig. 5.3 shows the expected changes in the numbers of an animal species that occur through time in any particular place of limited area and food supply. The numbers initially increase rapidly, during the 'growth phase', and then stabilize at an 'equilibrium phase' characterized by more or less constant numbers. Fig. 5.3 is called a population growth, or *logistic*, curve.

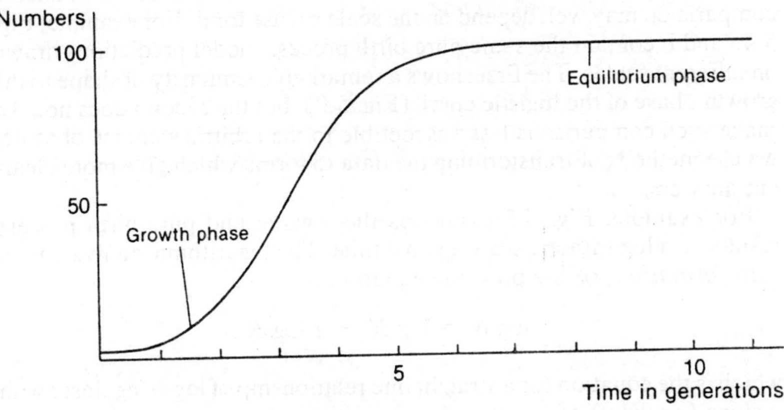

Fig. 5.3 The population growth or logistic curve.

All pest problems arise from the relationship between the pest's logistic growth curve and the economic threshold, rather arbitrarily set in the ways described by P.T. Walker in Chapter 4. Although some sorts of pests (for example many vectors, and pests of tree crops) exist in their equilibrium phase, others (such as pests of seasonal or annual crops) are often only in their growth phase. Each phase of population development is the product of a rather delicate balance between the rates of increase and decrease referred to in Fig. 5.2. In the growth phase the birth processes predominate, whilst in the equilibrium phase they are exactly balanced by natural mortalities of various kinds.

In understanding the nature of this balance between birth and death rates we often rely on population models, simple mathematical descriptions that express the essence of any process or relationship.

For example, a hypothetical population that doubles in size each generation will show the following changes in numbers:

Generation no.	0	1	2	3	4	5	
Population size	2	4	8	16	32	64	Reproduction rate = 2
or, in symbols	N_0	N_1	N_2	N_3	N_4	N_5	Reproduction rate = R

Clearly $N_1 = N_oR$
$N_2 = N_1R$
$N_3 = N_2R$ etc.

Substituting the first equation into the second
$$N_2 = (N_oR)R$$
$$= N_oR^2$$

and in general $N_t = N_oR^t$ where t is the generation number.

This equation is the general model of a *pure birth process*. To find out if a real population, such as that in Fig. 5.3, ever behaves as if it were subject to a pure birth process alone, we must compare its changes with those predicted by a pure birth process model. Our conclusions about such a comparision may well depend on the scale we use for it. For example, Fig. 5.4a and b contain the same pure birth process model predictions drawn on different scales. The first shows a remarkable similarity of shape to the growth phase of the logistic curve (Fig. 5.3), but the second does not. To make such comparisons less susceptible to the arbitrary choice of scale, we use methods of transforming the data to forms which give more clear-cut answers.

For example, Fig. 5.5 expresses the logistic and pure birth process results on a logarithmic scale against time. The logarithmic equivalent, or *transformation*, of the previous equation is

$$\log N_t = \log N_o + t \log R$$

which is the equation for a straight line relationship of $\log N_t$ against t with a slope (gradient) of $\log R$. All pure birth processes give a straight-line relationship after such a transformation, and comparisons with field data are therefore easier.

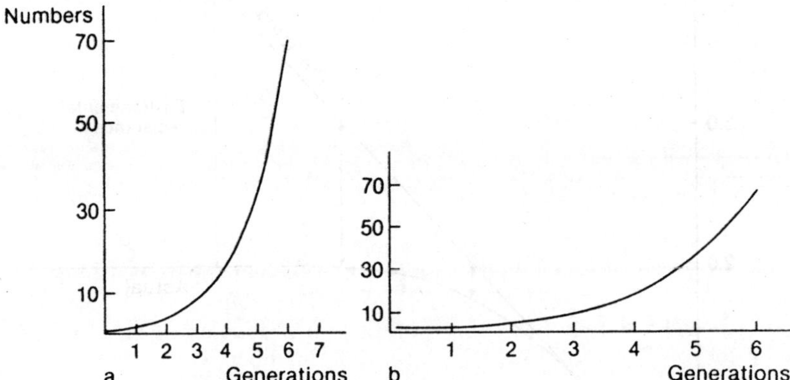

Fig. 5.4 A pure birth process, involving a doubling of numbers each generation, is drawn on two different arithmetic scales for comparison with the growth phase of the logistic curve of fig. 5.3.

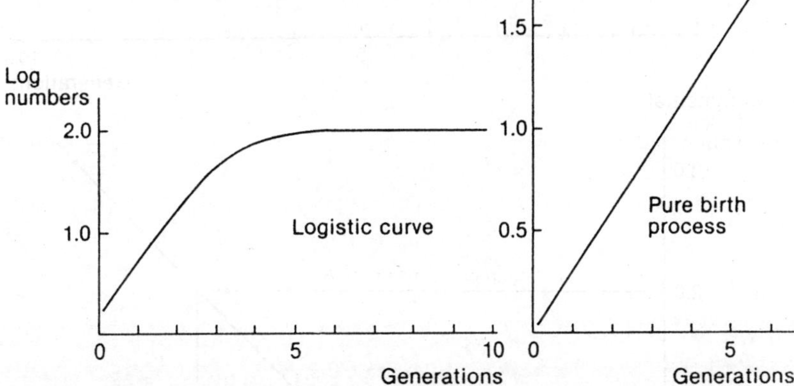

Fig. 5.5 Logarithmic transformation of figs 5.3 and 5.4 renders the pure birth process and the growth phase of the logistic curve both linear.

Fig. 5.5 shows that whereas the initial growth phase of the population appears to conform to the predictions of the pure birth process model, this is not the case later on. The straight line begins to bend round to the horizontal (equilibrium) position. Since the slope of the line is determined by the rate of increase, it is obvious that this rate must have decreased from an initial high level to zero as the population increased. This decrease is caused by environmental resistance to further population growth, which can be quantified by comparing the population's actual performance with the increase it would have shown if it had continued to grow at its initial rate. Fig. 5.6 shows how this is done, the 'predicted line' of Fig. 5.6a being

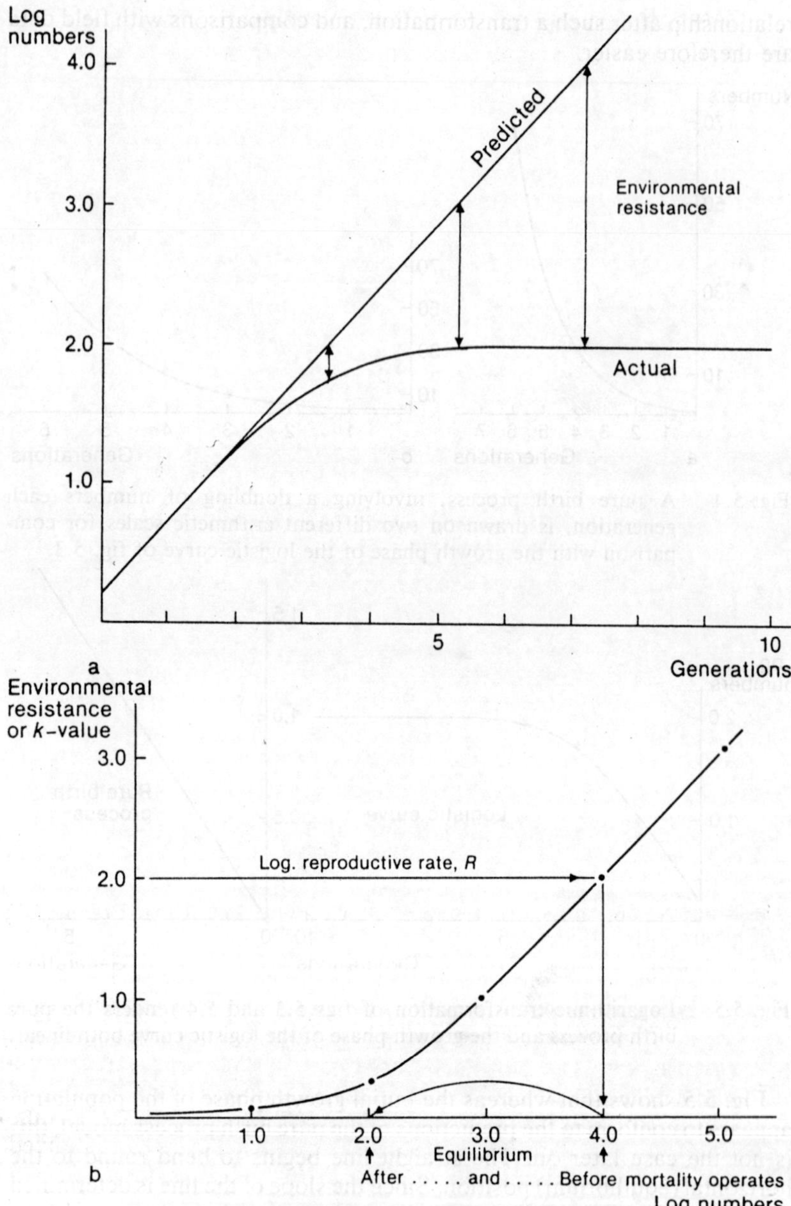

Fig. 5.6 (a) The prediction of a pure birth process model is compared with the logistic curve. (b) Environmental resistance to growth increases as predicted density rises and is plotted as a k-value against log. density. The equilibrium density reached by this population depends on its reproductive rate; in this example it is at a log density of 4.0 *before* the mortality acts and 2.0 *after* it acts.

a continuation of the growth phase line of the logistic curve in Fig. 5.5. At any predicted log density the difference between the predicted and actual lines (i.e. the resistance to growth) is read off and plotted against density (Fig. 5.6b). Thus:

Environmental resistance = log (predicted) – log (actual), a quantity which is conventionally called the *k-value* (for 'killing value').

The same result would have been obtained had we counted out different numbers of eggs of the species concerned (the initial density) and allowed them to develop to the adult stage (the final density). In this case:

Environmetal resistance =k-value
=log (initial density) − log (final density)

Fig. 5.6b shows that the resistance increases as population density increases, i.e. it is an example of a *density-dependent mortality.*

Thus by applying a rather simplistic model, involving only a pure birth process, to a field result where both birth and death processes were involved, we have managed to abstract the underlying density-dependent mortality that is responsible for establishing the equilibrium level. Now we turn to predicting equilibrium levels from such density-dependent mortality graphs.

How many? How stable? And why?

Imagine a population with a fixed growth rate and affected only by the density-dependent mortality of Fig. 5.6b. If the population exists at a very low density, in each generation the reproductive rate exceeds the mortality rate, given by Fig. 5.6b, and so the population increases. If it were artificially raised to a very high level the population falls because mortality (which is now high) exceeds the reproductive rate. The population thus rises or falls until it reaches a density where the birth rate is just balanced by the death rate.

A simple population model that incorporates both a birth and a death process might be of the form

$$N_{t+1} = N_t Rs$$

where N is the number of individuals in successive generations t and $t{+}1$, R is the reproductive rate and s is the generation survival rate.

Again this model is more conveniently expressed in terms of logarithms:

$$\log N_{(t+1)} = \log N_{(t)} + \log R + \log s$$

It is a simple matter to show that the following relationship exists between generation survival rates and the equivalent k-values:

$$k\text{-value} = -\log s$$

and therefore:

$$\log N_{(t+1)} = \log N_{(t)} + \log R - k\text{-value}$$

To find out how the numbers of individuals change over time (i.e. to 'run' the model) it is therefore only necessary first to add the log rate of increase to the log of the initial numbers (effectively giving the numbers of offspring successfully produced), and then to subtract the *k*-value for the mortality that acts on them (which can be read off from graphs such as Fig. 5.6b). This gives the number of adults surviving to the next generation.

At equilibrium the numbers of individuals remain the same, and so

$$\log N_{(t+1)} = \log N_{(t)}$$

and the only time this can happen is when

$$\log R = k\text{-value}$$

as shown graphically in Fig. 5.6b.

It follows that if we know for any species the type and degree of density-dependence and the rate of increase, we can predict its equilibrium level by the simple graphical method presented in Fig. 5.6b. In each case the equilibrium level predicted is of the production *before* the operation of the density-dependent mortality in each generation. The population level *after* the mortality is obtained by subtraction of the mortality from the initial predicted level; this is also shown in the figure.

In itself a prediction of equilibrium levels is important enough, since it can be compared with the economic threshold to find out if the insect is likely to be a pest (equilibrium above the economic threshold) or not. Furthermore, graphs such as Fig. 5.6b can also tell us about the stability of the equilibrium, and whether or not it is likely to be easily disturbed by our applied mortality agents. This important point can be illustrated by Fig. 5.7 which shows that the same reduction in fertility (as may be achieved by control measures such as sterile male releases) applied to three different pest species, each experiencing different natural density-related mortalities, has different effects on the final population size. In the first case the final population size is decreased, although to a lesser extent than the applied reduction in fertility; in the second case there is no change in final population size; and in the third case there is an *increase* in final population size. Thus the effect of any particular control measure depends entirely on the type of natural density-dependent process operating. In all cases in Fig. 5.7 the final population size changes by a smaller amount than is suggested by the effect on fertility (for example, in Fig. 5.7a log fertility is reduced by one half, but the log final population size by only one tenth).

Mechanisms such as these explain the commonly observed phenomenon that insecticides that kill off 90 per cent or more of the insects to which they are applied in the laboratory, do not reduce the population levels of the same insects in the field by the same amount. (Technical reasons for such

Fig. 5.7 (*opposite*) Different species are affected by the same control measures in ways that are entirely determined by their characteristic density-dependent mortalities. The figure shows how final population size may be decreased, unaffected, or increased by the same reduction in fertility.

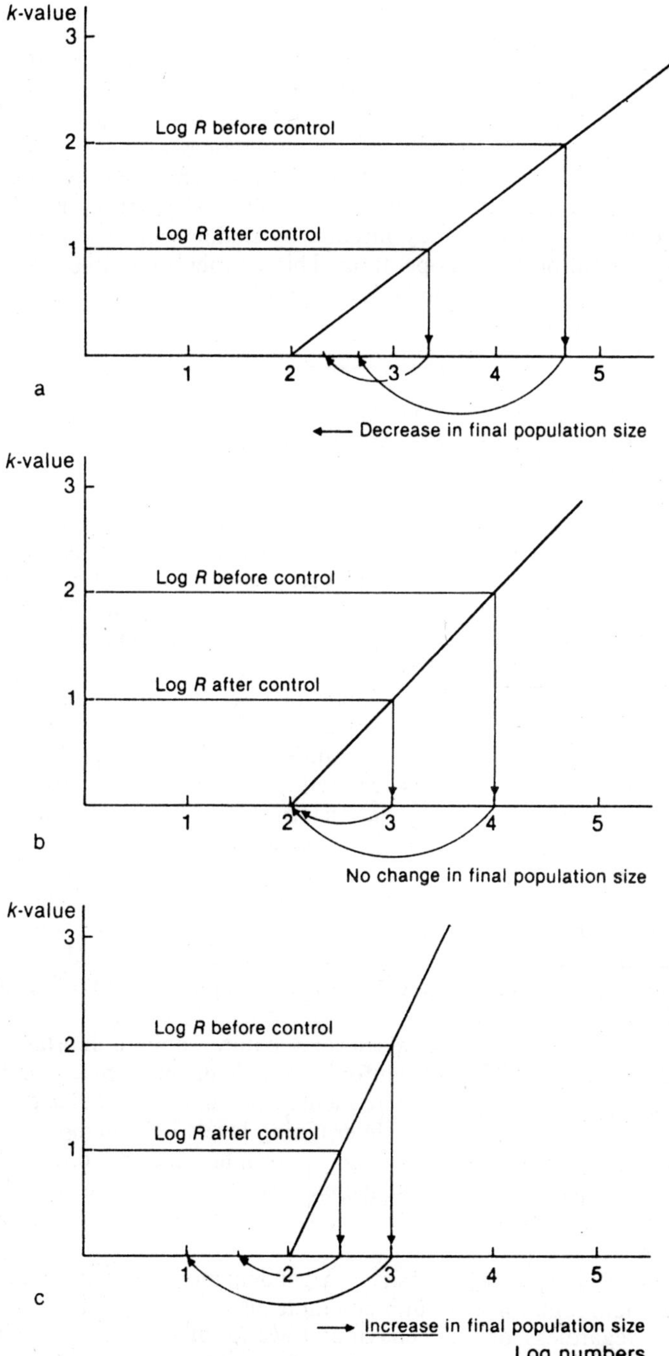
k-value
3
2
1
Log R before control
Log R after control
1
2
3
4
5
a
← Decrease in final population size
k-value
3
2
1
Log R before control
Log R after control
1
2
3
4
5
b
No change in final population size
k-value
3
2
1
Log R before control
Log R after control
1
2
3
4
5
c
→ Increase in final population size
Log numbers

failures are generally given, such as failure of the insecticide to reach the target area; whilst this is sometime true, the ecological processes just discussed are *always* operating and so will *always* have some effect.)

To conclude this section we can state that, in general, density-dependent mortalities *resist* changes in population size, whilst control measures are attempts to *change* population size. These two processes must therefore interact whenever we manage a pest population. Furthermore we can only tell how they interact by finding out which sort of density-dependent process is operating. This is where life-table studies, discussed later on, make an important contribution, but we must first consider other life-history mortalities.

The role of other life-history mortalities

The biotic mortalities that formed the theme of the foregoing pages are not the only ones to act. The effects of weather (e.g temperature, rainfall) are often of dramatic importance to insect populations. Acting alone, however, they cannot regulate because they do not adjust their severity according to population size. They are said to be *density-independent* in their action, and are often described as *abiotic*.

As weather-induced mortalities reduce the rates of increase, they must have some effect on population size by acting on the interplay of rates of increase and density-dependent mortalities outlined in the previous section. When equilibrium levels are predicted as previously described, the rate of increase used is the population fertility (the potential number of offspring produced *per individual*) *minus* the effects of all other (i.e. abiotic) life-history mortalities. For example, if potential fertility is 100 offspring per individual, and all abiotic mortalities combine to inflict 90 per cent losses, the effective rate of increase of the population is ten-fold. The log of this number (i.e. 1.0) is therefore used for the prediction of equilibrium level from the k-value/log density relationship of the species' density-dependent mortality.

Whilst spending little time discussing natural abiotic mortalities, because their action is rather straightforward, it should be emphasised that they are the type of mortality most frequently encountered in the field. It is not that they are unimportant in determining levels of abundance of a species, but rather that they are not likely to be interfered with one way or the other by man-made control measures. The same is not the case with density-dependent mortalities, to which more attention must consequently be paid.

We do not, therefore, take density-independent mortalities for granted, but regard them rather like the unpredictable gifts of kindly grandparents that arrive regardless of our behaviour towards them, whilst density-dependent mortalities are the potential gifts of friends whose kindness rather depends on our behaviour towards them.

Integrating life-history data – key-factor analysis

We now have some idea of how populations are regulated, the vital role of density-dependence, and the usefulness of logarithmic transformations. Our understanding of each of these is not an end in itself, but a means to the end of interpreting population changes detected by our sampling programme in the field. Our aim at each step is to understand why population *numbers* change from generation to generation and we do this by examining changes in life-history *mortalities* which must clearly hold the key to such changes. If in one generation the total mortality is higher than average, then the population density will be lower than average, and *vice versa*.

Usually our sampling programme allows us to identify a number of periods in the life-cycle of the species we are studying. For each of these periods we can calculate a survival rate, or a k-value, for the submortality taking place. The periods concerned need not necessarily correspond exactly with the developmental stages, egg, larva, pupa and adult (although for clarity this will be the case in our example), often they are determined by the convenience of sampling. A sticky trap that samples very young larvae drifting with the wind onto a crop may be followed in the sampling sequence by a soil surface trap that samples mature larvae leaving the crop to pupate in the soil. Such a period covers virtually all of the larval instars, but neither the total period of the first nor the whole of the last instar, both periods when considerable losses may occur.

A simple set of census data, called a *life-table*, is given in Table 5.1. Notice that the entire lifespan is covered by the census techniques used. The first part of the table gives the untreated data and calculates the survival rate both for each life-stage in turn and for the entire period of egg$\longrightarrow$adult. When the results are expressed in this way it is impossible to examine directly the contribution of each individual survival rate to the total (which is in fact the product of the individual values). When, however, the data are converted to logarithms and when, by subtraction, the k-values are found for each submortality, these are found to add up to the total generation mortality, also expressed as a k-value (conventionally as K).

$$K = \text{generation mortality} = k_1 + k_2 + k_3 \text{ etc.}$$

This relationship allows us to examine over a series of generations both the variability of total mortality, and therefore of population size, and also the contribution to this mortality of each submortality in turn. When this is done graphically, as in Fig. 5.8, it is often simple to pick out one

Table 5.1 An example of a life-table for a single generation of an insect pest. Sample data are given in bold type, and all other results are calculated from them.

Stage	Estimated number	Survival rate	log numbers	k-value
Eggs	**1000**		3.00	
		0.5		0.31 k_1
Larvae	**500**		2.69	
		0.2		0.69 k_2
Pupae	**100**		2.00	
		0.1		1.00 k_3
Adults	**10**		1.00	

Generation survival = 10/1000 = 0.01;
Generation mortality, K = 3.00 − 1.00 = 2.00 = $k_1 + k_2 + k_3$

submortality that contributes most to changes in total mortality (k_0 in Fig. 5.8). Because it is the key to changes in total mortality, it is called the *key-factor*, and the whole exercise is called *graphical key-factor analysis*.

In all complete life-table studies some assumption has to be made about population fertility, that is the maximum number of eggs that can be produced per individual, because this allows us to 'join up' the end of one generation with the start of the next. Maximum fertility can be found either from laboratory studies, or can be guessed from the field data (errors in its estimation do not affect the identification of the key-factor). The difference between the predicted number of eggs present and the actual number counted forms the first life-history submortality (k_0) in each generation, and is often an important source of loss.

$$\log(\text{potential eggs})_{t+1} = \log(\text{adults})_t + \log(\text{maximum fertility/individual})$$

and $k_0 = \log(\text{potential eggs}) - \log(\text{actual eggs})$
where t and $t+1$ refer to successive generations.

The identification of the key-factor, which allows us to concentrate our attentions on the most important life-history submortality (not always the most obvious one), is only the first step in the analysis of life-table data. It gives us the correct perspective on our sample data. The next step is to examine the density relationships of each life-history submortality in turn, by plotting the k-value for the mortality in each generation against the logarithm of the population on which it acts. This allows us to pick out the important density-dependent relationships whose underlying causes we must proceed to investigate.

As emphasised previously, density-dependence of some sort is always present, but it is not necessarily easily demonstrated. It may be obscured by the inclusion of density-independent components acting at more or less the same time of the life-cycle. Its detection will also be difficult when environmental resources such as food, which (through for example, competition) elicit density-dependence, change in quantity from one

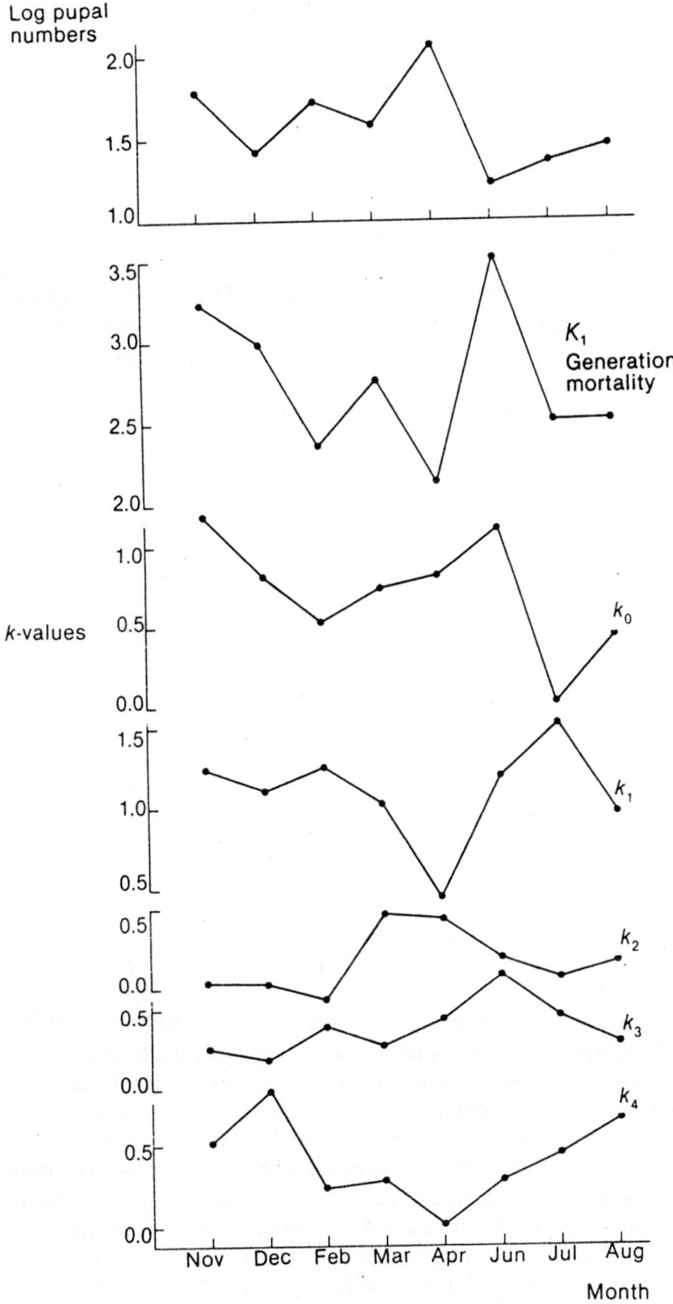

Fig. 5.8 Graphical key-factor analysis of the mosquito, *Aedes aegypti* in Bangkok. The analysis is initially presented as an example of the technique, and later forms the basis of a population model. k_0, the difference between the potential and actual number of eggs present, is the key factor (other details in Table 5.2).

generation to the next. When this happens rather more subtle tests have to be applied, such as the expression of density not per unit area, but per unit of resource, which must therefore also be measured. Such a lack of ease of demonstration does not imply a lack of importance, since applied control measures will have the same effect on density-dependent mortalities regardless of our awareness of their existence!

It is often useful to discover the causes of density-independent mortalities, especially when they are key-factors. The level of mortality can be correlated with environmental variables such as meteorological conditions at the appropriate time in the life-cycle (for example, amount of rainfall or sunshine), or during the weeks or months preceding this (such as the number of days of drought). The most useful correlation can then be used to predict the level of mortality during later population modelling.

Problems and how to avoid them

Graphical key-factor analysis takes a very short time to explain, but a whole lifetime to learn. At each stage one acquires more knowledge of what the technique can and cannot do. As a tool in the right hands it is invaluable, but it is likely only to confuse or mislead those who apply it without thinking. One common pitfall is to ignore accidentally an important source of loss at the beginning or end of the life-cycle. For example, the adult insects that emerge in a field or plantation are not necessarily those that lay the eggs sampled there later. Immigration and emigration often take place before oviposition and these losses (unless appreciated and therefore estimated) will be included in the first life-history mortality, k_0, whose calculation was outlined previously. k_0 thus often has a number of components (for example, reduction of fertility from the maximum). Because it is not directly measured in quite the same way as other losses, k_0 is often forgotten by field workers when they analyse their data. A useful check to avoid this mistake is to compare the log maximum fertility with the average totai generation mortality, K. If the population is at approximate equilibrium these two quantities will be about equal, for the reasons explained in previous sections. If they are not, it is likely that some life-history mortality has been accidentally omitted.

Another common error is to assume that the key-factor should always be density-dependent. In fact this is only rarely the case (although it may be more common among certain types of pests than among insects in general). Since density-dependent mortalities oppose changes in population size they are, as a general rule, unlikely to be key-factors which, by definition, are responsible for changes in density. Key-factors are usually density-independent, such as weather related mortalities about which we can do rather little. But it should never be forgotten that the aim of much biological control is to replace the natural key-factor by the mortality caused by the newly introduced predator or parasite. The natural mortality that was the key-factor of course continues to operate, but becomes relatively less important in the life-cycle. Sensible sampling before, during and after control programmes will allow us to see whether this objective has been achieved and, if not, whether it is worth continuing. Many biological control programmes require occasional 'inoculations' of additional natural enemies, and therefore imply the maintenance of rearing facilities for them. Whether or not these are cost-effective may well depend on whether the parasites or predators become the new key-factor, or approach it in importance. If they do not, although they will reduce the variability of the pest numbers, they will not reduce the equilibrium level by a large amount.

Occasionally it is difficult to identify the key-factor by the graphical method alone; occasionally we need to make comparisons of the same species either before and after control, or in two different places. These problems are all simplified by applying a regression technique to the plotted data. If each life-history mortality is plotted in turn against total generation mortality for a series of generations, the key-factor is, by definition, that submortality that gives the highest slope. It is a simple matter to calculate the slope of each submortality in turn and to choose which is the key-factor. The technique also helps to rank the submortalities in decreasing order of importance (i.e. decreasing slopes) and therefore to compare analyses at different times or places.

Key-factor analysis, which was designed as a technique for the study of an insect with one generation a year and with synchronized life-stages, has now been applied to a range of animals as varied as mosquitoes in Thailand and buffaloes on the Serengeti Plains of Tanzania. Each application involves new problems, but also provides new and often startling insights into the population dynamics of the animals concerned.

Population modelling

Once a key-factor analysis is complete and once all the density relationships are appreciated, it becomes possible to construct a predictive model of population change.

The first step in model building is to list in sequence the mortalities that occur. Density-independent mortalities can be included using their average observed value, or can be calculated from some useful predictive variable, as discussed above. Density-dependent mortalities are included as the regression equations (k-value/log density) that best describe them. Finally population fertility is included as its observed maximum value.

Each mortality is allowed to act in turn on an initial population and after only a few generations the natural equilibrium levels will be established. All these calculations are more simply done in logarithms; antilogarithms can be taken when necessary. Many small programmable calculators nowadays have the capacity to run quite complex population models and there is therefore no need to depend on large computers far from the field site.

Finally, once the model is running correctly we can investigate the effectiveness of alternative population manipulations using the various means at our disposal. We can thus answer the sorts of questions posed at the start of this section.

Case study – mosquito control

In this final section we give an imaginary case study based on a set of real life-tables for the mosquito *Aedes aegypti*. We seek to choose an effective control strategy from the two alternatives of applying an insecticide to kill off 90 per cent of the adults before they lay their eggs, or a larvicide to kill off 90 per cent of the mature larvae in the breeding sites. The objective is to reduce the number of emerging adult mosquitoes as much as possible, since these are able to transmit disease fairly soon after emergence.

Ae. aegypti was studied in Bangkok by regularly counting the eggs, larvae and pupae present in water-jars left uncovered for a period of 48 hours. This brief period of exposure to the ovipositing females was chosen to avoid the problem of many different life-stages occuring together, making counting and interpretation more difficult.

Using this protocol the life-tables shown in Table 5.2 were compiled each month during the period October – August with the exceptions of January and May. The maximum observed rate of increase between the

Table 5.2 Life-tables for the mosquito, *Aedes aegypti*, in Bangkok. The original sample data, in bold type, come from Southwood *et al.* (1972) and all the other quantities are calculated from them.

Month	Stage	Number	log	k-value
October	Pupae	**228.0**	2.36	
November	Potential eggs	100 320	5.00	
				1.19 k_0
	Eggs	**6 503**	3.81	
				1.23 k_1
	Larvae 1 + 2	**380.9**	2.58	
				0.03 k_2
	Larvae 3	**357.0**	2.55	
				0.25 k_3
	Larvae 4	**197.9**	2.30	
				0.53 k_4
	Pupae	**59.3**	1.77	
				K = 3.23
December	Potential eggs	26 092	4.41	
				0.81 k_0
	Eggs	**4 025**	3.60	
				1.12 k_1
	Larvae 1 + 2	**302.8**	2.48	
				0.03 k_2
	Larvae 3	**281.6**	2.45	
				0.18 k_3
	Larvae 4	**184.6**	2.27	
				0.85 k_4
	Pupae	**26.6**	1.42	
				K = 2.99

Month	Stage	Number	log	k-value
February	Potential eggs	11 704	4.06	0.52 k_0
	Eggs	**3 499**	3.54	1.26 k_1
	Larvae 1 + 2	**190.2**	2.28	−0.06 k_2
	Larvae 3	**216.8**	2.34	0.40 k_3
	Larvae 4	**87.3**	1.94	0.23 k_4
	Pupae	**51.1**	1.71	
				K = 2.35
March	Potential eggs	22 484	4.35	0.73 k_0
	Eggs	**4 141**	3.62	1.03 k_1
	Larvae 1 + 2	**384.7**	2.59	0.47 k_2
	Larvae 3	**131.2**	2.12	0.26 k_3
	Larvae 4	**72.9**	1.86	0.27 k_4
	Pupae	**38.5**	1.59	
				K = 2.76
April	Potential eggs	16 940	4.23	0.82 k_0
	Eggs	**2 554**	3.41	0.44 k_1
	Larvae 1 + 2	**936.7**	2.97	0.44 k_2
	Larvae 3	**336.8**	2.53	0.43 k_3
	Larvae 4	**125.0**	2.10	0.02 k_4
	Pupae	**121.5**	2.08	
				K = 2.15
June	Potential eggs	53 460	4.73	1.12 k_0
	Eggs	**4 044**	3.61	1.20 k_1
	Larvae 1 + 2	**256.1**	2.41	0.19 k_2
	Larvae 3	**157.2**	2.22	0.72 k_3
	Larvae 4	**31.9**	1.50	0.28 k_4
	Pupae	**16.7**	1.22	
				K = 3.51
July	Potential eggs	7 348	3.87	0.01 k_0
	Eggs	**7 280**	3.86	1.51 k_1
	Larvae 1 + 2	**222.9**	2.35	0.07 k_2
	Larvae 3	**190.9**	2.28	0.47 k_3
	Larvae 4	**64.7**	1.81	0.45 k_4
	Pupae	**22.2**	1.35	
				K = 2.51
August	Potential eggs	9 768	3.99	0.43 k_0
	Eggs	**3 646**	3.56	0.97 k_1
	Larvae 1 + 2	**388.9**	2.59	0.17 k_2
	Larvae 3	**260.2**	2.42	0.31 k_3
	Larvae 4	**129.1**	2.11	0.65 k_4
	Pupae	**28.9**	1.46	
				K = 2.53

Regression equations

$k_0 = 0.536K - 0.77$
$k_1 = 0.276K - 0.34$
$k_2 = -0.087K - 0.41$
$k_3 = 0.073K - 0.18$
$k_4 = 0.203K - 0.15$

pupae of one month and the eggs of the next was 436, and so the completed life-table has used a number slightly in excess of this (440) in order to calculate the potential number of eggs present (by multiplication with the number of pupae of the previous month). Since mosquitoes can pass through more than one generation a month, this figure represents something more than straightforward reproduction and includes all processes leading to the net increase in the intervening period (including additional entire generations). The data are converted to logarithms in the table, and k-values and total generation mortalities calculated. The graphical key-factor analysis has already been given as Fig. 5.8 and k_0 identified as the key-factor. This is confirmed by the regression test mentioned earlier (details in Table 5.2), which shows that k_1 is in second place, followed by k_4; the other two mortalities seem to have a negligible effect on changes in K. With a rapidly breeding species such as this one, capable of many generations a year, we would predict a seasonal shift in the relative importance of each submortality, and therefore perhaps a seasonal sequence of key-factors.

Fig. 5.9 shows each submortality plotted against the log of the density on which it acts. k_0, k_1 and k_2 all appear to be density-dependent, but with different slopes, whilst k_3 and k_4 are variable and density-independent. This is therefore an example where the key-factor is density-dependent.

The simplest population model for this situation is one where k_0, k_1 and k_2 are applied in sequence, using the regression equations given in Fig. 5.9, whilst k_3 and k_4 are included as their average values (Table 5.3). With a rate of increase of 440, as used in the analysis, this model very quickly reaches a stable equilibrium at a log density of 1.57 (antilog = 37.2) pupae per water-jar, close to the observed log value of 1.58 (= 37.6) pupae. Variations around this predicted equilibrium could be introduced easily by including some of the variability present in k_3 and k_4, but this is not necessary in the present context.

Proceeding from this model description of the natural situation we can investigate the effects of the alternative control strategies (Table 5.3). A 90 per cent mortality is equivalent to a k-value of 1.0, and when this is assumed to be applied to the adults it appears between the log pupal numbers of one generation and reproduction giving rise to the next. The effect on the number of pupae, and therefore probably on the number of adults emerging, is negligible because the reduction in the number of eggs oviposited is later compensated for by a similar reduction in the density-dependent mortalities k_0, k_1 and k_2.

When, however, the same mortality is applied by means of a larvicide operating after the density-dependent mortalities, each generation we achieve a significant reduction in the numbers of adults emerging.

The general rule that emerges from this and similar studies is that it is best to apply control measures *after* the operation of density-dependent mortalities and *before* the emergence of the harmful stage each generation. Very often the mortalities coincide with the harmful stage, and this makes control rather difficult. In this case models can be invaluable in deciding

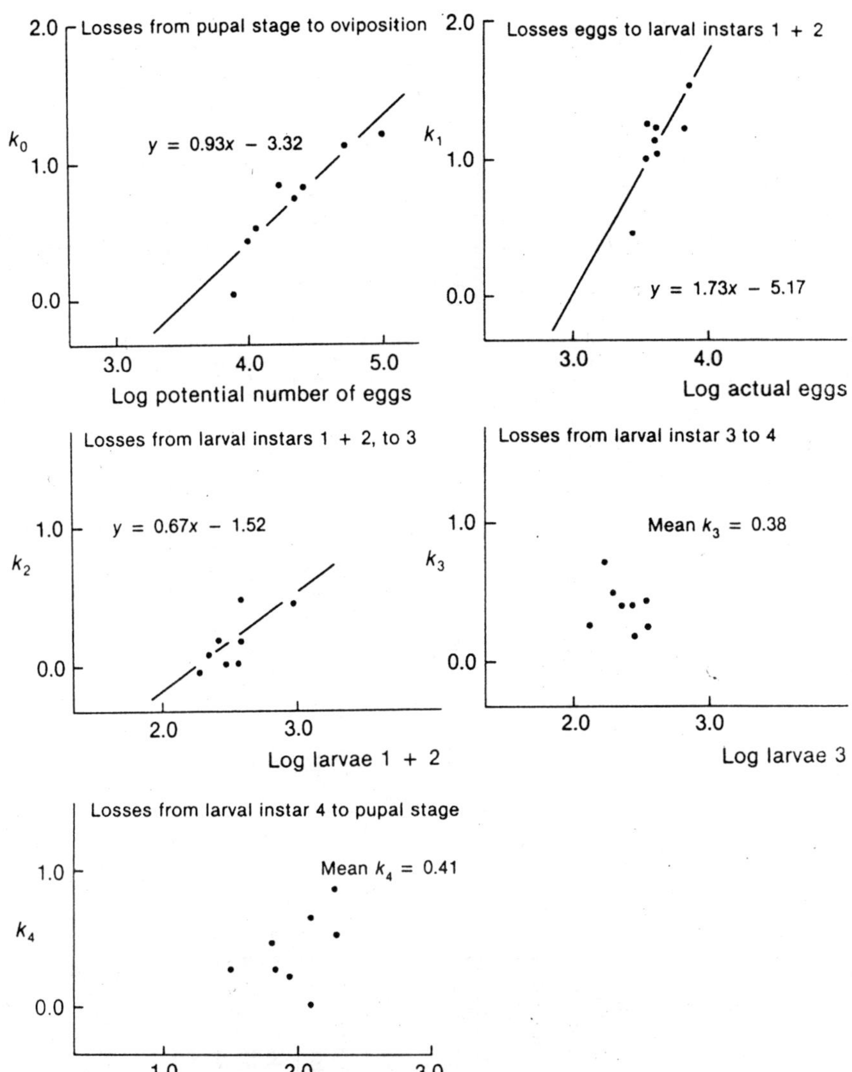

Fig. 5.9 The relationship between each submortality of the life-table of *Aedes aegypti* and the log of density on which it acts. k_0, k_1 and k_2 are density-dependent; k_3 and k_4 are density-independent.

between alternative strategies.

We must not expect initially that our answers will always be right. But the approach we have adopted is more rigorous and more likely to lead quickly to effective integrated strategies than the 'trial and error' methods most often applied in the past. Shall we be the architects of future control strategies, or shall we continue, as at present, to be their victims?

Table 5.3 Population models for the mosquito, *Aedes aegypti*, developed from the life-tables of Table 5.2 and the graphical key-factor analysis of Figs. 5.8 and 5.9. The object of insecticidal control (cases 2 and 3) is to reduce the density of emerging adults (bold type). All figures are of log densities. Gen = Generation number

Case 1: Natural regulation – no applied mortality

Instructions		Add log 440 (= 2.64)		subtract density-dependent mortalities						subtract density-independent mortalities $k_3 + k_4 = 0.79$			
				k_0		k_1		k_2		k_3		k_4	
Example	Pupae	Emerging adults	Potential eggs		Actual eggs		Larvae 1 + 2		Larvae 3		Larvae 4		Pupae
Gen 1	2.36	**2.36**	5.00	1.33	3.67	1.18	2.49	0.15	2.34				1.55
Gen 2	1.55	**1.55**	4.19	0.58	3.61	1.08	2.53	0.18	2.35				1.56
Gen 3	1.56	**1.56**	4.20	0.59	3.61	1.08	2.53	0.18	2.35				1.56
Gen 4	1.56	**1.56**											

Case 2: Insecticide application kills 90% of adults before reproduction

Example	Pupae	Emerging adults	Survivors	Potential eggs	k_0	Actual eggs	k_1	Larvae 1 + 2	k_2	Larvae 3	k_3	Larvae 4	k_4	Pupae
Additional instructions		subtract k-value for a 90% mortality = 1.0												
Gen 1	2.36	**2.36**	1.36	4.00	0.40	3.60	1.06	2.54	0.18	2.36				1.57
Gen 2	1.57	**1.57**	0.57	3.21	0	3.21	0.38	2.83	0.37	2.46				1.67
Gen 3	1.67	**1.67**	0.67	3.31	0	3.31	0.56	2.75	0.32	2.43				1.64
Gen 4	1.64	**1.64**	0.64	3.28	0	3.28	0.50	2.78	0.34	2.44				1.65
Gen 5	1.65	**1.65** etc.												

Case 3: Insecticide application kills 90% of late instar larvae

Example	Pupae	Emerging adults	Potential eggs	k_0	Actual eggs	k_1	Larvae 1 + 2	k_2	Larvae 3	k_{3+4}	Larvae 4	Survivors	Pupae
Additional instructions												subtract k-value for a 90% mortality = 1.0	
Gen 1	2.36	**2.36**	5.00	1.33	3.67	1.18	2.49	0.15	2.34	0.79	1.55	0.55	0.55
Gen 2	0.55	**0.55**	3.19	0	3.19	0.35	2.84	0.38	2.46	0.79	1.67	0.67	0.67
Gen 3	0.67	**0.67**	3.31	0	3.31	0.56	2.75	0.32	2.43	0.79	1.64	0.64	0.64
Gen 4	0.64	**0.64**	3.28	0	3.28	0.50	2.78	0.34	2.44	0.79	1.65	0.65	0.65
Gen 5	0.65	**0.65** etc.											

Recommended reading

Begon, M. and Mortimer, M. (1981) *Population Ecology: A Unified Study of Animals and Plants*, Blackwell Scientific Publications, Oxford, 192 pp.

Elliott, J.M. (1977). *Some Methods for the Statistical Analysis of Samples of Benthic Invertebrates,* Freshwater Biological Association Scientific Publication, No. **25**, 160 pp.

Krebs, C.J. (1978) *Ecology: The Experimental Analysis of Distribution and Abundance*, Harper and Row, New York, Evanston, San Francisco and London, 678 pp.

Rogers, D. (1977) 'Tsetse population dynamics and distribution: a new analytical approach.' *Journal of Animal Ecology,* **48**, 825–49.

Service, M.W. (1977) 'Mortalities of the immature stages of species B of the *Anopheles gambiae* complex in Kenya: comparison between rice fields and temporary pools, identification of predators, and effects of insecticidal spraying.' *Journal of Medical Entomology*, **13**, 535–45.

Siegel, S. (1956) *Nonparametric Statistics for the Behavioral Sciences*, McGraw-Hill, New York, Toronto and London, 312 pp.

Snedecor, G.W. and Cochran, W.G. (1967) *Statistical Methods*, 6th ed., Iowa State University Press, Iowa, 593 pp.

Southwood, T.R.E. (1978) *Ecological Methods: With Particular Reference to the Study of Insect Populations*, Chapman and Hall, London, 524 pp.

Southwood, T.R.E., Murdie, G., Yasuno, M., Tonn, R.J. and Reader, P.M. (1972) 'Studies on the life budget of *Aedes aegypti* in Wat Samphaya, Bangkok, Thailand'. *Bulletin of the World Health Organization*, **46**, 211–26.

Varley, G.C., Gradwell, G.R. and Hassell, M.P. (1973) *Insect Population Ecology: An Analytical Approach*, Blackwell Scientific Publications, Oxford, 212 pp.

Williamson, M. (1972) *The Analysis of Biological Populations*, Edward Arnold, London, 180 pp.

6 Management of plant pests

This chapter considers management strategies for selected crop plant pests. Two groups of crop plants were selected for treatment: food crops, such as rice, maize and legumes; and cash crops such as cotton, cocoa and coffee. For each crop an outline of the major pests and their damage is given and the various methods available for the management of these pests are described.

For most tropical crops systematic integrated pest management schemes have not been developed due mainly to incomplete ecological and behavioural data on the plant–pest interrelationships. However, where this has been developed, for example, in cotton, maize and rice, it has been demonstrated that integrated pest management provides an ecologically sound, cheap and reliable system for dealing with pest problems. The various undesirable side effects associated with heavy pesticide application have been largely eliminated; in particular, the problem of environmental pollution has been reduced to the barest minimum. Pesticide usage in most tropical countries is still comparatively small but there are strong indications that this will increase rapidly because of the high demand for self-sufficiency in food production for the various countries. Insecticides for crop pest control maintain a primary role in pesticide usage in the tropics, but the time has come when appropriate integrated pest management programmes must be developed for crop pests.

The most important elements in the development of integrated pest management are:

(a) Adequate knowledge through research of the ecology and behaviour of the pests in relation to the phenology of the crop;
(b) Reliable systems for monitoring and assessment of pest populations and pest damage together with the determination of economic thresholds;
(c) The availability of materials and other resources to manipulate pest populations at a time when they are needed.

The pest management systems described in this chapter draw attention to the variety of protection methods which are actually is use, or are available for the selected crop pest situations discussed. The aim is to emphasize that alternatives to wholesale pesticide applications for crop protection do in many cases exist, and that crop protection specialists and farmers need to move away from total dependence on insecticides for the protection of their crops.

Management of coffee pests

R.O. Abasa

Coffee will grow anywhere in the tropics except in deserts and places liable to freezing temperatures during any part of the year. Once it is set in plantations and survives, the coffee plant is almost permanent and lives for many years, flowering and fruiting. There are many coffee plants which are still yielding beans 50 years after they were established in plantations. Because of this semi-permanent ecosystem, the pests associated with coffee have a relatively steady food source for many years with short breaks during 'stumping' – that is cutting down the multi-stemmed coffee 'tree', usually called the 'vertical', to within 30 cm of the ground. Pest infestation starts at the nursery level and continues into the field where there are other pests already present to attack the crop. The variety of pests in plantations is further complicated, first by the existence of alternative wild host plants which are a source of infestation, and secondly by new insects becoming adapted to the coffee plant. There are, for example, cases in which caterpillars associated originally with weeds between coffee plants, turned to feeding on coffee leaves, and which eventually became pests of considerable importance.

Changes in cultural practices have also sometimes resulted in high incidences of pests. In Kenya early Arabica coffee (*Coffea arabica*) plantations at high elevations had shade trees to protect coffee from hailstones and cold. These shade trees tended to favour certain coffee pests, such as *Hypothenemus hampei*, *Antestiopsis* spp., scale insects like *Planococcus citri*, the leaf miner, *Leucoptera caffeina* and *Eucosma nereidopa*. The introduction of mulch which increased the moisture content of the soil has also been associated with increased incidence of leaf miner (*Leucoptera* spp.) infestations in Kenya. The application of copper to promote leaf retention has also been associated with leaf miner infestations. It is important, therefore, to weigh the advantages against the disadvantages of any change in cultural practices in coffee plantations. Early attempts at control of coffee pests depended entirely on the application of insecticides but this practice has produced undesirable side effects. For example, the destruction of the natural enemies of some minor pests enabled these pests to develop into major pests in the absence of their natural enemies. The giant looper, *Ascotis selenaria reciprocaria*,and the green looper, *Epigynopteryx coffeae*, (probably a geographical form of *E. strictigramma*) in Kenyan coffee estates are examples of previously minor pests that became major pests for a time as a result of irresponsible use of parathion to control leaf miners, *Leucoptera* spp. Because of this problem with chemical control, attention was focused on the practice of integrated pest management for many coffee pests wherever possible. The success of an integrated pest management programme depends largely on a thorough knowledge of the biotic environment or ecosystem of the target pest, and recognition of the economic threshold of the pest (see p. 75).

Through research and development some major coffee pests have been successfully controlled by the adoption of integrated management schemes. The control of both the grape mealybug, *Pseudococcus maritimus*, and the citrus mealybug, *Planococcus citri*, in South Africa has been achieved by encouraging predators and parasites and by controlling ants (which attack the natural enemies) with barriers or poisons. Control of the common coffee mealybug, *Planococcus kenyae*, has been obtained in Kenya by importation, breeding and dissemination of its natural parasite, *Anagyrus* spp., combined with restrictions on the application of residual insecticides. These are examples of some of the major successes in pest management in Africa which have had little adverse effect on the environment.

Pest management

The number of coffee pests listed by Le Pelley in 1968 is extremely large, but included were potential and minor pests which in an undisturbed ecosystem are unlikely to cause economic loss to coffee growers. A small number, however, are sometimes important pests. The danger lies in the unpredictability of weather and allied natural factors which often help keep their numbers in check. The grower cannot rely entirely on natural control factors against these pests, and the only solution is to apply insecticides selectively to supplement the action of parasites and predators.

The proper management of the agro-ecosystem, by sanitary cultural methods such as prompt stripping of unwanted sucker growth, amelioration of fertilizer treatment, and pruning, together with selective spraying and augmentation of natural enemies, has been used in East Africa, particularly in Kenya, to control some major coffee pests. A few examples are described here.

Coccoidean (mealybug) pests

White waxy scale, *Gascardia brevicauda*

This is probably indigenous to Kenya but it occurs also in Uganda and Angola where it is a pest of both Arabica and Robusta (*Coffea canephora* = *C. robusta*) coffee. At one stage it was such a destructive pest that parasites were imported from Australia to control it. A careful study revealed that heavy infestations originated from frequent use of poison baits for the control of shield bugs, *Antestiopsis* spp. The crawler stages of *G. brevicauda* occur on leaves while the older scales move to the shoots. The scale, only occasionally attended by ants, had many indigenous parasites and predators which usually kept its population below the economic threshold. The cessation of poison baits (against *Antestiopsis* spp.), coupled in cases of minor outbreaks with the removal of badly infested branches, successfully reduced *G. brevicauda* to a minor pest. The cut-off branches are laid on the ground within the plantation to allow parasites to emerge. The use of white oil (a refined grade of petroleum oil) and malathion as 'full cover sprays' is recommended only for heavily infested trees. Attendant ants are stopped by spraying or painting a band, at least 15 cm wide, of dieldrin around the stumps of the trees. These measures have been so effective that few serious outbreaks of *G. brevicauda* have been reported in recent years.

Green scale, *Coccus alpinus*

This scale insect has been reported from Kenya, Zaire, Tanzania, and Uganda on mature Arabica coffee and on other host plants in Eritrea and Ethiopia. Previously limited to young coffee and seedlings in nurseries in Kenya, it was later (1958) found on old coffee. The scales have a preference for leaves, particularly along the main leaf veins, and for the tips of green shoots where they settle and remain immobile while they feed. Consequently, the upper surfaces of the leaves often have spots of sticky transparent honeydew secreted by the scales; at other times the leaves are covered with sooty mould growing on the honeydew. Apart from the mould which reduces the photosynthetic capacity of the leaves, the sucking of the scale has the additional effect of weakening the tree and this may lead to death, especially in nursery plants. Mould growth is characteristic of infested trees in the absence of ants attending to the scales. Attendant ants feed on the honeydew and interfere with parasites and predators of the scales. When so protected the scales multiply abundantly and greater damage is inflicted on the tree.

Chemical control using the organophosphate compound azinphos-methyl to spray green scale insects is recommended only when the infestation is very high and even then, only badly infested trees are treated. In most cases when infestation is average or low, dieldrin-banding to keep off attendant ants is recommended to allow parasites and predators to establish themselves and clean up the infestation. Encouragement of natural enemies and occasional application of insecticides have thus reduced green scale to a minor pest.

Brown scale, *Saissetia coffeae*

This is another important pest which is widespread on coffee, tea, guava and citrus throughout the tropics and in some subtropical areas.

The mobile nymphal stages which are green in colour, and the immobile dark brown adults feed on leaves, shoots and green coffee berries. Though a minor pest in East Africa, the brown scale is elsewhere occasionally a severe pest, especially on weak unhealthy trees. It has been reported to be harmful to coffee trees in Réunion and of economic importance in the Dominican Republic. Except when attended by ants, brown scale is attacked by many parasites and predators. A rarely mentioned fungus, *Cephalosporium lecanii*, was reported in 1917 to parasitize the pest heavily in Puerto Rico; it was also recorded in Cuba in 1929. There is, however, no record of this fungus having been exploited for biological control.

Based on an ecological understanding of the pest it has been possible to keep it under control in Kenyan coffee plantations by manipulating the ecosystem. The cultural approach is to provide optimum quantities of mulch and fertilizer to the coffee trees, and any attendant ants are warded off the trees by dieldrin–methylene blue banding. (Methylene blue is not insecticidal but assists the grower in identifying which trees have been banded.) Badly infested branches are cut off and left on the ground for parasites to emerge. If these practices alone do not give the desired results then white oil is recommended as a drenching spray to infested trees when young stages of the scale are abundant. The spray will not kill the mature females and a second application may be necessary a month later. In Indo-China liquid paraffin emulsified with rice flour or soap and applied as a spray has been found effective.

Star scale or yellow-fringed scale, *Asterolecanium coffeae*

This scale insect has been recorded in Tanzania, Uganda, Kenya and Zaire. In Kenya it is a minor pest of Arabica coffee at high altitudes (above 1524 m) but a sporadically serious pest at lower altitudes. Coffee trees exposed to dusty conditions, particularly near dusty roads, are most liable to infestation. This is because the dust inhibits the activity of the parasites thus allowing unchecked multiplication of the pest. Infestation is usually on the underside of green branches, near a node and also on the trunk. The affected branches are often elbowed at the nodes with pits formed in the green bark on the inside of the bends; the leaves from such nodes droop and die. In a heavy infestation the internodes beyond the elbow are elongated and whip-like. Not only is the crop yield seriously affected but the entire tree may die under heavy infestation. In addition to several ladybird beetles which prey on the young crawlers, there are two primary parasites, both encyrtids, which attack the scales.

Heavy infestation by this scale on Kenyan coffee is rare because of a combination of cultural methods. Along short stretches of earth roads (e.g. murram or laterite roads), it is recommended that roads are treated with old sump-oil or some other waste products to reduce dust. This method, however, is not used very much because it is cheaper to plant living dust barriers, such as napier grass, *Pennisetum purpureum*, or passion fruit vine, *Passiflora edulis*, on *Commiphora* poles between the road and the coffee. Infested trees should be pruned, stripping off most of the crop, and the useless heads cut from multiple-stem trees. Application of optimum quantities of mulch and nitrogen to infested trees is also recommended. To reduce infestations on the stem, the bark is painted with tar oil whose ovicidal properties are well known.

The tendency of *Asterolecanium coffeae* to infest coffee trees exposed to dust and to concentrate on branches and stems is mirrored in that of *Cerococcus catenarius* in Brazil. There is no record of parasites attacking this insect.

Root mealybug, *Planococcus citri*

This is one of the commoner mealybugs reported from nearly all coffee-growing countries of the world, including those around the Mediterranean basin, in South Africa and in some parts of the southern USA. It has a wide host-range including cotton, cocoa, coffee and citrus trees. It is an important pest of coffee in Java, India, Guatemala, Puerto Rico and East Africa. Apparently there are two forms or varieties of the same pest: an aerial form reported from Java, Guatemala, Puerto Rico and Vietnam; and a root form from India, Uganda and Kenya. It is not, however, unusual to find the two forms co-existing. These mealybugs attack *Coffea arabica*, *C. canephora* (= *robusta*) and *C. liberica*. The aerial form occurs on the twigs, fruit and branches, causing the immature berries to become deformed or drop to the ground. The root form, as the name suggests, is confined to the roots. In heavy infestations the roots often become stunted and, on older trees, encased in a crust of greenish-white fungal mycelium of *Polyporus coffeae*, under which the bugs live. Seedlings and very young trees are often free of the fungus. The fungus itself does not in any way impair the health of the plant. Obviously the mealybugs on the roots are injurious to the plants and characteristic symptoms of infestation are wilting and yellowing of the leaves as if affected by drought. The trees may die as a result.

Although the aerial form does not produce large quantities of honeydew, it is known to be attended by ants which protect the mealybugs from predators and parasites and are suspected to disseminate the pest. The root form has also been reported to be attended by ants in Brazil, Puerto Rico, Uganda and Kenya. The association with ants by providing protection from natural enemies leads to increased reproduction of the mealybug.

The root mealybug, although a minor pest of both Arabica and Robusta coffee in Kenya, is potentially serious and hence control has to be undertaken. Investigations have indicated a distinct correlation between nutrients and the incidence of attack by the root mealybug complex. It was experimentally demonstrated that coffee grown in soils deficient in potassium, with little exchangeable calcium and a $pH<5$, were invariably susceptible to infestation.

The ant–mealybug association on infested, but otherwise healthy, trees is controlled by application of ground sprays of aldrin. In addition, any replacement or 'gapping' trees in infested plantations have 55 g of 2.5 per cent aldrin dust mixed thoroughly with the soil in the planting hole. There has been no record of parasites or predators associated with the mealybug in Kenya but nevertheless there is reason to believe they exist.

Common coffee mealybug, *Planococcus kenyae*

This pest, which is sometimes called the Kenyan mealybug, was originally identified as *P. citri* and later as *P. lilacinus* before its true identity was established. It is of limited distribution in East Africa. It was inadvertently introduced into Kenya from Uganda during the First World War, and reported from Nigeria in 1954 but this has not been confirmed. Whereas in Uganda it was a pest of no consequence because of the presence of a large complex of hymenopterous parasites, it was free of these in Kenya. Consequently, it multiplied profusely to become the worst destructive pest ever experienced on Kenyan coffee between 1923, when the first epidemic was reported, and 1951 when it had at last been reduced to a minor pest.

The coffee mealybug has many host plants but it has been reported to cause economic damage on only *Coffea* spp., *Passiflora* spp. (passion fruit or grenadillas), *Dioscorea* spp. (yams) and *Cajanus cajan* (pigeon peas). In a severe infestation of coffee, masses of mealybugs are found either between clusters of berries or flower buds or on the sucker tips. The upper surfaces of leaves carry spots of sticky transparent honeydew or become covered with a crust of sooty mould growing on the honeydew. Whenever a severe and widespread infestation occurs, the mealybug is always attended by ants, *Pheidole punctulata*. By interfering with natural enemies of the pest the ants allow the mealybug to flourish abundantly.

Following the failure of insecticides to control the mealybug, efforts were directed towards importation into Kenya of natural enemies. Initial mis-identification of the mealybug pest led to years of frustration during which the wrong parasites and predators were imported and tried out to no avail. When the costly mis-identification of the mealybug was corrected it was realized that the pest existed in Uganda where it was of a low pest status because of a complex of primary and secondary parasites. From 9 primary parasite species imported and bred in insectaries, only 5 were liberated and of these only 2 – an *Anagyrus* species near *kivuensis* and *A. beneficans* – became established. These parasites attacked mealybugs on both coffee and indigenous shrubs and herbs. However, it was later realized that whenever a coffee mealybug infestation was accompanied by *Coccus alpinus*, the parasite *A.* nr. *kivuensis* was a less effective control agent because parasites of *C. alpinus* , are hyperparasites of *A.* nr. *kivuensis*. It was necessary, therefore, to remove unwanted sucker growth (upright shoots growing low down on the trunk of the tree) since these were often heavily attacked by *C. alpinus*, and were thus a source of this scale insect.

The establishment of ants, *Pheidole punctulata*, has to be prevented by banding the stem of the trees with 1 per cent dieldrin in methylene blue, thus allowing the parasites to clean up the infestation. Badly infested trees or suckers have to be sprayed, to the point of 'run-off' with 60 per cent diazinon emulsion concentrate (150 ml in 200 litres of water).

Coleopteran pests

Six species of beetles (two cerambycid, one bostrichid, one curculionid, one scolytid and one tenebrionid) infest coffee in Kenya. By means other than wholesale dependence on insecticides, it has been possible to effectively reduce three of these from major to minor pest status.

White borer, *Anthores leuconotus*

This beetle is widely distributed in Africa south of latitude 5 °N. It breeds in all species of coffee and is particularly serious in Kenya on Arabica coffee grown below 1524 m. Among its alternative host plants are a number of rubiaceous trees and shrubs. This pest is indigenous to Kenya.

The adult beetle causes negligible damage on coffee but the larva severely damages young trees (1–2 years old) to the point that they are nearly always killed. Older trees (3–4 years old) wilt and turn yellow and do not produce any crop. The young larvae bore under the bark of the tree downwards from the point of insertion of the egg. Most of the damage at this stage is in the form of ring-barking. As they continue downwards towards the ground they penetrate the wood of the tree at the junction of a lateral root with the stem. Pupation occurs within the trunk. Unfortunately few parasites have been recorded.

The control of white borers is largely aimed at both ovipositing and emerging adults as well as the larvae actively boring downwards. The trunks of trees, from ground level to a height of 45 cm, are banded just before the onset of the rains with a mixture of dieldrin (500 ml of 18 per cent emulsion concentrate (=miscible liquid) in 20 litres of water) and methylene blue dye for identification. Adult beetles touching the sprayed bark during oviposition or emergence, and larvae eating or touching the treated bark, pick up a lethal dose of the insecticide. This selective treatment is supplemented by the removal of wild host plants. A related beetle, *Bixadus sierricola*, is a severe pest in the whole of West and Central Africa and parts of Uganda. It has a similar life-cycle to that of *A.*

leuconotus, but fortunately it has a large number of natural enemies which keep the population below the economic threshold.

Dusty brown beetle, *Gonocephalum simplex*

This beetle is a minor, sporadic pest in East Africa, Malawi and Zaire. The larvae live in the soil and feed on vegetable debris and roots. It is the adult beetles which cause the damage. They also live in the soil and decaying leaves under the shade of coffee trees from where they emerge at night to attack young coffee trees. They feed on the bark around the collar, gnaw the bark of branches and twigs and cut off the stalks of young berries at night. They are particularly destructive in plantations lacking adequate mulch and weeds. Only two carabid beetles, *Carabomorphus brachycerus* and *Scarites madagascariensis*, are known to prey on the pest in Kenya.

The control of the brown beetle takes advantage of the behaviour of the adults to leave, feed and then return to their habitations near and around the bases of coffee trees. It is recommended to apply aldrin baits selectively or spray dieldrin round the base of each infested tree. Baits are most effective when applied in the late afternoon or early evening so as to be available to beetles moving out to feed at night. In heavily infested plantations, new plantings should receive 60 g of 2.5 per cent aldrin dust mixed with the soil in each planting hole. When properly carried out these practices give excellent control.

Berry borer, *Hypothenemus (= Stephanoderes) hampei*

This pest occurs in many parts of the world's chief coffee-producing areas and is responsible for great losses if left uncontrolled. It is endemic in Central Africa and a severe pest of Robusta everywhere, and Arabica grown at low latitudes in Tanzania and Uganda. It is currently a pest of minor importance on Arabica coffee in Kenya. Damage is caused by ovipositing beetles and by the developing larvae. The female beetle enters the coffee berry by a circular hole made in the tips of large green or ripe berries, in which it tunnels and lays eggs. The developing larva feeds on the coffee bean until it pupates inside the bean. After emergence, adult females usually fly away but males remain within the bean. Infestations are carried over the post-harvest season by breeding in over-ripe berries on the tree or on the ground. Females live through the period feeding on immature berries causing many of them to fall prematurely. Probably the greatest damage is inflicted on the beans, which become unusuable or else produce very low quality coffee. Only three parasites, *Prorops nasuta*, *Cephalonomia stephanoderis* and *Heterospilus coffeicola*, have been recorded. A number of fungal species, including *Beauveria bassiana* and *Spicaria javanica*, have been reported attacking the beetle in Brazil, Java and West and Central Africa but they do not seem to have been evaluated as biological control agents.

Effective control of the berry borer takes advantage of the fact that they are most abundant in plantations heavily shaded by shade trees and where coffee trees are inadequately pruned, together with the fact that the beetle carries over from one crop to the next in over-ripe berries either on the tree or the ground. It is recommended that ripe berries should be picked every two weeks during the peak season and that no ripe or dried berries remain on the tree or the ground. Infested berries should be destroyed by burning, burying deep in the soil or by rapidly drying on trays so as to stop reinfestation. Complete stripping of the old crop just before flowering is recommended as this destroys any beetles that are still in the plantation.

When efficiently applied these cultural measures alone have resulted in satisfactory control of the berry borer. However, there have been cases when such procedures have appeared to be inadequate. In such situations insecticides have to be used as a supplement. Dieldrin 18 per cent emulsion concentrate at the rate of 625 ml in 200 litres of water, sprayed on each infested tree twice at 3-weekly intervals, has given excellent control in Kenya.

Partial management success with stemborers

Insect borers have presented probably the greatest challenge to pest management in coffee plantations. Many reasons could be advanced to explain this. Firstly, the ovipositing females have no marked swarming season and many of them are active throughout the year. Secondly, infestations become apparent only after the damaging larvae have already bored into the branch or stem and at this time they are not reached easily by insecticides. For these and many other reasons the control of borers has been mostly partial, as the following examples illustrate.

The black borer, *Apate monachus*, occurs widely in Africa and has been reported in Cuba and Puerto Rico. The adults tunnel into the coffee tree from ground level to the tip of the stem including the branches. Females lay their eggs after the plant or portion of it has died. The stem is so weakened by the tunnelling that it breaks easily even in a gentle wind. Attempts to control the borer have included: (i) burning all dead wood which may contain larvae; (ii) destruction of infested coffee and alternate host plants; (iii) hand-collection of adults; (iv) light-trap catches; (v) extraction of larvae from the tunnels using a wire; and (vi) injection of petrol or carbon disulphide into the borings to kill the larvae. Larval extraction by wire has the disadvantage of causing damage to the tree because chiselling is necessary to allow the insertion of the wire. Besides this it is inefficient, expensive and time consuming. The above methods may be practical and effective on small farms, but they are very costly in large plantations.

The coffee stemborer, *Xylotrechus quadripes*, is an occasional severe pest of coffee in southern India and south-east Asia. Eggs are laid in crevices in the bark in which the larva initially tunnels before tunnelling the wood. The beetle is a pest of *C. arabica* although it will also attack *C. canephora* and other wild host plants. The tunnelling which may extend to the roots causes extensive damage to the plant often resulting in death of 7–8 year old trees. Partial control has been achieved by providing shade and optimum nutrition to the trees, removal of both infested trees and the dry scaly bark to limit oviposition sites, and the application of insecticides to the stem and thick primary branches to kill adult beetles during the flight season.

Another stemborer, *Neonitocris princeps*, is a pest of Excelsa, Arabica and Robusta coffee in West Africa and parts of Central Africa. The fertilized female lays eggs on the main branches and the young larva having penetrated the bark into the wood bores a gallery through the branch into the trunk and down into the principal roots. Pupation takes place within the principal gallery. Tunnelling weakens the stem which as a result easily breaks or may die altogether. Cultural control measures include the removal of branches bearing symptoms of attack and hand-collection of adults. Whereas these measures can effectively control the pest in small coffee plots, they are almost impractical in large plantations. In the latter it is recommended to introduce carbon disulphide or carbon tetrachloride into the borings and then to seal the holes with clay. The problem here is that identification of infested trees by careful examination of all trees is an expensive exercise in both labour and time.

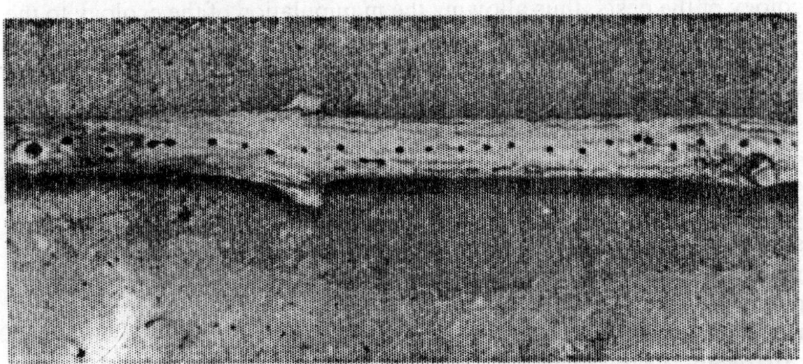

Fig. 6.1 A coffee branch showing holes made by the yellow-headed borer, *Dirphya nigricornis.* (Courtesy of P.T. Walker.)

Dirphya nigricornis, the yellow-headed coffee borer (Fig. 6.1) is a minor pest on Arabica coffee growing areas in Kenya, although severe localized infestations are reported from small-scale growers in other Kenyan localities, such as the Taita Hills and large plantations in Kiambu. It is not a pest of significance in Malawi and Tanzania. Control measures advocated to growers in Kenya include the cutting off of infested primary branches before the larva reaches the vertical (i.e. main stem), burning old heads to destroy any larvae or pupae within, and destruction of larvae using a wire inserted through the lowest hole through which frass (i.e. excreta) is ejected. Small quantities of dieldrin 18 per cent emulsion concentrate (= miscible liquid) diluted 1:100 with kerosene and injected through the lowest frass-ejection hole, is more reliable than killing larvae by inserting a wire. These measures, and the inclusion of hand-collection of adults for destruction, have not been entirely successful and alternative methods are under investigation.

Beetle borers are not the only problem that besets the coffee grower. Lepidopteran borers such as *Xyleutes armstrongi*, of considerable importance in Sierre Leone and Zaire, and *Zeuzera coffeae*, a severe pest in India, Indonesia, Papua New Guinea and neighbouring countries, are equally difficult to control. The use of attractants, such as in pheromone traps, for both forecasting pest outbreaks and pest control, the introduction of suitable parasites, and the destruction of alternative wild host plants all require further investigation.

Conclusions

The successful control strategies of the coffee pests that have been described are based on a thorough understanding of the biology and ecology of the pests, thus allowing the manipulation of the ecology to the disadvantage of the pests. Insecticides are not altogether prohibited, but their use is minimized and they are applied selectively so as to greatly reduce environmental damage. Correct identification of a pest is essential, especially when biological control is to be considered. Incorrect identification, as happened with the common coffee mealybug, *Planococcus kenyae*, in Kenya, can lead to frustrations and expensive importation of parasites and predators not specific to the pest. Indigenous pests are usually easier to manage through encouragement and augmentation of their natural enemies, possibly combined with careful selective insecticidal applications, than introduced ones which often require more extensive insecticidal treatments.

In recent years there has been a definite shift in Kenyan coffee plantations towards using insecticides with short-lived residual activity, and metabolites which are easily biodegraded. Anti-feedants, such as triphenyl hydroxide, which inhibit the feeding of leaf-eating caterpillars have also been used. They have been applied effectively to control lepidopteran giant loopers, *Ascotis selenaria reciprocaria*, and green loopers, *Epigynopteryx coffeae*.

Above all, the success of any pest management programme is highly dependent on the co-operation of the coffee planter. Legal enforcement alone will not convince the planter to follow instructions from the extension officer.

Management of cotton pests

R. Kumar

Although cotton is a tropical plant, commercial cotton production is not restricted to the tropics; indeed the greater part of the world's crop is produced in the temperate zones. Cotton is one of the world's more important crops supplying the basic raw material for the textile industry and assuming greater importance in comparison to synthetic fibres during an energy crisis.

The Asiatic or Old World cottons, *Gossypium arboreum* and *G. herbaceum*, which originated in Sind, India formed the bulk of cotton produced for centuries. During the last 150 years, these species have largely given way to *G. hirsutum* (American 'Upland' cotton) which originated, and was first established, in what is now southern Mexico and Central America. Another species, *G. barbadense* (Egyptian varieties), with its extra long, fine fibre is in demand by manufacturers of sewing thread and by fine weavers throughout the world, and is largely confined to Egypt and Sudan.

Actively growing cotton seems ill-equipped to defend itself against most pests. No less than 1326 species of insects have been recorded on this crop; 482 from Africa south of the Sahara. Fortunately, most of these insects are of no economic significance but the few (10–12 species) which are serious pests can devastate the crop, some in a matter of days. With proper insect pest management, however, yields of seed cotton can be increased from 224–448 kg/ha to 1680–2240 kg/ha.

Old World pests

Insect pests of cotton in the Old World comprise the following:-

Bollworms

These are the larvae of certain lepidopterous species; bollworms are probably the most important pests all over the world. One of the most destructive genera is *Heliothis* with species found throughout the Old World. *H. armigera* is frequently called the American bollworm and is closely related to *H. zea* and *H. virescens* of the western hemisphere. The larvae of this bollworm can damage cotton as soon as flower bud formation ('squaring') commences. They are more mobile than other bollworms and feed with much of the body exposed, whereas other bollworms feed inside the fruit, moving out of it only if unable to complete development there.

They normally hollow out a bud completely before attacking the next one. Each larva can damage up to 12 fruits. All species of bollworms not only cause damage by direct feeding but also provide entry points for secondary infections which result in boll rot.

The pink bollworm is the larva of *Pectinophora gossypiella* which tunnels inside fully mature bolls, feeding on developing seeds in the bolls. The damage is often well advanced before their presence is realized. Fully grown larvae may enter diapause but in the equatorial regions (7°30′N to 7°30′S) they survive on alternative malvaceous plants.

Earias spp. include the spiny and spotted bollworms of Africa and Asia. They feed on the growing shoots of young plants and bore downwards inside the stem; late season populations attack the buds and bolls as does *Heliothis*. There is no diapause in the life-cycle and they feed on many malvaceous plants. Red bollworms include two species of *Diparopsis*: *D. castanea* in southern parts of Africa; and *D. watersi* (Sudan bollworm) in northern parts. The young larvae may consume flowers, attack flower buds and bolls which are then shed. The larvae pupate within 8 cm of soil and towards the end of the season pupae undergo diapause.

The false codling moth, *Cryptophlebia leucotreta*, is of economic importance in many parts of tropical Africa, especially in Ghana, Nigeria, Zaire, Uganda. The larvae prefer partly ripened bolls with soft tissues and tend to feed mainly on large green bolls. They cause boll rot which is a major cause of crop losses in Ghana.

Other lepidopterous pests include the cotton leafworm, *Spodoptera littoralis*, cotton leaf roller, *Sylepta derogata*, and the cotton semi-looper, *Cosmophila flava*.

Sucking pests

Cotton stainers (*Dysdercus* spp.) These insects, though distributed widely, are of economic importance on cotton only in Africa. During feeding on green bolls they transmit a fungus, *Nematospora gossypii*, which causes discoloration of the fibres. Similar damage results from the feeding of stink bugs, e.g. *Nezara* spp. and iridescent Calidae spp.

Jassids (*Empoasca* spp.) Nymphs and adults feed on small veins of the cotton leaves injecting phytotoxic saliva into the food-distributing vessels. The toxic saliva reaches the margins of the leaves and causes the development of 'hopperburn'. In the Sudan it has been shown that in a season of heavy attack by jassids, application of nitrogen fertilizers to the crop prolonged the infestation. *Empoasca fascialis* and *E. lybica* in Africa, and *E. devastans* in India, are particularly important.

Aphids The cotton aphid, *Aphis gossypii*, not only damages cotton by its feeding but secretes honeydew on the open bolls. Black sooty mould grows on this secretion and this stains the lint, thereby lowering the lint quality.

Whiteflies The cotton whitefly, *Bemisia tabaci*, causes debilitation of the plants because of leaf feeding by the flattened, immobile, scale-like nymphs and the active adults. A sooty mould develops on the secreted honeydew and this affects the quality of the lint. The pest is particularly important in areas with high temperature and humidity and low rainfall – notably the Sudan and parts of the Indian subcontinent. It is also found in other African and Asian countries.

Lygus bugs These insects attack young and tender vegetative growth of the cotton plant, puncturing plant tissues. This causes the leaves to become ragged and the flower buds to be rapidly shed.

Mirid bugs The mirid bug, *Helopeltis schoutedeni*, damages leaves by sucking young green tissues. Heavy attacks on bolls cause them to open prematurely and the contents to be exposed to secondary infestation by bacteria and fungi.

Spider mites These arthropods, especially the *Tetranychus* spp., have become a problem in most cotton growing areas. Severe leaf shedding and consequent loss of crop follow feeding by the pests.

New World pests

The pest fauna of most South and Central American cotton growing areas is in many ways similar to that of the USA. Here the most destructive pests are: boll weevils (*Anthonomus grandis*); bollworms (*Heliothis zea*); tobacco budworms (*Heliothis virescens*); spider mites (*Tetranychus* spp.); pink bollworms (*Pectinophora gossypiella*); aphids (*Aphis medicaginis* and *A. gossypii*); thrips (*Frankliniella* spp.); cabbage loopers (*Trichoplusia ni*); leaf perforators (*Alabama argillacea*); armyworms (*Spodoptera* spp.); and the whitefly (*Bemisia tabaci*).

There is an intimate relationship between the phenology of the cotton plant and the insects attacking it. Figure 6.2 illustrates the relationship which must be taken into account when considering strategies for pest management.

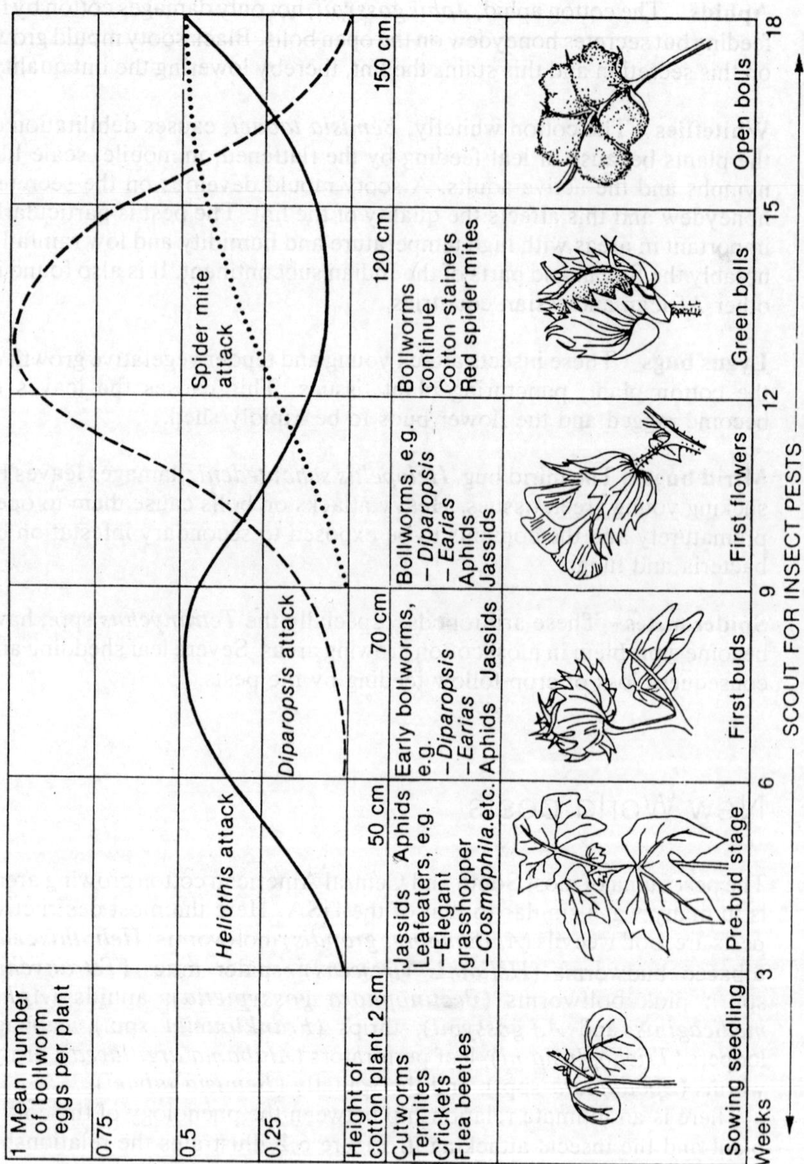

Fig. 6.2 Typical pest attack timing, and build-up in relation to the growth stages of cotton. (Modified from Tunstall and Mathews (1972).)

Management of cotton pests

The cotton plant requires attention throughout its life and strategies for pest management have been well studied and documented in the literature.

Cultural practices

The history of pest control in cotton has shown that several cultural practices – such as enforcement of a closed season and destruction of crop residues – cannot be bettered by modern insecticide technology. However, they were abandoned in favour of synthetic insecticides. In the USA during the 'pre-calcium arsenate period', farmers were advised to follow some 18 comprehensive practices at various intervals. With the advent of modern insecticides these practices were discontinued, but it now seems that the methods available some 50 years ago will form the core of pest management programmes currently being developed in the USA.

Correct sowing date

Within the constraints of environmental factors such as temperature, rainfall or irrigation, there may be a date *before* which cotton must not be planted for entomological reasons. Planting at such a date would avoid the emerging populations of diapausing bollworms, such as the pink bollworm; the dates would vary according to the local environmental conditions and must be worked out for each situation. Cotton should then be planted over a large area simultaneously to avoid build-up and movement of pests through cotton of varying ages. Where the flea beetle, *Podagrica puncticollis*, which is a pest in most parts of the world and the harvester termite, *Hodotermes mossambicus*, which is a nuisance in some parts of Africa, are a problem, the crop must be protected by dressing seeds with a suitable insecticide. In the Sudan, early-sown cotton gives a better crop but is liable to flea beetle and thrips damage, but these are managed by all available protection methods.

Fertilizer applications

Work in the Sudan and elsewhere has emphasized the need for correct fertilizer application. Over-fertilization with nitrogen increases the severity of jassid and whitefly attack as well as prolonging the period of pest control. Under-fertilization, however, offsets the gains of successful chemical control of the pests.

Destruction of crop residues and weeds

Many cotton pests tend to survive in crop residues and weeds to infest the new crop in the following season. This is especially true of the pink bollworm and the Sudan bollworm in many parts of the world. The practice of a closed season for cotton was therefore developed to eliminate this source of primary pest infestation. This remains a primary defence against several cotton pests and reduces the incidence of viral and bacterial diseases. As soon as possible after harvest, all the previous season's cotton plants must be destroyed, preferably by burning. This practice of phytosanitation is decreed by legislation in the Sudan. Before starting to plant the next year's crop, seedlings growing from diseased bolls or seeds are uprooted, and shoots and old root stumps are also removed. The enforcement of a closed season helps to break the food-cycle of many cotton pests. In China, hand-collection and burning of infested bolls and stalks is practised.

Where irrigation facilities are available, flooding the fields after harvest is an effective method of destroying the resting (diapausing) pupae of bollworms such as the pink and Sudan bollworms. Recent work in the USA suggests that irrigation of firmly compacted soils does not result in the high mortality of bollworms and tobacco budworms as does irrigation of loose soils.

It should be remembered that the management of an insect pest requires a knowledge of its ecology and of its host crop. With this knowledge, known methods can be adopted for its control as well as devising new ones.

Use of trap crops and the hand-picking of pests

In California manipulating alternative crop hosts of certain cotton pests such as *Lygus* spp. has been reported to have given successful control. The preferred plants are alfalfa (*Medicago sativa*), and safflower (*Carthamus tinctorius*) which constitutes a reservoir crop. On the latter crop just two well-timed insecticidal treatments kill *Lygus* spp. in large numbers.

Hand-picking of eggs, larvae and insect-infested parts of plants was widely practised before the advent of modern insecticides, and still has a definite place in labour-intensive agriculture of the Third World tropics. For example, hand-picking of egg masses as a control measure of the cotton leafworm, *Spodoptera littoralis*, is practised in Egypt.

Use of resistant varieties

Crop resistance is probably the most promising way of checking pest and disease attack, and much progress has been made by geneticists and plant breeders. In the USA, the gossypol content in cotton varieties has been correlated with resistance to certain insect pests and plant pathogens. Considerable effort has been employed in increasing the gossypol content of the plants, especially of the flower bud (square), which is the feeding site of *Heliothis virescens*. Recently, it has been shown that the activity of gossypol is supplemented by the antibiotic activities of 3 other compounds present in the flower buds.

Currently attention has also been focused on the development of early maturing cotton capable of escaping late season pests. Such short-season cotton has the added advantage to growers of requiring considerably smaller amounts of inputs such as fertilizers, irrigation and insecticides. Recent work in the USA has shown that short-season cottons may be harvested 22 days earlier, thereby disrupting the regular breeding and feeding patterns of many pests. Large scale plantation of such varieties has obvious advantages and requires further research. Development of multi-pest resistant lines of cotton has reached experimental stages. The process of development of such cultures is admittedly slow but such research should continue side by side with the available short-term methods of combating pests.

Use of biological control agents

The use of predators, parasites and pathogens such as viruses, bacteria and fungi, in natural control to help keep the pests in check is an important component of integrated control. Cotton pests have been shown to possess a rich and complex fauna of natural enemies. However, their precise role in the regulation of pest populations in any cotton ecosystem is poorly understood. Until recently little attempt was made to use these enemies in the management of pests in the cotton agro-ecosystem. Although a number of countries have pests of foreign origin, little attention has been directed to searching for their enemies in their countries of origin.

With the development of pesticide resistance some effort is now being directed to using natural control to suppress pest populations in cotton fields. Experiments in the USA are currently in progress using microbial insecticides such as *Bacillus thuringiensis*, nuclear polyhedrosis viruses (NPVs), egg parasites, *Trichogramma* spp., either alone or in conjunction with insecticides. The *Trichogramma* are, however, highly susceptible to insecticides and additional research in determining the most effective ways of using such enemies in different situations is urgently required.

Use of pheromones and hormones

During the last 20 years considerable research has been carried out in the use of chemicals which mediate vital life processes. Sex pheromones are important in insects seeking mates. Among cotton pests they have been extensively studied in the pink bollworm moth (Gossyplure) and boll weevil (Grandlure). A number of attempts have been made to use these chemicals to control the respective pests by mass trapping and disruption of mating communication. During 1976 and 1977 feasibility studies of area-wide trapping of male pink bollworm in a cotton insect pest management programme in the USA have given encouraging results.

Pheromones of the red bollworm, *Diparopsis castanea*, have been studied in Malawi, while the disruption of communication between adult male and female cotton leafworm, *Spodoptera littoralis*, has been reported by spraying micro-encapsulated formulations of a pheromone inhibitor in north-west Crete.

In experiments in 1972, in the USA, sterile male boll weevils, with their wing covers glued together, were released on cotton plants to provide a natural source of pheromone to attract over-wintered native populations to insecticide-treated strips of cotton crops where they were killed. Despite these experiments and the value of pheromones in surveys and monitoring, no spectacular results have been achieved in controlling cotton pests with these chemicals. However, they still have a role to play in integrated control schemes for the management of cotton pests.

The use of hormones in cotton pest control has been limited to experiments using the insect growth regulator, diflubenzuron, which inhibits chitin synthesis thus disrupting the formation of insect cuticle. Field trials suggest that the chemical is effective in suppresing boll weevil populations provided the entire cotton areas are treated to avoid influx of weevils from untreated areas. Further trials in this field are required.

Genetic control

The use of genetically impaired pests to limit reproduction and survival of their own species in natural populations is another tool in the arsenal of the entomologist. However, the only known use of this method in cotton pest management is in San Joaquin Valley in California. Here the pink bollworm has not as yet become established, but there remains the danger that it might spread into the valley from nearby heavily infected areas. Because of this threat sterile moths have been released regularly since 1968, and during 1975–77 about 1–1.5 million sterile moths were released daily during the cotton growing season. These releases are believed to have helped suppress breeding of the pest in the San Joaquin Valley, from immigrant moths known to enter the valley during the cotton growing season.

Chemical control of cotton pests

With so many different important pests attacking cotton, it has been necessary to use broad-spectrum insecticides in their control. Altogether, 25–30 per cent of the total amount of insecticides used on crops are applied to cotton. Till the development of pest resistance to insecticides, chemicals gave spectacular results and were used routinely as an insurance policy. It is such a large-scale use of insecticides (5–9 kg/ha in California in 1974 and as much as 75 kg/ha in Guatemala) in the cotton agro-ecosystem which is responsible not only for the development of resistance but also for the emergence of pests which were formerly unimportant species.

The history of cotton pest control using insecticides is especially well illustrated by the countries of Central America. Cotton forms the backbone of the economies of Nicaragua, Guatemala and El Salvador and, together with Honduras (where only a small number of hectares are planted), they produce over 2 per cent of the world's cotton. From 1949 to 1965 there were sustained increases in cotton yields due to the use of pesticides. But thereafter, as pest resistance to insecticides developed, cotton yields declined and each country suffered a severe economic crisis during 1965–70. In Nicaragua alone, it was estimated that as a result of an annual expenditure of $US10 million on the importation of pesticides, there was environmental damage estimated at $US200 million per annum. Losses included:

(a) Presence of residues in terrestrial and aquatic biota;
(b) The rejection of beef exports by buyers due to excessive DDT residues;
(c) Human poisonings and deaths due to pesticide toxicity;
(d) Significant increase in the incidence of malaria in the cotton growing areas due to the development of resistance to DDT by the malaria mosquito vector;
(e) Failure of farmers to grow crops such as beans and maize in cotton growing areas because of aggravation of pest problems due to the heavy use of pesticides on cotton.

After these disasters, an integrated pest management scheme shown in Table 6.1 was developed by research entomologists from the University of California, USA.

It is clear that reduction in the usage of insecticides along with non-chemical control measures is effective in controlling cotton pests, and is consistent with human concern with environmental pollution.

Experience in Central Africa has shown that satisfactory results in using pesticides are obtained when spraying is based on weekly scouting for cotton pests. Small-scale farmers, who grow 90 per cent of the cotton crop in Zambia, are trained to do their own scouting. This enables the correct timing of pesticide applications and prevents unnecessary spraying. Ultra-low-volume spraying is being used in areas of water shortage, and training is being imparted to farmers in the handling and use of insecticides. By scouting, the use of pesticides is kept to a minimum and allows them to be integrated with the use of jassid-resistant varieties and a closed season. The pink bollworm is a minor pest as a result of the closed season and the number of over-wintering Sudan bollworm is also reduced.

In Africa, cotton entomology is best developed in the Sudan where spray levels against the pests are determined by pest counts. The effects of application of nitrogen fertilizers on pest populations are regularly monitored. Despite excessive reliance on insecticides, cotton growing is still profitable in the Sudan, but the time has come for vigorous efforts to be directed to developing pest management programmes similar to those of Central America (Table 6.1).

Table 6.1 A summary of chemical pesticide usage for cotton pests in Central America (L.A. Falcon, personal communication)

Pest problems	Pest control programmes		
	Chemical approach	Integrated control approach	
		Earlier stage of development	Advanced stage
Boll weevil	15–20 applications, 4-day intervals when plants 21 days old	5–10 applications, based on need as determined by scouting	Use trap crop, hand-pick or apply chemical pesticide only to trap crops
Cotton leafworm	5–10 applications during first 100 days of cotton plant growth	0–3 applications of selective pathogens (*Bacillus thuringiensis*) as determined by scouting	Same as for earlier stage of development
Bollworm	10–20 applications, especially during boll formation period	5–10 applications as determined by scouting	0–5 applications of selective virus as determined by scouting
Other pests	5–10 applications, when present	0–5 applications as determined by scouting	0
Total applications	25–50	8–25	0–5
Kg of pesticide (a.i.) per metric ton of cotton produced	50–100 kg	25–50 kg	0–1
Cotton production/ha	2–5 bales, actual	2.5–5.5 bales, actual	3–6 bales, estimated

Conclusion

In conclusion, the complex cotton agro-ecosystem provides an ideal opportunity for entomologists to develop pest management systems. Experience in cotton entomology clearly shows that non-chemical methods of pest control, such as cultural practices and the use of resistant varieties, are indispensable when used in conjunction with insecticides in managing cotton pests. Further, there is scope for the incorporation in modern discoveries such as pheromones, hormones, and genetic manipulation in the pest management programme. Far more research is required on cotton pests with a view to developing suitable pest management schemes, especially the Third World tropics. There is also the urgent need to educate farmers and officials in the controlled use of pesticides.

Cotton pest management in Egypt

H.S. Salama

Cotton has dominated Egyptian agriculture for the last 10 years and is synonymous with super-high quality, long-staple cotton the world over. In 1978, the Egyptian cotton growing area was nearly 600 000 ha and the yield average 0.8 tonnes/ha of lint cotton; thus a total of about 0.5 million tonnes was produced that year. Because of the importance of Egyptian cotton and the great effort made by the Egyptian government, and other bodies, to introduce practical management strategies, a separate account on Egyptian cotton is merited.

Cotton (*Gossypium barbadense*) in Egypt is attacked by various pests during the different stages of development. The most important insect pests are thrips (*Thrips tabaci*), aphids (*Aphis gossypii*), cotton leafworm (*Spodoptera littoralis*), corn earworm or American bollworm (*Heliothis armigera*), pink bollworm (*Pectinophora gossypiella*), spiny bollworm (*Earias insulana*) and the bug *Oxycarenus hyalipennis*. Nematodes, spider mites and the plant diseases, *Rhizoctonia solani* and *Fusarium* wilt, are of lesser economic importance.

S. littoralis and *P. gossypiella* are the major pests on Egyptian cotton and they damage the crop between June and August each year. Special measures are taken to manage these pests.

Management of cotton pests

Chemical control

Until the late 1940s cotton pest control in Egypt was confined to legislative, mechanical, cultural and biological control; chemical control was largely limited to dusting cotton fields. Emphasis on the control of cotton pests with chemicals started with DDT in 1950, followed by a DDT–HCH–sulphur mixture in 1956, and by toxaphene initially at 0.6 kg/ha. Toxaphene failed in 1961 and was replaced by carbaryl, a carbamate insecticide, and by the organophosphate, trichlorphon. From 1965, various organochlorines, and mixtures of organophosphates and organochlorines including the cyclodienes were introduced to control the cotton leafworm.

In 1977, appreciable amounts of pyrethroids were used against cotton pests. For achieving a good efficiency of the currently used insecticides, some moulting inhibitors such as Dimilin are now being combined with organophosphorous compounds and carbamates for cotton pest control. Aerial spraying of cotton fields was introduced in the late 1960s and by 1972, the cotton area contracted for aerial spraying was 202 500 ha; by 1978, it was in excess of 0.5 million ha. Cotton is the major pesticide-consuming crop in Egypt. It accounts for about 70 per cent of all pesticides used in the country. The average annual cost of pesticides used in Egypt (excluding the cost of ground and aerial spraying) increased from $US25 million in 1974 to more than $US60 million in 1978. Many investigations have been made in Egypt in an attempt to apply other control methods and to reduce insecticidal applications. No single method, however, other than insecticides has been found to control the cotton pests at a reaslistic and acceptable level.

From the beginning of 1972, the agricultural co-operatives in the Egyptian villages became the sole agents for the chemical control of cotton pests. None of the pesticides recommended against pests are sold directly to the growers but only through co-operatives, and pesticides are used only under the supervision of agricultural officers and more or less on a collective basis. Half of the costs of pest control operations are shared at the end of the season by all growers in the area according to the hectarage of each, and the other half is met by the government.

Integrated control

Each year a national committee for pest control sets up an annual programme and recommendations for the next year, based on the current year's tests and observations. The Ministry of Agriculture, universities, national research centres and other appropriate institutes are represented on the committee which is presided over by the Minister of Agriculture.

Once the programme is approved by the committee, it is implemented by the Ministry of Agriculture and Co-operatives. The general principles which form a system for cotton pest management can be summarized as follows:

(a) The use of pesticides if necessary must be an integral part of a pest management system.

(b) Spraying cotton seedlings as a treatment against infestation by early season pests (e.g. thrips, aphids and occasionally spider mites and certain 'greasy' cutworms) is bad for beneficial insects and results in subsequent attack by *Heliothis* and *Spodoptera*. Accordingly, cultural control methods are stressed, and the earlier recommendations for a general protective spray against thrips has been discontinued. Instead, thrips populations are monitored and cotton fields are not sprayed unless the economic damage threshold of 8–12 insects per seedling is reached. By this procedure, the area of cotton sprayed against early season pests has been reduced from 400 000 ha in 1972 and to under 20 000 ha in 1979.

(c) The cotton leafworm breeds during the winter and spring in clover fields. Of the egg masses deposited in cotton fields during late May and June, 90 per cent are produced by moths emerging from pupae in clover fields. So it is required by law that the last irrigations of clover should not be later than 10 May to maintain dryness of the soil, thus causing mortality or malformation of at least 50 per cent of the pupae and moths. It had also been recommended that during this last irrigation of clover 30 litres of oil should be added to the irrigation water to kill the larvae and pupae of cotton leafworms. This procedure considerably reduced the level of infestation on cotton by first-generation pests. This general protective chemical spraying of all clover fields in May has, however, been discontinued because of its extremely harmful effect on beneficial insects.

(d) Direct control of the cotton leafworm in cotton fields starts in late May or early June. From late May to the end of June, all egg masses deposited by female moths on the cotton leaves are hand-picked by organized groups of young labourers. About 2 million children are employed during the period June to August. These children carefully search the rows of cotton plants for egg masses (300–500 eggs each) once every 3 days. It is estimated that children pick and ultimately destroy, by burning, about 50–80 per cent of the deposited egg masses. This method of mechanical control is environmentally acceptable since it assists natural biological control of cotton pests by preserving a population of beneficial insects during a very crucial period of the cotton. The pattern of abundance of most insect predators and parasites in the field indicates a rising population in May to a peak in June, then declining rapidly in July. When plants are large and hand-picking of egg masses is no longer efficient, or when leafworm populations are high, insecticides are applied as deemed necessary. For example, chemical control of cotton leafworm is delayed as long as possible until it can be carried out at the same time as the regular programme of sprays against bollworm infestations. However, in cases of emergency, for example, when the number of eggs are too large for the pickers to cope with and a threatening number of eggs hatch, treatment with insecticide is allowed.

(e) Periodic spraying against bollworm infestation begins when the economic damage threshold is reached in the bolls of cotton plants nearest to villages. This threshold is taken as a 10 per cent infestation rate of the green bolls sampled at random from cotton plants. The main sources of pink bollworm infestation of the new crop are the diapausing larvae within the dry bolls of the previous season still attached to cotton stalks piled on the roofs of village houses. Therefore, a project was implemented during the last season which aimed at collecting and burning these bolls before adult moths emerged. This eliminated much infestation. Moths emerging from these bolls in June and July deposit their eggs on the flower buds and newly formed green bolls of plants adjacent to the village boundaries – within a distance of 150 m to the north-west and 250 m to the south-east. This initial attack starts the first generation of the new season's infestation which spreads rapidly into the fields further away from the villages.

Bollworm infestation in Egypt is therefore more of a socio-economic than an entomological problem. Had another cheap source of fuel been available to the farmers and were they willing to change their age-old custom of relying on dry cotton stalks as a source of fuel in their homes, the control of pink bollworm in Egypt would have been much simpler.

The adoption of these management strategies in combination with only 3 insecticidal spray applications during the last 3 years has led to the production of a cotton crop with comparatively good yields.

Management of cocoa pests

R. Kumar and A. Youdeowei

The cocoa tree, *Theobroma cacao*, is a native of South America from where it has been introduced to various parts of the tropical forest belt since the sixteenth century. It seems to thrive best on moist porous soils and the young plant needs shade, protection from winds and high temperatures – conditions which are provided by tropical rain forests. Currently Africa, especially West Africa, accounts for about 75 per cent of world cocoa production.

Cocoa has evolved under 'pest pressure', escaping from natural enemies in its homeland only to be confronted in its new environments with many different pests and diseases. On a world scale, no other crop seems so prone to pests and diseases of serious consequences. This may be partly because cocoa thrives in the wet humid tropics in close association with forests which already harbour various pests and pathogens. Further, as a crop rich in proteins, carbohydrates and minerals, attacks by a variety of pests are perhaps not unexpected.

Although over 1500 different insect species are known to attack cocoa, with different continents having their own pest spectra, most are unimportant and only the following deserve attention.

Pest fauna of cocoa

Mirids (capsid bugs)
Some members of the Bryocorinae (Heteroptera:Miridae) are known to feed on cocoa wherever it is grown. These include the black, *Distantiella theobroma*, brown, *Sahlbergella singularis*, and glossy, *Bryocoropsis laticollis*, mirids in West Africa; cocoa mosquito-bugs, *Helopeltis* spp., in Africa, south-east Asia and Papua New Guinea; *Boxiopsis madagascariensis* in Madagascar; and *Monalonion* spp. in South and Central America. It is believed that these insects are attracted by methyl purines present in the plant. Mirids feed by inserting their long, needle-like mouthparts into the plant and sucking the juices. During the process the feeding points become water-soaked areas of damaged tissues, which later provide entrance points for a parasitic fungus – *Calonectria rigidiuscula* in West Africa (Fig. 6.3). This fungus often causes extensive die-back of branches. Light appears to be an important factor in influencing the damage by mirids. Anything which damages the canopy encourages mirid attack as light is then able to reach ground level, leading to the production of basal shoots on which the mirids flourish. Attack of the young growing branches causes severe damage to the trees (Fig. 6.4). Mature-pod feeding by mirids is seldom of severe consequence although *Monalonion* and *Helopeltis* can be destructive, and here also the damage is attributable largely to the action of pathogenic fungi – *Gelosporium* in Papua New Guinea and *Phytophthora* in Sri Lanka. Mirids feeding on young pods cause cherelle wilting and crop loss. Although crop losses due to mirids have been estimated at about 20 per cent of the total annual production in West Africa, adverse environmental conditions such as severe dry seasons and the northern harmattan wind also contribute to this crop loss.

Other heteropteran bugs
A number of shield bugs (Pentatomidae) attack cocoa – *Antiteuchus* in Central and South America and Trinidad, and *Bathycoelia thalassina* in West Africa. The latter causes the loss of about 18 per cent of Ghana's cocoa crop. Coreid bugs (Coreidae) such as *Pseudotheraptus devastans*

Fig. 6.3 Cocoa pods damaged by cocoa mirids in West Africa. (From P.F. Entwistle (1972).)

in West Africa and *Amblypelta* spp. in Papua New Guinea and the Solomon Islands may sometimes be important.

Mealybugs

These small, usually flat insects (Homoptera:Pseudococcidae) are generally covered with white secretions and are found wherever cocoa is grown. However, it is in West Africa that these bugs, *Planococcoides njalensis, Planococcos citri* and *Ferrisia* (= *Ferrisiana*) *virgata*, through their transmission of deadly virus diseases such as cocoa swollen shoot virus (CSSV), are of major importance to the cocoa industry. Trees infected by virulent strains of CSSV die and also provide foci for the spread of the disease. Trees with mild strains of CSSV develop a damaged canopy which encourages mirids to increase.

Fig. 6.4 Mirid damage to cocoa trees in Ghana. The trees have lost most of their leaves leaving exposed bare projecting branches. (From G.A.R. Wood (1975).)

Caterpillars and borers

Larvae of moths such as *Earias biplaga* (cocoa bollworm), *Anomi : leona* and *Characoma stictigrapta* (Noctuidae) attack young cocoa in Africa (Fig. 6.5). Larvae of the cocoa-pod borer *Acrocercops crai ierella* (Lithocolletidae) are very serious pests of pods in Saba, Philippi ies and Indonesia and also in Papua New Guinea. Pod damage by this pes t varies from 5–100 per cent depending upon the variety of cocoa and the locality in south-east Asia.

Ambrosia beetles, *Xyleborus ferrugineus*, (Coleoptera: Scolytidae) attack cocoa in the New World. The injury inflicted by the beetles enhances the effect of the vascular wilt fungus, *Ceratocystis fimbriata*. The weevil borers, *Pantorhytes* spp., have become increasingly important in Saba, Papua New Guinea and on the Bismarck and Solomon Islands. The longhorn beetles, *Glenea* spp., (Cerambycidae) are common in the Old World tropics and may be important pests in Java, Papua New Guinea and New Britain.

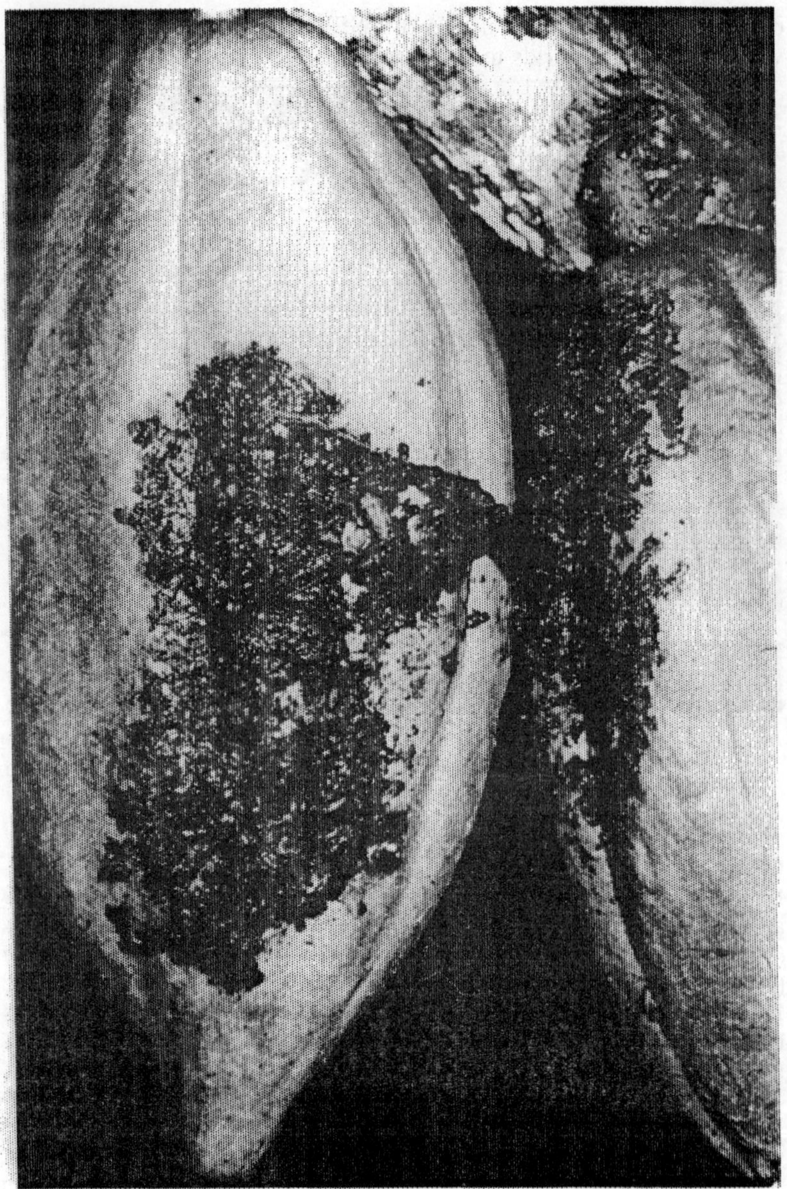

Fig. 6.5 Cocoa pods damaged by the moth borer, *Characoma strictigrapta*. (From P.F. Entwistle (1972).)

Ants

Leaf-cutting ants, *Atta* and *Acromyrmex* spp., are important in Central and South America and Trinidad. The *Azteca* group of ants tend sap-sucking bugs (aphids, whitefly, scale insects etc.). Some species, for example *Azteca paraensis* var. *bondari* in Brazil, bite young terminal shoots to extract mucilage for nest building thereby causing terminal defoliation. In West Africa ants such as *Crematogaster striatular* and *Pheidole* spp. tend mealybugs which transmit swollen shoot disease (CSSV). There is some evidence that *Crematogaster* ants can cause damage to cocoa pods, although this is not serious.

The background fauna

A wide variety of insects, for example grasshoppers (Orthoptera), stick and leaf insects (Phasmida), beetles (Coleoptera), thrips (Thysanoptera), mantids and cockroaches (Dictyoptera), springtails (Collembola), termites (Isoptera), ant-lions (Neuroptera), moths and butterflies (Lepidoptera), flies (Diptera), ants, bees and wasps (Hymenoptera), are found on cocoa farms. In addition to these invertebrates, vertebrates such as monkeys cause considerable damage to pods in Sierra Leone, while rodents such as rats and squirrels are of great importance in south-east Asia and in West Africa. Parrots also damage cocoa pods in Indonesia, causing up to 15 per cent crop losses. With so many of the invertebrates and vertebrates trying to subsist on one crop alone, it is amazing that any cocoa pods survive at all! Although most insects on cocoa are of little economic significance, changes in pest status occur from time to time. This may be due to crop and ecological changes caused by water stress, over-insolation and insecticide applications (see pp. 41, 192, 194, 195).

Insects and disease transmission

Black pod disease, caused by the fungus *Phytophthora palmivora*, has been reported from 59 cocoa growing countries. Global losses amount to as much as 10 per cent of world production. Under conditions of high rainfall and humidity this disease causes pod rot, but the losses due to this disease are variable and unpredictable, making it one of the major factors influencing world cocoa prices. Evidence regarding the role of invertebrates in the dissemination of *P. palmivora* has been steadily mounting in recent years. In West Africa, radioisotope tagging has implicated a variety of invertebrates in the spread of this disease, though atmospheric humidity is considered probably the most important factor affecting the incidence of black pod disease. The fungus can infect flower cushions, leaves, shoots, roots and young seedlings. The other important disease is cocoa swollen shoot virus (CSSV) which is transmitted by mealybugs in West Africa. So, here are two confirmed cases of the association of insects with the spread of disease in cocoa.

Management of cocoa pests

The cocoa farm forms a complex and heterogenous ecosystem – with large variations in the degree of canopy density, the number and species of shade trees present, the depth of litter, amount of dead wood and in the nature of the neighbouring vegetation – both from farm to farm and within the same farm throughout the annual cycle. Thus cocoa provides niches for a large and diverse fauna. It has been suggested that since this faunal diversity equals and resembles that of primary forest, the cocoa farm should be regarded ecologically as a type of forest – 'the cocoa forest'.

The insect fauna of cocoa has been studied in different parts of the world – in South America, Melanesia, as well as in West Africa – and has always been shown to be extremely diverse. Recent surveys in West Africa indicate that the magnitude of this diversity exceeds all previous estimates. With the great diversity of the background fauna, pest and disease management in the cocoa-ecosystem offers interesting possibilities. About 30–40 per cent of the cocoa crop in Africa and elsewhere may be lost due to the destruction and debilitating effects of pests and diseases. It must be remembered, however, that losses attributable to pests and diseases are often compounded by the trees suffering from old age, bad management, poor nutrition, and competition with weeds.

The cocoa-ecosystem is an example of a semi-permanent ecosystem. A cocoa tree lives for at least 80 years and so the system persists almost indefinitely unless destroyed by external agencies. However, the cocoa-ecosystem is in a constant state of flux. The dynamics involve not only the gross morphology of individual trees, but also the quantity and quality of its biotic and abiotic environment. Qualitative changes in the cocoa-ecosystem are associated with changes in the stage of development of the crop. In young cocoa-ecosystems, sap-sucking insects (aphids, psyllids, mealybugs and thrips) and defoliators (caterpillars) constitute the dominant pests. As the system matures, dominant insect species are mirids, shield bugs, pod-boring caterpillars, psyllids, ants and grasshoppers. The management systems adopted for cocoa pests must therefore be sufficiently dynamic to respond to changes in the relative importance of pest species. Moreover, the interaction between the effect of seasonal climate on host phenology and the dominance and abundance of the various pest species must be carefully considered. For example, there are particular months during which cocoa mirids appear in large numbers, and if insecticidal treatments are to be used, they must be used only during these months.

A successful cocoa pest management scheme involves operations during both plantation establishment and plantation maintenance, as well as a well co-ordinated 'Pest Surveillance System'.

Plantation establishment

The main objective is to eliminate all possible sources of pest incidence, and some of these are described below.

Wild hosts of cocoa pests When plantations are established under forest shade, the shade trees should be carefully examined, and all plants identified as wild hosts of cocoa pests must be removed. For example, *Fagara macrophylla* is a known alternative host plant for cocoa mirids in West Africa, *Nephelium* and *Cynometra* are host plants of the cocoa-pod borer in south-east Asia. The continued presence of such plants in the vicinity of cocoa creates continuous foci of pest infestation on the cocoa crop.

Shade trees Shade trees planted during the establishment of the cocoa farms must be regularly inspected for the presence of cocoa pests. *Leucaena* and *Gliricidia* tend to harbour defoliators, mealybugs and pathogens which later affect the cocoa trees adversely.

Sanitation All dead and decaying vegetation must be collected and buried within new cocoa plantations. All planting material transported from one place to another should be inspected carefully for the presence of adult insect pests, their eggs or immature stages, before they are used. Stemborers, mealybugs, and larvae of cocoa-pod borers are readily spread with pods or rooted cuttings which are used as new planting material.

Plantation maintenance

Routine maintenance is essential not only for the health of the cocoa tree but also for the control of pests and diseases. For this purpose the following maintenance practices must be followed routinely to form an integral part of pest management in cocoa farms.

Weeding Weeds are particularly important during the early years of the establishment of cocoa. They also grow rapidly whenever there is a break in the canopy. Weeds compete with the crop for minerals and water and other essential requirements, so their removal means that more raw materials will become available for the growth and yield of cocoa. Removal of weeds also helps to circulate air and lessens the relative humidity within the farm with a consequent reduction of the incidence of black pod disease.

Pruning This involves regular removal of low branches and all basal chupons (cocoa stems growing vertically upwards) so as to give a closed, umbrella-shaped canopy. All unwanted growth should be pruned ruthlessly. Formation of a closed canopy is a good assurance against mirid attack which tends to occur wherever there is a break in the cocoa canopy. Pruning assists in obtaining a tree structure convenient for harvesting, spraying and other agronomic operations.

Shade management and intercropping In Africa deciduous trees do not provide as satisfactory a shade regime as could be obtained by an economically orientated policy of intercropping, such as practised in south-east Asia, notably Malaysia, and the Philippines where cocoa is interplanted with coconuts or rubber. Emphasis should be on introducing trees which support a fauna of predatory ants, such as *Oecophylla*. Deciduous trees shed their leaves during the dry sunny season when cocoa needs most protection; consequently the use of such trees should be avoided. The importance of shade manipulation as a pest control measure has not been appreciated sufficiently. In West Africa shade removal leads to a large build-up of ground-nesting ants of the genus *Camponotus* which are active predators of some cocoa pests, but further thinning of shade trees and breaks in canopy reduce populations of predatory ants such as *Oecophylla*. Field trials have shown that in addition to preserving predators, shade removal can also increase the yield of Amelonado cocoa. Whenever large gaps occur in the canopy they should be immediately closed by planting fast growing trees like bananas, plantains or *Gliricidia*.

Plantation sanitation Removal of epiphytes, climbers, mistletoes (in West Africa, Indonesia, and Costa Rica), and the removal of dead wood from the trees and the ground should be carried out routinely. Dead, decayed and rotting wood should be removed regularly to prevent the establishment of termites and foci for the development of fungi and wood-boring insects. Dead pods, whether resulting from black pod or not, should be buried and allowed to rot at the bases of the cocoa trees as a mulch.

Continual cropping of cocoa leads to the removal of potassium from the soil because cocoa pods contain 0.8 per cent dry weight of potash. Old pods and broken pod husks should be buried and allowed to rot within the plantation in order to return some potassium to the soil. Old pods left on the trees, besides being a source of black pod sporangia, are used together with dead twigs, as nest sites by *Pheidole* and *Crematogaster* ants in West Africa. Good farm sanitation is therefore of prime importance in limiting the effects of cocoa pests.

Pest surveillance system An efficient pest monitoring or surveillance system is a vital component of an integrated pest management scheme for cocoa pests. The system involves the regular use of a team of field staff who have been trained in the pattern of growth of the cocoa trees and in the recognition of all forms of insect pests and their damage to cocoa. The objectives of accurate pest surveillance are:

(a) To monitor the growth and development of the cocoa crop in relation to pest incidence;

(b) To monitor populations of pests and evaluate the extent of their damage to cocoa;

(c) To provide information which will be used for making early decisions on the control of insect pests.

The procedures for pest surveillance and control are as follows:

(a) Perform routine inspection of the cocoa trees throughout the year for pest incidence and damage; mark those trees with pests and showing damage and spray selectively with an appropriate insecticide;

(b) Evaluate the success of the insecticidal treatment and repeat if necessary;

(c) If pest control is achieved, then continue pest surveillance throughout the year, in order to detect fresh sources of pest infestation.

Cocoa pest management methods used in the tropics

Chemical control

The use of organochlorine, organophosphate or carbamate insecticidal sprays is the quickest method available in West Africa and south-east Asia for the control of major cocoa pests. Recommendations for a spray regime in West Africa were based on field experiments in Ghana, Nigeria and the Ivory Coast. Cocoa plantations are usually sprayed with appropriate insecticides when the pest population has appeared on the crop and is causing considerable damage. Cocoa is sprayed against cocoa mirids in August, September, October and December when the pest populations approach their maximum numbers.

Chemical control of mirids (Fig. 6.6) is most beneficial for a cocoa tree. Even seriously affected trees whether young or old, have a remarkable power of recovery as soon as they are properly treated. After effective spraying the canopy is soon restored, cocoa production commences and chemical treatments can be drastically reduced. Furthermore, the cost of such treatment is equivalent to just a few dozen kilogrammes of dried cocoa per hectare. While it is true that the use of chemicals in cocoa farms has resulted in the development of insecticide resistance in cocoa mirids and in the upsurgence of secondary pests such as *Bathycoelia thalassina* in Ghana and *Characoma stictigrapta* in Nigeria, rational chemical treatment still offers the best hope for the control of mirids. It has been shown that in 3 years, by effective mirid control, the potential production of a cocoa plantation can be doubled.

Fig. 6.6 Chemical control of cocoa mirids in Nigeria: spraying mature cocoa trees with 'Gammalin-20' (HCH) from a pneumatic knapsack sprayer. (A. Youdeowei).

Physical and cultural control

Not all major pests of cocoa have been successfully controlled by spraying with insecticides. In the south-east Asian countries of Malaysia, Philippines and Indonesia the cocoa-pod borer moth, *Acrocercops cramerella*, has not been successfully controlled by the use of chemicals. Two main methods have been adopted to deal with this major pest problem.

Rampassan This is a system of removing all cocoa pods from every cocoa tree in the plantation in order to break the breeding cycle of the cocoa-pod borer. The system has not been particularly successful because of the continuous podding characteristics of the new Amazon cocoa varieties which have been planted, and also because certain pods within the cocoa canopy are missed during the rampassan operation. Thus, some foci of pod borer infestation are left in the plantation to continue the moth breeding cycle. A more drastic method which has been adopted for controlling this pest is the complete cutting out of the cocoa trees, similar to cutting out trees infested with the cocoa swollen shoot virus in West Africa. By removal of all trees and replanting after about 10 years, it has been possible to eliminate this pest from parts of Java in Indonesia.

Fig. 6.7a

Fig. 6.7b Bagging cocoa pods: (a) young trees with pods covered with paper bags to protect them from attack by *Acrocercops cramerella*, the cocoa-pod borer, in Indonesia; (b) trunk of an old cocoa tree in Indonesia showing pods covered with plastic bags to protect them against *A. cramerella*. The use of plastic bags is preferred to paper ones because the former do not tear so easily. This method of cocoa pest management is now used successfully in Indonesia and the Philippines. (A. Youdeowei).

Bagging This method of control involves enclosing 3-month-old cherelles in paper or polythene bags (400 X 200 mm) open at the free end, so as to protect the pods from oviposition by adult cocoa-pod borer moths (Fig. 6.7a and b. If the pods are bagged at the appropriate time, this method gives up to 90 per cent protection of the crop and is comparatively cheaper than unreliable chemical control for this pest. It is, however, only practicable for small cocoa plantations of less than 100 ha. The method is now widely practised in Indonesia and the Philippines.

Control of mealybugs and swollen shoot (CSSV) in Ghana

The difficulty in controlling mealybugs lies in the fact that their many natural enemies already keep pest populations at low levels. It is the ability of the low numbers of mealybugs to cause considerable damage which defies attempts in using natural enemies to control them. Therefore attempts at propagating exotic and indigenous parasites against them have had no effect in reducing the incidence of swollen shoot disease (CSSV). Since 1946 swollen shoot disease in Ghana and Nigeria has been controlled by cutting out the infected trees. In Ghana, authorised agencies have been carrying out a campaign of location and destruction of such trees. The enormity of the task is frightening – over 160 million trees have been removed so far. The problem has not only been to do the job. It has been equally difficult to persuade the farmers, first to have their cocoa trees cut down when they may still be carrying a crop and, secondly to replant the land with cocoa.

Biological control

In Java, the black ant, *Dolichoderus bituberculatus*, has been used to deter *Helopeltis* on cocoa. This ant protects trees not by predation on mirids but simply by disturbing them. In Cameroun, the farmers artificially propagate the ant, *Wasmannia auropunctata*, to keep cocoa trees mirid-free. In Papua New Guinea, the ground-nesting ant, *Anoplolepis longipes*, is also known to keep cocoa free from mirids. *A. longipes* is also known to restrict infestation of weevil borers in Papua New Guinea and neighbouring cocoa growing areas. Success has been reported from the Ivory Coast in rearing egg parasites (*Trichogramma* spp.) of *Earias biplaga* on a commercial scale, and releases have been made in the cocoa-ecosystem.

Prospects for biological control of cocoa pests

Parasitism by the braconid wasp, *Leiophron sahlbergellae,* of 3rd- and 4th- instar larvae of *Sahlbergella singularis* has been found to vary from 15–41 per cent (mean 25%) in Ghana and 6–21 per cent in Nigeria. In Ghana, parasitism of *Distantiella theobroma* by this parasite is negligible, being of the order of 0.1 per cent and it is in turn attacked by a hyperparasite. Egg parasitism by a scelionid wasp, *Telenomus* sp., both in *D. theobroma* and *S. singularis* is normally below 10 per cent.

As *D. theobroma* is a major pest in Ghana and it is virtually without a parasite, the possibility of introducing an enemy against the bug from south-east Asia has been considered. Due, however, to the heavy financial investment required for pursuing such a programme, no progress has been made in this direction. Large-scale multiplication and release of *L. sahlbergellae* have not been practicable in view of the difficulties in continuously rearing appreciable numbers of any species of cocoa mirids.

An undetermined but highly lethal bacterial pathogen of *S. singularis*, discovered in Nigeria, has been found to be equally effective against *D. theobroma* and *Helopeltis* pests, but not *Bryocoropsis laticollis* in Ghana. Experimental spraying with a suspension of this bacterium resulted in a 32 per cent kill of mirid populations, compared with unsprayed trees which acted as a control. Difficulties in maintaining and building up viable cultures of this pathogen have limited its use.

As established by radiotracer and other studies, mirid bugs in West Africa are attacked by a wide range of predators. These include mantids, reduviids, crickets, long-horned grasshoppers and ants. From mirid survival studies, these organisms have been found to cause 3.5–9.0 per cent reduction in mirid populations. The only specific negative association was found to exist between the mirid, *Distaniella theobroma*, and the ant, *Oecophylla longinoda*, which was estimated to reduce mirid populations by 15–20 per cent. The idea of manipulating the environment of a facultative predator so that it may heavily predate its prey (cocoa pests in this case) has been developed in Ghana. But attempts at the artificial spread of *Oecophylla* in Ghana, have not been successful. Colonies of this ant tend to be stable and may occupy a nesting and foraging territory over several years without venturing into mirid infested trees. Also, this ant is absent from areas of broken canopy where the mirids thrive.

Several authorities have emphasized that ants exercise a major influence on the composition of the fauna in tropical ecosystems. Recently some workers in West Africa have shown that major pests and diseases of cocoa are influenced by the distribution of some dominant ant species. They have suggested that manipulation of ant distribution could form the basis of an integrated control scheme for cocoa pests and diseases. However, considerable work is required in elucidating the characteristics of interactions between different ants before manipulation of ant distribution can be introduced as a viable method for cocoa protection.

Use of resistant cocoa varieties

Recently a project supported by British Technical Aid has been working at Tafo, Ghana, to produce cocoa hybrids resistant against swollen shoot virus. The team has developed a technique for screening progenies from which selected strains go forward into field tests. The programme aims at selecting replacements for Amelonado cocoa which is at present used as the male parent in the seed gardens of Amazon female clones which provide the current seedlings for farm planting. Some of the inter-Amazon progenies which were consistently more resistant in screening tests than the equivalent Amazon–Amelonado hybrids have been planted in field trials for further assessment. The complexity of the programme may be judged from the fact that it will take 8–13 years before the resistant material can be released to the farmers for planting.

Conclusion

In conclusion, cocoa should be treated as a horticultural crop and for its proper health, routine maintenance and use of various methods of keeping pests and diseases in check are essential. Considerable research is required in developing novel management practices suitable for particular situations. Other methods of management need to be introduced such as: the use of barrier crops; various insect traps and baits to lure away pests; introduction of suitable natural enemies from one continent to another; development of resistant varieties by exploiting the fast-disappearing varieties of wild cocoa in South America; and use of relatively new techniques (such as genetic manipulation of pests). All these methods should be developed into an integrated programme, in addition to using chemicals to control cocoa pests and diseases. Finally, educating the farmers in pest management methodology through efficient extension services is also required.

Management of sugarcane pests

R. Kumar

Sugarcane, *Saccharum officinarum*, is essentially a plant of the tropics and subtropics requiring large amounts of water, very fertile soils for high tonnages, and a hot dry spell for ripening the juice. It is a perennial plant that permits the harvesting of successive crops (ratoons) without the expense of replanting. It is said to produce more food per hectare than any other crop. Its origin is thought to have been somewhere in south east Asia but currently the major areas of production in the tropics are East and South Africa, Barbados, Brazil, Cuba, Guyana, Hawaii, India, Mauritius, and Puerto Rico. Large sugarcane plantations are currently under development in West Africa.

Over 1500 species of insects have been listed as feeding on the sugarcane plant, with a distinct fauna of indigenous and introduced species in each geographic region or island on which it is grown. Very few pests are cosmopolitan and these are economically unimportant. Lack of phytosanitary precautions have resulted in the transport of some serious pests from one country to another. Further, in many instances, insects on grasses, maize and sorghum have switched over to feeding on sugarcane.

Pest fauna

Defoliators

A variety of locusts and grasshoppers (at least 60 species of Acridoidea) and caterpillars (Lepidoptera), mainly the cutworms, can damage sugarcane severely. Leaf miners and leaf rollers (Lepidoptera) also occur but these are usually unimportant. Among the defoliators the following migratory plague and locust species deserve attention:

(a) The African migratory locust, *Locusta migratoria migratorioides*, in Africa south of the Sahara (but not reaching Natal cane growing areas) – especially destructive outbreaks occur in East and West Africa;
(b) The red locust, *Nomadacris septemfasciata*, in most of East, Central and South Africa, being notably destructive in Natal, Mozambique and Mauritius;
(c) The oriental migratory locust, *Locusta migratoria manilensis*, in the Philippines, Malaysia and Taiwan;
(d) The Central and South American locust, *Schistocerca paranensis*, in Nicaragua, Venezuela and Mexico.

These plague locusts cause major losses to stands by defoliation and destruction of the growing points, leading ultimately to reduction of sugar yields. Defoliation, whether by cutworms or locusts, delays cane development until a new foliage canopy has been formed.

Vectors of diseases

At least 5 different virus diseases of sugarcane are transmitted by homopteran insect vectors. Virus diseases affect the tonnage produced, mainly due to poor growth of the infected plants. However, reduced germination from infected seeds may also contribute to losses. Aphid species (Aphididae) have been implicated in the spread of sugarcane mosaic virus disease which has been considered to be one of the world's most important sugarcane diseases. Apart from Mauritius and Guyana, it is found nearly everywhere sugarcane is grown. The aphid vector species include *Acyrthosiphon pisum*, *Carolinaia cyperi*, *Dactynotus ambrosiae*, *Hysteroneura setariae*, *Nasonovia lactucae*, *Schizaphis granarium* and *Rhopalosiphum maidis*. Three species, *Longiunguis indosacchari*, *L. sacchari* and *R. maidis*, transmit grassy shoot disease, which is believed to be confined to India.

Species such as *Longiunguis sacchari* in India and *Sipha flava* in Puerto Rico may cause economic damage by direct feeding.

Leafhoppers (Fulgoroidea) are associated with the transmission of Fiji and streak diseases. Three species of Delphacidae, *Perkinsiella saccharicida*, *P. vastatrix* and *P. vitiensis*, transmit the former disease in the Pacific region, south-east Asia and Madagascar. *Cicadulina mbila* (Cicadellidae) transmits streak disease which, although indigenous to Africa, has been recorded in India, Madeira and Réunion. A mealybug, *Pseudococcus saccharifolii* (Pseudococcidae), is said to transmit spike disease in India, believed to be caused by a virus.

General sap-feeders

A large number of sap-feeding Homoptera, although not transmitting any diseases, may damage sugarcane by sap removal and by loss of photosynthetic tissue. This may lead to stunted growth, loss of yield and poor juice quality. There are 66 species and subspecies of froghoppers or 'spittle bugs' (Homoptera:Cercopoidea) known to feed on sugarcane in the tropics and subtropics but they are of economic importance only in the New World. Species of *Aeneolamia* are a serious pest of canes in Mexico, Venezuela and Trinidad. The feeding by adults on border parenchyma of leaves causes removal of chloroplasts resulting in 'froghopper blight' from leaf necrosis, reduced growth and thin stunted internodes. Toxic effects of salivary secretion may also be involved with the 'blight'.

More than 150 species of planthoppers (Fulgoroidea) have been found on sugarcane but only a few are important. These include *Pyrilla perpusilla* in Sri Lanka and India, and *Saccharosydne saccharivora*, a major pest in Jamaica. Attack by these pests may retard the plant growth and affect the quality of the sugar juice. Injury by the insects whether during feeding or oviposition may provide easy entry to red rot fungus. The formation of sooty mould on liquid excreta left on the leaves by *S. saccharivora* also reduces photosynthesis. Feeding by *Numicia viridis* on canes in South Africa causes necrosis of the tip and edges of the leaf. The central whorl becomes soft and flabby and the juice has lower sucrose and higher glucose content than normal.

There are 35 species of scale insects (Homoptera) known to attack sugarcane. Important among them are the soft scales of the genus *Pulvinaria* (Coccidae) and armoured scales of the genera *Melanaspis* and *Aulacaspis* (Diaspididae). Infestation by *Aulacaspis tegalensis* may reduce tonnage and the amount of sucrose in the juice by as much as 35 per cent.

At least 30 species of mealybugs (Pseudococcidae) have been recorded on sugarcane. Heavy infestations affect germination, growth, yield, increased infestation by micro-organisms and difficulty in filtration and clarification of juice in the factory.

Moth borers

The larvae of about 50 species of moths Lepidoptera have been found to bore into sugarcane attacking stalks at any stage of growth and causing enormous losses to sugarcane industries throughout the world. They may be shootborers, *Chilo infuscatellus* (Pyralidae), *Diatraea* spp. (Pyralidae) and *Sesamia* spp. (Noctuidae); top borers, *Chilo tumidicostalis, C. infuscatellus* and *Scirpophaga nivella* (Pyralidae); or internode borers, *Diatraea* spp., *Chilo auricilius*, *C. tumidicostalis* and *Eldana saccharina* (Pyralidae) (see Fig. 6.8). Some species may also attack roots. Moth borers in the New World are mostly *Diatraea* spp., while *Chilo* and *Sesamia* predominate in the Old World. According to taxonomists, *Diatraea* and *Chilo* are really a monophyletic group but are kept as distinct genera for practical purposes. *Sesamia* and *Eldana* species which normally attack maize in Africa are currently gaining notoriety in attacking sugarcane in East and West Africa.

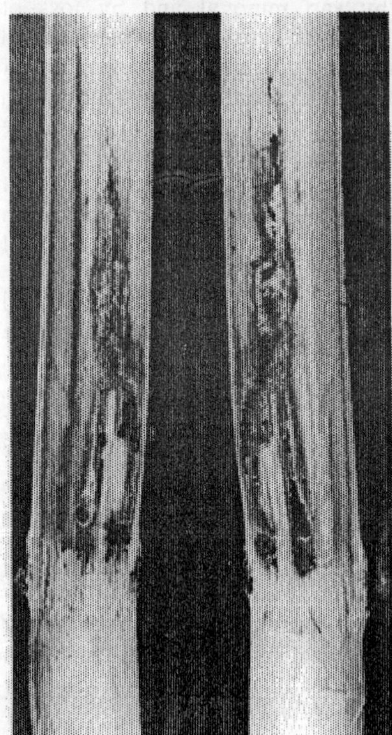

Fig. 6.8a

Fig. 6.8b

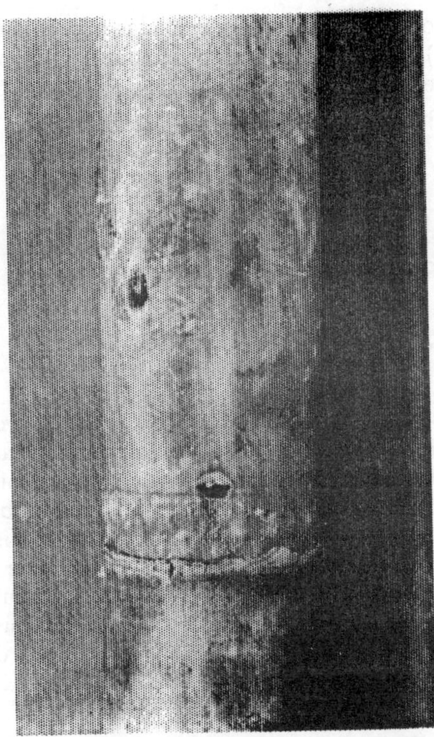

Fig. 6.8c Example of sugarcane moth borers: (a) damage caused by larvae to the stems internally; (b) pupal stage of a borer; (c) exit holes of escaping adult moths. (R. Kumar).

Borer infestation may reduce cane tonnage and drastically decrease the sucrose content of the affected canes as well as increasing the acidity of the juice and the amount of impurities such as ash and gum. To these should be added financial losses due to the higher cost of processing damaged canes in the factory.

Soil insects

Insects also damage sugarcane in the soil. These include small soil-inhabiting arthropods such as springtails (Collembola), bristletails (Thysanura), centipedes (Myriapoda), as well as other small animals. Termites (Isoptera) are important pests only in India, China, Central Africa and the Caribbean and their effect is particularly marked during drought conditions.

White grubs (beetle larvae of Melolonthidae, Rutelidae, Dynastidae) are pests of sugarcane in many countries notably Australia, India, Mauritius, Philippines and Puerto Rico and in several African countries. They feed mainly on the living roots. In severe infestations as much as 80 per cent of the crop may be lost.

In addition, other insects such as cicadas (Cicadidae), ground pearls (Margarodidae), mealybugs (Pseudococcidae), wireworms (Elateridae), ants (Formicidae), negro bugs (Cydnidae) and mole crickets (Gryllotalpidae), attack sugarcane roots, sometimes killing the tissue or sucking the plant sap.

Management of sugarcane pests

Defoliators

Large swarms of locusts are now controlled by the co-ordinated actions of national and international organizations which employ aerial applications of insecticides on the flying insects. Biological control of locusts is not yet a practical proposition as these insects are highly mobile, occur in vast numbers and are irregular in occurrence. Natural enemies require time to multiply and seldom migrate with their pests. The most promising approach to their control appears to be a regular surveillance of the outbreak areas. These are areas where locust populations build up prior to mass migration, and it is here that such populations can be destroyed before increasing to large numbers.

Vectors of diseases

The use of insecticides to control vectors of sugarcane diseases has seldom been successful. Even when vector populations are reduced greatly, the survivors and incoming populations continue to spread the disease. The importance of controlling weeds, grasses and preferred hosts which harbour the virus or its vectors must be stressed. The use of seed-cane free from diseases has also proved successful. However, best results have been obtained by growing resistant sugarcane varieties which have reduced the incidence of virus diseases to negligible levels.

General sap-feeders

Rotation of sugarcane with other crops and reduction of ratooning (i.e. springing up from roots of previous crop) have been suggested to control the 'froghopper blight'. But frequent replanting of seed-canes is a costly operation and the use of insecticides has been found more convenient and cheaper by growers. Sugarcane varieties resistant to froghoppers are yet to be developed and further research in this field is likely to prove rewarding. A large number of natural enemies of froghoppers are known but they are considered to be far less adapted to the ecological conditions of sugarcane fields than the pests themselves.

Little success has been achieved in the control of *Pyrilla perpusilla* by biological control methods. Labour-intensive practices such as the collection of adults by handnets, and hand-collections of egg masses have been used in India to lessen the incidence of the pest. Insecticide spraying is also practised. Attempts at developing cane varieties resistant to *P. perpusilla* have been disappointing.

Although entomophagous fungi are known to play an important role in the natural control of *Saccharosydne saccharivora* in Jamaica, biological control alone in controlling this pest has been unsuccessful. Insecticides such as malathion are used in keeping these pests in check. No commercially acceptable resistant cane varieties against the pest are available. Introduction of the egg predator, *Tytthus (Cyrtorhinus) mundulus*, against *Numicia viridis* in South Africa has given encouraging results.

However, a major contribution to the pest management programme of planthoppers is agronomic. It is now known that for an outbreak of leafhoppers such as *S. saccharivora, Aeneolamia* or *Perkinsiella* to occur, there must be a retardation of normal protein synthesis in the plant. The most important factor responsible for this appears to be adverse soil conditions at certain stages in the growth of the crop. These include periods of inadequate water supply to the roots, factors preventing deep rooting of the plant, and so on. There is therefore the need to develop quick and reliable methods of detecting the first signs of metabolic disorders in the crop. It is also necessary to study the pest behaviour under conditions adverse to growth. With information on controlling soil moisture and the response of pests to such treatments, it should be possible to apply the necessary remedial action to control the pests.

Sudden increases in the populations of scale insects have been largely attributed to the use of susceptible sugarcane varieties, drought and absence of effective natural enemies. Scale insects are attacked by a number of natural enemies such as Hymenoptera, coccinellid beetles and predatory mites. But no systematic attempt has been made to use such agents in pest control. The use of healthy seed-canes for planting has been found to reduce the incidence of diaspid scales. Organophosphate insecticides with a contact action have been used successfully in killing the crawlers, but not the eggs and fixed stages because of the protective scale over the insect's body.

The control of mealybugs on standing cane by insecticidal treatment has been considered to be uneconomical. However, cultural practices such as selection of uninfested seed-canes for planting and destruction of crop residues, such as cane trash and stubble after harvest (Fig. 6.9), has been found to reduce or avoid infestation and losses. Known alternative hosts of the insect pests should not be allowed to grow near cane fields.

Fig. 6.9 After harvesting, illustrated here, has been completed, the removal and destruction of sugarcane stubble and other crop residues should be encouraged as this reduces the refuges for sugarcane pests such as moth borers and sap-feeding insects. (R. Kumar).

Moth borers

Effective insecticidal control of these pests is expensive and except in parts of the USA, chemical control is seldom intensively practised. For a judicious use of insecticides in controlling infestation, adequate knowledge of the economic threshold of the borer is required (see p. 75). Continuous monitoring of the incidence of infestation before spraying any insecticides is essential. By this method management of pest populations that are injuring the stalks is permitted, regular insecticide applications are precluded and biotic and abiotic factors are taken into account in arriving at treatment schedules.

Biological control of borers has consisted mostly of introduction and establishment of exotic parasites and the rearing and release of indigenous egg parasites, especially *Trichogramma* species. The effectiveness of *Trichogramma* in controlling pest populations is in doubt because most mass rearing and releases of *Trichogramma* seem to have been done rather thoughtlessly, without adequate ecological studies on the pests and the parasites. In view of the adaptability of *Trichogramma* and its vast genetic potential, breeding of suitable *Trichogramma* strains to suit particular environmental conditions has been suggested as a largely unexplored field of research in the control of borer pests.

The introduction of hymenopterous and dipterous parasites (mostly tachinids) of borer pests of canes has so far met with comparatively little success. The recent importation and release of these parasites from India against *Sesamia* and *Eldana* species in Ghana has been without any effect. No field recoveries have been made. These releases seem to have been made without adequate ecological studies of the pest and parasites.

Labour-intensive cultural practices such as early planting, flooding of fields and destruction of crop residues have been mostly discontinued in the USA. But in many tropical countries with relatively cheap labour, such practices have a definite place in sugarcane production. Delayed planting in some countries has been of value, but such periods need to be worked out for each situation. In Ghana, ratooning seems to propagate the incidence of *Sesamia calamistis* and *Eldana saccharina*. Growers would have to decide whether losses from these borers outweigh seed-cane planting costs every season.

Much literature exists on the differences in resistance to borer attack between sugarcane varieties. But the contribution of resistant varieties to the management of these pests has been as yet very slight. More research is required into developing varieties which are resistant to borer attack and have the desirable economic characteristics.

In India, hand-collection of moths and their egg masses coupled with light-trapping have not generally been very effective. The use of sex attractants (pheromones) to trap or confuse the pests is in a developmental stage in a few tropical borer species, while the use of sterile male techniques to suppress the borer populations has yet to be attempted on a commercial scale. Further research against borers using bacteria and viruses is also very necessary.

Soil insects

Insecticides have been used effectively in controlling the 'small soil arthropods'. Organochlorines, whether in treating seed-canes before planting or as insecticides worked into the soil, have given highly satisfactory results in the control of termites. Destruction of termite mounds has also been helpful in controlling termite infestations. On the whole, insecticides are now used as a preventive measure against soil pests and are effective in low dosages against white grubs (i.e. beetle larvae). Although some side effects, such as destruction of the natural enemies of pests and phytotoxicity, are a hazard with insecticidal treatments, these effects can be avoided or minimized by sensible application. Here mention should, however, be made of the white grub problem in Mauritius and Hawaii. On these islands these pests have been introduced from other lands, but have been controlled by importation and release of exotic natural enemies, alone or in combination with resistant crop varieties.

Conclusions

In considering the management of sugarcane pests, the great variability of the fauna from region to region and island to island must be recognized. Management strategies must be developed to deal with each particular pest situation. More basic research is required on sugarcane pests to enable a systematic development of integrated pest management schemes for sugarcane pests in the tropics.

Management of rice pests

R.C. Saxena

Rice is the staple food of about 60 per cent of the world population and is grown on about 140 million hectares. Of the world's rice, 90 per cent is produced and consumed in Asia and forms an important part of the regional economy. Rice production will have to increase further especially in most developing countries to keep pace with the population growth.

Rice is grown under extremes of environmental conditions, from the humid tropics to semi-arid, warm temperature climates; from latitudes 53°N to 40°S, and from sea level to altitudes of 3000 m. It is grown either from seeds – broadcast or drilled – or by transplanting under diverse landforms and water regimes; as an upland rainfed crop or as a lowland, bunded, irrigated or rainfed crop. On slopes, rice is cultivated in terraces, and in low lying areas deep-water rice may be grown in 5–6 m of standing water. Temperature is a critical factor in rice cultivation. The optimum temperature is from 25-33 °C for tropical rices and from 18-33 °C is suitable for rice grown in temperate areas; lower temperature regimes induce sterility, particularly at the flowering stage. Thus, only one crop a year can be grown in regions with cool winters, but in warm areas up to three crops are common.

Rice belongs to the grass family and the genus *Oryza*. There are about 20 *Oryza* species, but nearly all cultivated rice belongs to *O. sativa*, which originated in Asia more than 5000 years ago. The relatively recent African cultivar, *O. glaberrima*, is restricted to West Africa's Niger valley and because of its inferior yield it is being replaced by *O. sativa*.

Insect pests of rice

In most of the Asian countries, rice is produced almost exclusively by

Fig. 6.10 Striped stemborer, *Chilo suppressalis*, moth resting on a leaf-blade of a rice plant. (Courtesy of IRRI.)

Fig. 6.11 Adults of the brown planthopper, *Nilaparvata lugens*, feeding at the base of rice plants. (Courtesy of IRRI.)

small-scale peasant farmers who have little or no hope of increasing their land-holding capacity. Any significant increase in rice production will, therefore, have to come from an increase in output per unit area and can be achieved only through the use of high-yielding varieties, improved cultural practices, such as irrigation, fertilization, and crop intensification, and appropriate crop protection practices against diseases and insect pests. Rice suffers mostly from insect damage in tropical countries where the peasant farmer has limited access to capital, insecticides, and other necessary inputs. The development and implementation of pest management strategies in rice cultivation is therefore an absolute necessity.

More than 100 insect species attack rice. Of these, about 20 are major pests (Figs. 6.10, 6.11). Table 6.2 gives the scientific and common names of these pests. Together they infest all parts of the plant, at all stages; a few transmit virus diseases. Even stored rice is attacked by a number of insect pests.

The warm and humid environment in which rice is grown is highly conducive to insect proliferation and activity. Also, heavily fertilized, high-tillering plants and the practice of growing the crop throughout the year favours the build-up of pest populations. It has been observed in the tropics that rice grown with modern technology often has more severe pest problems than the poorly managed peasant farmers' fields planted with traditional varieties.

Changing status of insect pests

Some insect pests have increased in severity in recent years. For example, brown planthoppers, once a minor pest, became the number one pest in south and south-east Asia in the early 1970s causing extensive feeding damage or 'hopperburn' in fields planted with the introduced high-yielding but hopper-susceptible varieties. The development of virulent biotypes of the brown planthoppers, following large-scale planting of hopper-resistant varieties, and the spread of the hopper-transmitted grassy stunt virus and the newly identified ragged stunt virus disease, to which most of the popular varieties are susceptible, has caused serious concern among scientists, administrators and rice farmers throughout tropical Asia. These developments have changed attitudes towards rice pest control. The need for early detection of changes in the status of insect biotypes is necessary to permit sufficient time for the development of pest-resistant varieties with new genes for resistance. Also, in most rice production programmes, exploitation of host plant resistance along with other pest management approaches is considered more desirable than sole reliance on the use of pest-resistant varieties.

The whitebacked planthopper, a minor pest so far, has become increasingly important in rice areas when brown planthopper-resistant varieties are being cultivated. Also, the rice leaf folder, formerly a minor pest, has increased in abundance with the introduction of high-yielding rice varieties. On the other hand, the importance of certain insect pests has been over-estimated. For example, the whorl maggot was considered to be an important rice pest, but studies conducted recently at the International Rice Research Institute (IRRI) in Manila, Philippines have shown that the gain in yield attributed to whorl maggot control, in fact resulted from a direct stimulation of the crop's growth through the use of the carbamate insecticide, carbofuran.

Causes for the changes in rice pest status are not well understood; probably agronomic practices aimed at obtaining high yields may contribute to pest severity. For example, improved fertility levels, irrigation, and the year-round cultivation of high-tillering varieties all favour the build-up of brown planthopper populations. Whatever the reasons, there is need for a pest management approach to prevent or delay pest occurrence, abundance and damage to the crop.

Table 6.2 Categories of major rice insect pests and their distribution

Category	Insect	Common name	Family	Distribution
Root feeders				
1.	*Lissorhoptrus oryzophilus*	Rice water weevil	Curculionidae	Americas
Internal stem feeders				
2.	*Chilo suppressalis*	Striped stemborer	Pyralidae	Asia, Australia
3.	*Diopsis thoracica*	Stalk-eyed borer	Diopsidae	Africa
4.	*Sesamia inferens*	Pink stemborer	Noctuidae	Asia
5.	*Tryporyza incertulas*	Yellow stemborer	Pyralidae	Asia
6.	*T. innotata*	White stemborer	Pyralidae	Asia, Australia
External stem feeders				
7.	*Scotinophora coarctata*	Black paddy bug	Pentatomidae	Asia
Foliage feeders – sucking type				
8.	*Baliothrips biformis*	Rice thrips	Thripidae	Asia
*9.	*Nephotettix virescens*	Green leafhopper	Deltocephalidae	Asia
†10.	*Nilaparvata lugens*	Brown planthopper	Delphacidae	Asia, Australia
11.	*Sogatella furcifera*	Whitebacked planthopper	Delphacidae	Asia, Australia
**12.	*Sogatodes orizicola*	Rice delphacid	Delphacidae	Americas
Foliage feeders – chewing type				
13.	*Cnaphalocrocis medinalis*	Leaf folder	Pyralidae	Asia, Australia
14.	*Dicladispa armigera*	Rice hispa	Chrysomelidae	Asia
15.	*Hydrellia sasakii*	Whorl maggot	Ephydridae	Asia
16.	*Mythimna separata*	Earhead cutworm	Noctuidae	Asia, Australia
17.	*Nymphula depunctalis*	Caseworm	Pyralidae	Africa, Asia, Australia, Americas
18.	*Orseolia oryzae*	Gall midge	Cecidomyiidae	Africa, Asia
19.	*Spodoptera mauritia*	Armyworm	Noctuidae	Africa, Asia, Australia
Grain feeders				
20.	*Leptocorisa* spp.	Rice bugs	Coreidae	Asia, Australia
21.	*Nezara viridula*	Green stink bug	Pentatomidae	Africa, Asia, Australia, Americas
22.	*Oebalus pugnax*	Rice stink bug	Pentatomidae	Americas

In each category insects are arranged in alphabetical order and not necessarily in order of their economic importance

* Vector of tungro virus

Yield losses caused by insect pests

Yield losses from rice pests in Asia, excluding mainland China, have been estimated to be 51.5 per cent, of which insects account for 31.5 per cent loss and diseases and weeds for the remaining 20 per cent. Losses from stemborers alone in tropical Asia range from 5–10 per cent (Figs. 6.12, 6.13). Depending upon the level of insect control, a 0.5–2.6 tonne/ha increase in yield has been obtained.

Fig. 6.12 'Dead hearts' caused by the larvae of the striped stemborer, *Chilo suppressalis*, on a stand of rice plants. (Courtesy of IRRI.)

Economic thresholds

Normally, plants can tolerate and recover from direct insect damage if the infestation is not high. It is thus important to know the pest density at which control measures should be applied to prevent an increasing pest population from causing economic loss. Tentative economic thresholds for some of the major rice insect pests have been established based on insect densities and level of damage (Table 6.3). Although refinement of these values is needed, even the existing threshold values are worth adopting in decision-making in pest control. Based on economic thresholds,

Fig. 6.13 A rice field on an IRRI farm in the Philippines completely damaged by the brown planthopper, *Nilaparvata lugens.* Excessive feeding by nymphs and adults results in 'hopperburn', or complete drying of the crop. (Courtesy of IRRI.)

the use of appropriate insecticides place pest control on a sound economic and ecological basis. However, economic thresholds are less useful when insects inflict heavy damage to the rice crop by transmitting viruses and other diseases.

Table 6.3 Economic threshold values for some important rice insect pests

Insect pest	Economic threshold*	
	Damage rating	Pest density
Brown planthopper	—	1 hopper per tiller or 10–20 per hill
Green leafhopper	—	5 hoppers per hill at vegetative stage, 10 per hill at flowering stage
Whitebacked planthopper	—	1–5 hoppers per hill
Rice bug	—	2 adults per m^2
Gall midge	5% galled leaves	—
Leaf folder	5–10% damaged leaves per hill at flowering stage	—

Whorl maggot†	9 (more than 50% of leaves damaged, some broken)	—
Yellow stemborer	5–10% 'dead hearts' at tillering and 2% 'white heads' at flowering stages	—

* Economic thresholds have been determined on the basis of either the damage rating or the pest density, depending on the pest species and the damage characteristics.

† Based on a 0–9 scale of damage; 0 = no damage, 9 = more than 50 per cent of leaves damaged.

Management of rice pests

Components for the development of rice insect pest management programmes

Components essential for a broad ecological approach to rice pest management include: pest sampling and surveillance techniques; forecasting; determination of economic thresholds; and chemical, biological and cultural methods, including the use of resistant varieties. Because of the numerous factors involved and the complexity of their interactions, the decision-making process in the implementation of appropriate control tactics and their final evaluation are important aspects in the development of a pest management system.

Sampling methods

Several different sampling techniques for rice pests have been used in different countries depending on the pest species and the location. Visual counting of nymphs and adults of hoppers is a simple but laborious sampling technique. The hoppers may be counted directly on the plants or after they have been dislodged on to the water surface by tapping the plants. Use of sweepnets is another commonly used sampling method, especially for leafhoppers. Occasionally, mouth aspirators are used to collect rice insects, especially nymphs; this method is not suited to large-scale sampling.

In recent years, motorized suction machines, such as 'Univac' and 'D-vac', have been used widely for sampling rice leafhoppers and planthoppers on rice hills, but these machines are inadequate for sampling insects on the water surface. The 'Farmcop' suction machine, designed recently at IRRI, has proved to be quite effective in sampling most pests and their natural enemies on the rice plant or the surrounding water surface. It consists of a modified insect aspirator attached to a car vacuum cleaner, operated by a portable car battery.

Other devices, such as yellow-pan oil–water traps, sticky traps and inexpensive kerosene light-traps with baffles are being used extensively for monitoring immigrant leafhoppers, planthoppers, and other rice insects. Pheromone traps as indicators of the activity of stemborer, leaf folder, caseworm, and gall midge adults are under evaluation in a few countries.

Surveillance and forecasting

Surveillance and forecasting pest outbreaks prepares the farmers to undertake prophylactic treatments. However, this area of research has not developed because of complexities involved in predictive monitoring. Short-term forecasting is being attempted in a few countries for a few rice insect pests. Forecasting the brown planthopper population levels is considered possible, particularly in temperate areas, such as Japan, where the hopper has distinct generations in a crop period. A 'Surveillance and Early Warning System' (SEWS) has been established in the Philippines as a basis for 'supervised crop protection'.

Forecasting can be most useful in the control of vector-borne virus diseases. Studies conducted at IRRI have demonstrated that tungro virus incidence can be predicted 2–3 months in advance by determining the percentage of virus-carrying green leafhoppers in rice fields. Light-trap catches and populations in rice stubble have been used in India to forecast stemborer infestation.

Chemical control and its implications

Intensive rice production systems still rely heavily on insecticides for insect control. Although the use of insecticides is not directly compatible with the ecological approach to pest management, their use becomes inevitable during heavy infestations and pest outbreaks. The use of selective, relatively less toxic insecticides, and improvements and modifications in application techniques, lessens their disruptive action and yet maintains their usefulness in pest management systems. Discovery of insecticides promoting the growth and yield of the rice plant, as well as being effective against insects, should prove to be a useful addition to insect pest management programmes.

Chemical control of rice insect pests is not always rewarding. Failure to apply the right insecticide at the proper time and at recommended dosages often leads to decreased benefits to the farmer. Insecticides applied without regard to economic thresholds and proper timing tend to decrease the cost/benefit ratio of the insecticide treatment.

Insecticide evaluation Insecticides effective against stemborers have been selected through field-screening programmes in Bangladesh, India, Korea and Thailand. Insecticides for gall midge control have been identified in India, where the insect is a major pest. Recently, insecticide screening has been intensified against the whorl maggot in the Philippines, India and Indonesia because it has become an important pest of irrigated, transplanted rice. Extensive laboratory and field trials with insecticides have been conducted against the brown planthopper which is currently the most important pest of rice in Asia.

Selective toxicity Ideally, an insecticide should affect only the target pest species, but most of the commercially available insecticides have broad-spectrum biological activity. A few insecticides with moderate to low toxicity to predatory spiders and mirid bugs, and yet effective against the brown planthopper, have been identified. Among 14 insecticides, the organophosphate, sumithion (fenitrothion), was reported to be the least toxic to *Lycosa* spiders. Spraying a rice crop with a carbamate insecticide, metalkamate, was found not to hamper parasitization of stemborer egg masses. Appropriate formulations and improved application techniques are reported to reduce toxicity to predatory spiders. Similar techniques have made fish culture possible in insecticide-treated paddy fields.

Insecticides should also be evaluated for secondary effects such as pest resurgence. A few insecticides have now been identified that control the brown planthopper without causing its resurgence.

Formulations Insecticides to control rice pests are commonly applied as foliar sprays in rice fields. While sprayable formulations are effective against foliage feeders, such as leaf folders, rice hispa, armyworms and cutworms, their application causes high mortalities of the natural enemies of pests. Granular formulations applied directly to paddy water or soil can effectively control a few insect pest species, and at the same time are easy to apply and are less deleterious to natural enemies than are foliar sprays.

Application methods The method of applying insecticide is crucial in determining the effectiveness of insecticides, cost of application, and ecological safety. Proper placement of foliar sprays can enhance the degree of control of certain pests. Thus, the brown planthopper is controlled more effectively if the insecticide spray is directed to the base of the rice plant rather than the canopy.

Soaking seedling roots in an appropriate insecticide solution, prior to transplanting, effectively controls early pests such as leafhoppers, planthoppers, and whorl maggots. Root-zone application of granular or liquid systemic insecticides has also proved effective. Studies conducted at IRRI demonstrated that one application of insecticide in the root-zone shortly after transplanting provided insect control and yields equal to, or superior to, 4 broadcast or foliar applications in a crop season. This technique was also less hazardous to parasites and predators and other non-target organisms; but root-zone placement of insecticides is laborious and requires good land preparation. Incorporation of carbofuran granules broadcast into the soil at the last harrowing has proved to be a much more practical and effective method for farmer's use.

Timing of application Insecticides applied at peak nymphal density in each generation adequately control brown planthopper infestations. Also, in areas where stemborers occur in distinct broods, application of insecticide in conformity with their brood emergence has been found to be more effective than calendar-based applications. Knowledge of the pest's local biology, and efficient monitoring of the pest's colonization and build-up, help immensely in proper timing of insecticide application.

Resistance to insecticides

Resistance to insecticides has been recorded for certain major rice pests, for example green leafhopper, brown planthopper, and striped stemborer in Japan where insecticides are intensively used. Reports of resistance to insecticides in tropical Asia are as yet few, but the intensification of rice production and increasing use of insecticides may lead to a situation similar to that existing in Japan. This tragedy in tropical countries can be averted by the implementation of integrated pest management programmes based on alternative control strategies and judicious insecticide application.

Induced insect resurgences

Resurgence of insect pests is another of the so-called 'hidden costs' in the use of insecticides. Extensive use of certain insecticides (DDT, HCH, methyl parathion) against the striped rice stemborer in Japan has been reported to induce the resurgence of these as well as other insects, such as leafhoppers and planthoppers. These outbreaks have been attributed to the destruction of natural enemies, especially spiders. Consistent resurgence of the brown planthopper after spraying with methyl parathion has been demonstrated in the tropics. In a 6-year study at IRRI, the average hopper-burn area was 94 per cent in the insecticide-treated plots, whereas in the untreated control plots it was only 18 per cent hopper-burned. It is possible that many of the brown planthopper outbreaks in farmer's fields throughout Asia are induced by insecticides, some of which are now known to induce resurgence. Resurgence of brown planthopper is more complicated than simply the destruction of natural enemies, and factors like the pest's enhanced feeding, reproductive stimulation, and shortened growth period probably contribute to this phenomenon.

Recent studies at IRRI have also shown that certain insecticides cause resurgence of the whitebacked planthoppers, a secondary pest. This fact complicates the chemical control of the whorl maggot as both whorl maggots and whitebacked planthoppers attack the crop shortly after transplanting. One of the insecticides recommended for whorl maggot control (diazinon) has been observed to stimulate whitebacked planthopper reproduction. Thus, more information than simply the biological efficacy of a candidate pesticide against the target pest is needed before recommending it.

Poisoning of humans, livestock and fish

Because of a lack of knowledge, most rice farmers do not protect themselves when applying insecticides on rice. A recent survey made in the Philippines showed that no farmer was aware of the contact toxicity of insecticides and individual cases of poisoning are seldom reported.

When pesticides are used to control rice pests, the problem of insecticide contamination may be very acute in rice fields where plants grow in standing water. Also, insecticide run-off can reach non-target organisms and almost immediately the consequences become disastrous to pisciculture. Farmers in Thailand for example, use paddies to raise fish as a regular source of food, but in the Philippines, this popular practice is almost abandoned now as insecticide use has increased. This is a loss of an important source of animal protein. The only agronomic practice which is not incompatible with fish culture within paddy rice fields is the use of insecticides as commonly applied. Minimal use of insecticides, proper timing of insecticide application, suitable formulations, appropriate application methods, or completely non-insecticidal control methods, are necessary to minimize hazards to fish culture and other valuable animals such as water buffalo, which in tropical countries are used for tillering paddy fields.

Use of host plant resistance

The planting of agronomically suitable varieties with natural resistance to pests now forms the foundation of pest management programmes for rice. Cultivation of insect-resistant varieties confers a number of advantages: (i) it provides inherent insect control to the crop thus emancipating the farmer from chronic dependence on insecticides; (ii) it does not impair the quality of the environment; and (iii) it is compatible with other pest control methods. When resistant varieties are planted over a relatively large hectarage, they cause a cumulative reduction in pest numbers, contrary to high pest infestation when susceptible varieties are grown.

The development of insect-resistant, high-yielding rice is considered a milestone in modern rice production technology. Their large-scale, rapid adoption by farmers has reduced insect pest problems and considerably augmented rice yields. Much of this progress has been made within the last decade. The rapid development and outstanding success of insect-resistant rice varieties can be attributed to several factors: (i) the occurrence of a vast genetic diversity in the world rice germplasm; (ii) willingness of national and international research systems to identify insect-resistant sources; (iii) support for inter-disciplinary co-operation among plant breeders, entomologists, pathologists and 'problem-area scientists' for incorporating this resistance into improved rices; and (iv) no extra cost is involved in growing resistant varieties once they are developed. Several insect-resistant rices are currently being grown in various Asian countries, although their resistance is limited to only a few major pests. Table 6.4 lists some insect-resistant varieties popular in India and the Philippines.

Table 6.4 Selected examples of resistant varieties recommended for management of insect pests of rice in the Philippines and India

Insect pest	Resistant varieties
Brown planthopper	
Philippine biotype 1	IR26, IR28, IR29, IR30, IR32, IR34, IR36, IR38, IR40, IR42
Philippine biotype 2 (currently predominant in Indonesia and the Philippines)	IR32, IR36, IR38, IR42
South Asian biotype (India and Sri Lanka)	Jyothi Bharti, Parijat, Shakti, Triveni, CR93-MR1523
Green leafhopper	IR8, IR20, IR26, IR28, IR29, IR30, IR32, IR34, IR36, IR38, IR40, IR42, Jagannath
Stemborers	IR20, IR26, IR30, IR36, IR40, IR42, Jaya, Ratna, Vijaya, Vikram, Triveni
Whorl maggot	IR40

Resistance to stemborers, leaf folders, whorl maggot pests

The level of resistance to these insect pests is not high, but enough to help prevent pest outbreaks, and when suitably combined with other control practices, the pest density can be maintained below the economic threshold.

Resistance to leafhoppers and planthoppers

During the last decade, a high level of hopper resistance has been bred into rice varieties. In about 1973, the brown planthopper in the Philippines, devastated rice fields which were well-irrigated, heavily fertilized, and planted with high-tillering varieties. Such a micro-environment favoured the pest, which inhabits the basal parts of the rice plant, and led to marked increases in its population in several areas, for example, around IRRI farms (Fig. 6.14). The first brown planthopper-resistant variety, IR26,

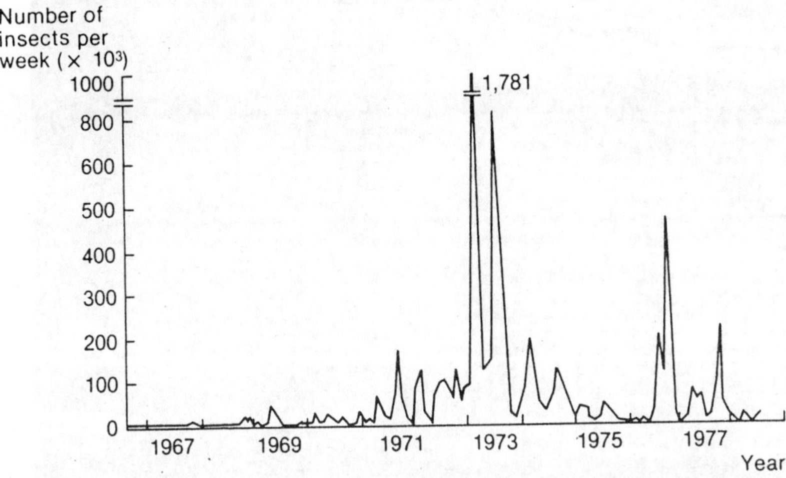

Fig. 6.14 Average number of brown planthoppers, *Nilaparvata lugens*, attracted to 3 light-traps at IRRI, Philippines. Intensive cultivation of resistant varieties is considered a major factor affecting the decline of brown planthopper numbers since mid-1973. (Courtesy of V.A. Dyck, IRRI.)

was released for cultivation in early 1973, but after 2 or 3 years of growing IR26 there were reports of widespread brown planthopper damage to this variety. At present, resistant varieties are being planted over 90 per cent of the irrigated rice area in the Philippines, and the subsequent decline in the pest's abundance may largely be attributed to the cultivation of these pest-resistant varieties. A slight increase in the pest population in 1976, demonstrated by high catches of insects in light-traps at IRRI farms may be due to the development of a new brown planthopper biotype which can survive on IR26, IR28, IR29, and sibling varieties (Fig. 6.15). Since then, introduction and cultivation of new rice varieties, such as IR32, IR36, IR38, and IR42, which carry a different source of brown planthopper resistance, has effectively stopped the brown planthopper incidence in the Philippines and made significant savings in the use of insecticides. In Indonesia the pest has been effectively controlled through large-scale cultivation of resistant rice varieties. However, brown planthopper populations in India and Sri Lanka belong to a different biotype to which the present IR varieties are susceptible. These countries have developed specific biotype resistant varieties for cultivation. Ptb 33 is a tall variety of *O. indica* from Kerala, India, and is resistant to all common biotypes of the brown planthopper. Several improved IRRI breeding lines now have this resistance.

Fig. 6.15 Rice variety IR32 having resistance to BPH biotype 2 (brown planthoppers) showed no visible damage when attacked by the planthoppers, whereas IR29 being susceptible to biotype 2 was 'hopperburned', as indicated here by its lighter colour. (Courtesy of E.A. Heinrichs, IRRI.)

The green leafhopper was only a minor pest of rice until a few years ago. Combining resistance to the pest and resistance to the pest-borne tungro virus in rice varieties became essential following epidemics of the tungro disease. Several commercial varieties now have this dual resistance, and their cultivation has reduced the tungro epidemic in many areas. However, new resistant varieties may still be needed if biotype selection, similar to the brown planthopper, occurs also in the green leafhopper.

Adequate levels of resistance to the whitebacked planthopper and the rice delphacid have also been bred into popular rice varieties.

Resistance to gall midge

This pest is serious in south and south-east Asia and a few African countries. Yield losses are total if infestation is high. Being an internal feeder, its control with chemicals is difficult, but the use of resistant varieties is practicable. Although biotypes among gall midges are reported to differ in different geographical areas, diverse sources of host plant resistance are available to counter this problem.

Multiple resistance

Because rice is attacked by a number of pest species, most of the breeding programmes are now developing varieties with multiple resistance to pests. For example, recent IR varieties, IR34, IR36, IR38, IR40,IR42, combine resistance to brown planthoppers, green leafhoppers, and stemborers.

Integration of varietal resistance with other control measures

Depending on its level, resistance can be used either as the principal method of control or can be integrated with other measures to develop an appropriate pest management strategy. For example, rice varieties resistant to the brown planthopper generally do not need insecticidal protection against this pest, but they do need protection from other pests to which they are susceptible. Applications of insecticides may be necessary to avert white-head damage even on stemborer-resistant varieties if borer infestation becomes severe at the flowering stage.

Resistant plant varieties do not adversely affect the natural enemies of rice insect pests. In fact, fewer insecticidal applications on resistant varieties may favour the survival of natural enemies. Experiments at IRRI demonstrated that a combination of varietal resistance and a predator gave better control of both the green leafhopper and the brown planthopper than either method alone.

Scope of resistant varieties in pest management

The use of insect-resistant rice varieties has been found to be highly practical for rice grown in tropical and subtropical Asia where infestations of a crop by about a dozen insect pest species is common. For the rice farmer in Asia it is the cheapest method of pest management. However, as most rice varieties resistant to pests have a narrow genetic base, their stability needs to be constantly monitored against the possible development of virulent biotypes of the pests. As a precaution, various rice improvement programmes at IRRI have been following a programme of sequential release of resistant rice varieties. If a variety resistant to a pest succumbs to a biotype, then new varieties with different genetic sources are released. Cumulative resistance, bred from a diverse gene pool, could be more lasting, but these efforts are still in their infancy. Integration of varietal resistance with other control measures is the most practical approach at present.

Anti-feedants

Insect damage to plants is primarily the result of their intense feeding activity. Anti-feedants which retard the pest's feeding activity on the treated rice plants would, therefore, be useful in controlling the pest because they are safe to the natural enemies that do not feed on the rice plant. Recent studies conducted at IRRI showed that oil from seed kernel of the Indian Neem tree, *Azadirachta indica*, significantly reduced feeding by the brown planthoppers and leaf folder pests on the treated rice plants. Neem oil disrupted the pest's normal growth and development and inhibited their ovipositional responses. Hatchability of leaf folder eggs was significantly reduced on Neem oil-treated rice plants, but the brown planthopper eggs which were laid within the leaf sheath tissue remained unaffected. In field trials, 50 per cent Neem oil sprayed at 4 litres/ha with an ultra-low-volume sprayer augmented the parasitization of the leaf folder larvae and pupae. Because Neem oil is relatively inexpensive (about US$1.00 per litre) and is plentiful in many rice-producing countries, research on its potential for controlling rice pests should be intensified.

Sex pheromones

Insects produce chemicals, called sex pheromones, which act as chemical messengers guiding insects to a mate. Their specificity renders them highly valuable for insect detection and surveillance in pest management programmes. Among rice insects, the occurrence of pheromones in the striped stemborer, the pink stemborer, and the rice leaf folder females has been demonstrated and similar studies are underway on the yellow stemborer and the caseworm, in collaboration with the Tropical Products Institute (TPI) in London. The major components of the pheromones of the striped stemborer and the rice leaf folder have been isolated and synthesized by TPI. The striped stemborer pheromones are two olefinic aldehydes, (Z)−11 hexadecenal (I) and (Z)−13 octadecenal (II) in a ratio of about 4.5:1. The attractiveness of a synthetic pheromone mixture to male striped stemborer moths has been demonstrated in several field trials. These pheromones have been used successfully for monitoring the striped borer moth activity on IRRI farms, and their proper use should permit efficient timing of insecticide treatments. To maximize the efficiency of the synthetic pheromones, various types of pheromone dispensers and traps are being evaluated.

Dispensed over a large area and released at a controlled rate, the synthetic pheromone may mask or neutralize the attraction of the natural pheromone produced by the females and may thus confuse the males during mate-finding. This confusion method has shown much promise in reducing the successful mating among striped borer moths and their infestation on experimental IRRI farms. The micro-encapsulated pheromone is dispensed over rice fields using ordinary spraying equipment. Attempts to utilize this method of rice pest control on peasant farmers' rice are under way.

Use of natural enemies: biological control

A well-developed complex of natural enemies of rice insect pests exists in unsprayed rice fields. Systematic study, conservation, augmentation, importation, and manipulation of natural enemies for rice pest control started only recently, partly because, in the past insecticides provided an easy and efficient insect pest control method. Even at the present level of insecticidal control, conservation of existing natural enemies, if not their

Table 6.5 Selected examples of parasites and predators of major rice insect pests

Natural enemy	Host	Host stage attacked
Parasites		
Anagrus flaveolus	Brown planthopper, Whitebacked planthopper	Egg
Apanteles chilonis	Striped stemborer	Egg
Gonatocerus sp.	Green leafhoppers	Egg
Neanastatus sp.	Gall midge	Larva
Oligoseta sp.	Rice leafhoppers, Planthoppers	Egg
Platygaster oryzae	Gall midge	Larva
Polygnotus sp.	Gall midge	Egg
Pseudogonatopus nudus	Brown planthopper, Whitebacked planthopper	Nymph, Adult
Telenomus rowani	Yellow stemborer	Egg
Tetrastichus schoenobii	Yellow stemborer	Egg
Trichogramma sp.	Rice leaf folder, Striped stemborer, Yellow stemborer	Egg
Tropobracon schoenobii	Yellow stemborer	Larva
Predators		
Callitrichia sp.	Brown planthopper	Nymph
Coccinella arcuata	Rice leaf folder	Egg
Cyrtorhinus lividipennis	Brown planthopper, Whitebacked planthopper Green leafhoppers	Egg, nymph, adult
Elenchus yasumatsui	Brown planthopper, Whitebacked planthopper	Nymph, Adult
Lycosa pseudoannulata	Brown planthopper	Nymph
Tetragnatha sp.	Rice leaf folder	Egg

actual manipulation, is necessary for successful rice insect pest management. Recent evidence shows that several natural enemies play a crucial role in rice pest control.

The natural enemies of rice insects include a range of parasites, predators and pathogens. The important parasites and predators of selected major rice insect pests are listed in Table 6.5.

Parasites

Several species of parasites of rice insect pests have been recorded in south and south-east Asia, but there is a need to determine the efficacy of the dominant parasite species. A fairly high level of natural parasitization (20–70%) of stemborer eggs by *Tropobracon schoenobii*, a braconid wasp, has been recorded in the Indian subcontinent, Korea, Malaysia, Philippines, Sri Lanka, and Thailand. Mass rearing and release of *Trichogramma* species have been attempted for biological control of stemborer and leaf folder pests of rice on mainland China. It is claimed that these operations, which cost about half that of chemical control, lead to 70–80 per cent parasitization of the eggs of the leaf folder.

A few parasites of the rice gall midge have been identified and their effects on pest abundance have been recorded. Thus, in India, larval parasitization by the gall midge, *Platygaster oryzae*, may be as high as 90–95 per cent. Similarly, in Indonesia, gall midge parasitization by *P. oryzae* and *Neanastatus* sp. has been reported to reach 96 per cent.

Recently, several parasites of rice planthoppers and leafhoppers have been identified. Egg parasitism may occasionally be as high as 80 per cent, with averages of 30–40 per cent. Since these egg parasities migrate into the rice fields soon after transplanting, their role in regulating the initial hopper population is very important. Parasites also attack nymphs and adults, but the level of parasitism is lower.

Predators

Predators are sometimes especially important as control agents as they attack more than one prey, devouring them or sucking their body fluids. Recent IRRI research stresses the importance of predators like the spider *Lycosa pseudoannulata*, the mirid bug *Cyrtorhinus lividipennis*, and the veliid bug *Microvelia atrolineata* in regulating the natural populations of the brown planthopper and green leafhoppers. The mirid bug is considered important because of its high mobility and high population density in the field, primarily during the first half of the crop period. *Microvelia* preys around the rice hills having high prey density and causing high prey mortalities. Its ability to survive long periods of starvation makes it a valuable biological control agent. Spiders are more abundant in upland than in lowland rice. Although they are general predators, a high spider population (peak at 40–50 days after transplanting) in the field effectively regulates hopper density. Ducks herded into rice fields are also reported to reduce the population size of the brown planthopper.

Pathogens

In a few cases *Bacillus thuringiensis* is sprayed to kill foliage feeders such as larvae of the rice leaf folder.

Conservation of natural enemies

Conservation of indigenous natural enemies of rice pests in farmers' fields is urgently needed, although their role in pest suppression has been recognized in only a few places in tropical Asia. Being indigenous they are already adapted to the host and its environment, and their conservation by favourable cultural practices is a better means of biological control than laborious and expensive attempts at mass breeding and inundative releases of indigenous or exotic species of natural enemies.

Conservation of natural enemies by selective chemicals or judicious use of pesticides, and by favourable cultural practices, is likely to give long-lasting control of rice insects.

Cultural control

Manipulation of cropping techniques to suppress pest populations has been practiced through the ages. But with the evolution of the pest management philosophy, interest has increased in cultural control measures. Cultural control can be practiced on a farm level and over a farming community.

Farm level cultural practices

The use of a seedbed for transplanting a hectare of rice concentrates the seedlings within a small area and allows easy inspection of developing pest problems. If insecticide treatment is warranted, the area to be treated is minimal. The close spacing within a seedbed is reported to reduce oviposition by the whorl maggot, while dry seedbeds control whorl maggot and caseworm which require standing water. The seedbed often serves as a trap crop for stemborers because egg masses are removed manually before transplanting.

The trap crop concept has been tested successfully for the control of the brown planthopper. A quarter of the total crop area planted as a trap crop

20 days prior to the main crop served to attract more colonizing hoppers on the trap crop than on the main crop. Control of the pest was easier, cheaper, and safer to the natural enemies, as the insecticide was applied only on the trap crop, and led to significantly higher yields in 'trapped fields' than in control plots without it.

Flooding rice fields eliminates many soil-borne insects such as root aphids, termites, and white grubs that occur in non-flooded fields. Limited control of armyworms, earhead cutting caterpillars and cutworms is also obtained by flooding infested paddies. Draining fields is practised to control whorl maggot and brown planthopper populations.

Heavy application of nitrogen fertilizer to high-yielding but pest-susceptible varieties aggravates the pest problem; however, the use of fertilizers cannot be reduced in favour of pest control. Using fertilizer-responsive insect-resistant varieties is the best approach to obtain optimal yields. Split applications of fertilizer over the crop growth period and the use of slow-release fertilizers are also beneficial.

Cultivation of early maturity varieties (100 v 150 days) reduces the threat of late season pests such as the brown planthopper and stemborers. Planting early or late, depending on the locaton, results in escape from highly seasonal pests such as the gall midge. Other escape mechanisms are ratooning (i.e. planting of successive crops) and transplanting of older seedlings.

Community level cultural practices

Some of the cultural operations need the co-operation of farmers in a village to suppress pests over a large area.

In many areas, farmers synchronize the planting within large contiguous farm units. This reduces the period of host availability as well as diluting an invading pest population over a large area so as not to overwhelm the natural enemies.

Rotation with non-rice crops breaks pest cycles, and tillage destroys volunteer rice (i.e. self-seeded plants) and other weeds. Sanitation or selective removal of alternative weed hosts of leafhopper virus vectors at the beginning of the rice growing season is widely practised in China. Continuous destruction of crop residues, such as stubble, on a large scale effectively reduced the rice stemborer menace in India.

Other cultural methods of choice in many areas are the hand removal of large insects, use of baits or traps, use of plant parts as repellents, or botanicals or other indigenous insecticides.

Implementation of pest management programmes

There is now a good appreciation and understanding of the need to implement rice insect pest management programmes in the tropics. The status of the development varies among countries, but there is sufficient technology available to implement programmes on at least a pilot basis. The following aspects should be included in the formulation of pilot projects.

Group action The pest management approach requires an interaction among farmers, extension workers, scientists and administrators. For optimal impact, pest management measures require a concerted and cooperative effort among farmers over a wide area.

Training Proper training and education of extension workers, technicians, and scientists, is the key to successful implementation of pest management programmes. Formal, degree-oriented education for the prospective rice scientist is needed; but of the greatest importance is the specialized, practical, field-oriented training of the extension staff, required for the transfer of technology from the research station to the farm level.

Monitoring This requires close scouting of rice fields sustained either by government organizations or by farmer co-operatives. Countries like China and Vietnam have regular scouts for monitoring rice fields and are motivated to perform well.

Input delivery system Availability of effective insecticides and credit for their purchase must be ensured by the chemical companies and government agencies, respectively. Seeds of insect- and disease-resistant varieties must be readily available to the farmers. This requires rice scientists and government agencies to work closely in the development of resistant varieties and for their seed multiplication.

Evaluation Evaluation of a programme permits identification of numerous constraints, including technological, political, and socio-economic factors, that may still need to be rectified by further scientific research. The formulation of realistic government policies, and a better understanding of the rice farmer's behaviour and background that limit the adoption of the new pest management technologies, need be taken into consideration.

Conclusions

Rice insect pest management seeks to employ and integrate a wide variety of naturally occurring control forces in the pest's environment. The complexity of the rice insect problem and the general socio-economic conditions of the rice farmer require that an integrated pest management approach be suitably adapted for practical, long-lasting control in the tropics. In this context, the identification of key pests, their economic status and seasonal abundance, the control methods available, and the economics of treatments, deserve attention before pest management programmes are designed. Research on sampling techniques, economic thresholds, and basic ecological studies must be further strengthened to provide the basic information needed in a pest management programme. Alternative methods to chemical control, such as repellents, anti-feedants, and botanical insecticides should be identified and tested. The development of strong farmer organizations is crucial to the success of pest management programmes. The transfer of pest management technology to farmer level will require an efficient network of extension teams which can translate into practice the concept of pest management.

Management of maize pests

P.T. Walker

The importance of maize

The production of maize reached 394 million tonnes in 1979, second only to wheat in world production, but in the tropics its importance as a staple food is far greater than this figure suggests, particularly in Africa and America. It is the most valuable crop in many countries, if both subsistence and marketing are considered. In Kenya, for example, maize is grown for subsistence on 70 per cent of the area of small farms and marketed for cattle feed, for export and processing. Its nutritive value is highly appreciated and its importance as an oil crop is increasing. Maize prefers temperatures of 10–30 °C with optimum growth between 24–30 °C, but rainfall is usually the main factor limiting distribution, the crop needing 45–60 cm during growth. Long- and short-term maturing varieties are developed for areas according to the length of the rainy season. New high-yielding hybrids and composites can increase yield up to 10 times if the recommended inputs are sufficient.

The main pests

Maize pests may be categorized according to the part of the plant attacked, some examples are given below.

At and below ground level pests
Beetles, adults and larvae: *Heteronychus, Schizonycha, Anomala, Gonocephalum, Diabrotica, Tanymecus.*
Crickets: *Gryllulus, Gryllotalpa.*
Ants and termites : *Hodotermes, Microtermes, Odontotermes.*
Rodents : *Rattus, Mus.*

Stem-feeding pests
Moth borer larvae : *Busseola, Chilo, Sesamia, Diatraea, Zeodiatraea, Eldana, Elasmopalpus, Ostrinia.*
Cutworms : *Agrotis, Euxoa.*
Rodents : *Rattus, Mus.*

Leaf-feeding pests
Locusts and grasshoppers : *Locusta, Schistocerca, Patanga, Zonocerus, Melanoplus, Hieroglyphus.*
Armyworms, etc. : *Spodoptera, Mythimna, Marasmia.*
Beetles, adults and larvae : *Epilachna.*
Sucking bugs, leafhoppers (including vectors) : *Cicadulina, Peregrinus, Blissus, Dalbulus, Pyrilla.*
Aphids : *Rhopalosiphum.*
Thrips : *Frankliniella.*
Mites : *Tetranychus.*

Ear and tassel pests
Beetles : *Carpophilus, Sitophilus.*
Stemborers : As above, under stem-feeding pests.
Earworms : *Heliothis..*
Aphids : *Rhopalosiphum.*
Bugs : *Nezara.*
Rodents : *Rattus, Mus.*
Birds : *Quelea, Ploceus.*

A wide range of these pests damage maize in different parts of the tropics.

Assessment of maize pests

To establish the status of a particular pest, and to ascertain whether there is any economic advantage in controlling it, the effect of the pest on yield must be known. There are different methods for assessing pests and the methods adopted will depend upon the kind of pest and the ecological situation.

Stemborers The incidence of stemborers is assessed as the percentage of plants attacked, the percentage of stem internodes bored, the number of exit holes, or number of live and dead insects present. The number of egg masses or plants bearing them can be used to forecast attack, and the use of pheromone traps for adult assessment is also possible. Light-trapping of adults is usually unreliable.

Armyworms Adults are assessed by light-trapping, and increasingly by pheromone trapping, specially to forecast outbreaks. The number of eggs or larvae per plant or per m^2 gives an estimate of the measure of plant attack.

Leaf and cob caterpillars These are assessed by their numbers, or on a rating scale; alternatively the degree of damage can be rated. Adults can be caught in light- or pheromone-traps (Fig. 4.8).

Aphids These can be counted per leaf, tiller or cob, or rated if infestations are high. Bugs, leafhoppers or thrips are sampled by sweep-netting or are shaken from plants or driven out by fumigation. Their populations can be monitored over a wide area by suction traps (Fig. 4.11) or by water or sticky traps (Fig. 4.7) which are often yellow.

Soil pests These are sampled by taking a standard volume of soil and sieving, or using more elaborate dry (Fig. 4.4) or wet (Fig. 4.5) extraction methods.

Birds and rodents These vertebrate pests can be trapped, counted or estimated on the basis of their damage to the maize crop.

Pest assessment should always be related to the decimal growth stage of the maize crop as illustrated in Fig. 6.16.

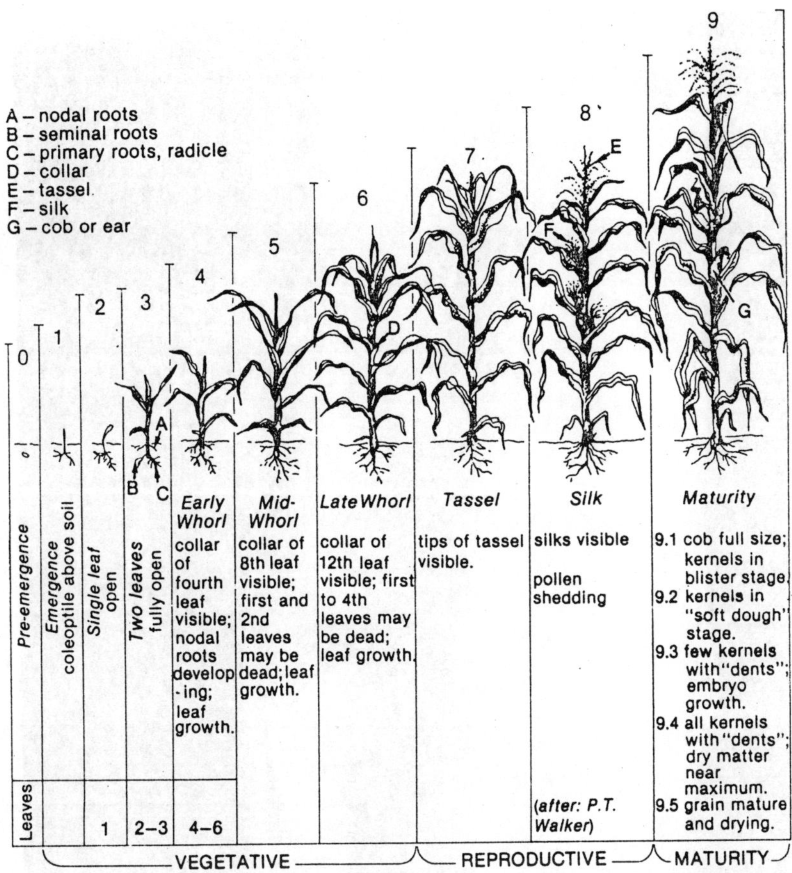

Fig. 6.16 The various growth stages of maize (From L. Chiarappa (1971).)

Crop losses

Loss of crop yield due to these pests may be caused in four ways: (i) by direct loss of plants; (ii) by the loss of an essential part of the maize plant, such as the growing point, roots, unripe or ripe grain; (iii) by the prevention of grain development by the interruption of the transport of water and nutrients when roots, stems (Fig. 6.17), or leaves (Fig. 6.18) are destroyed; or (iv) by the indirect introduction of pathogenic fungi, bacteria and viruses or by secondary pests after primary damage.

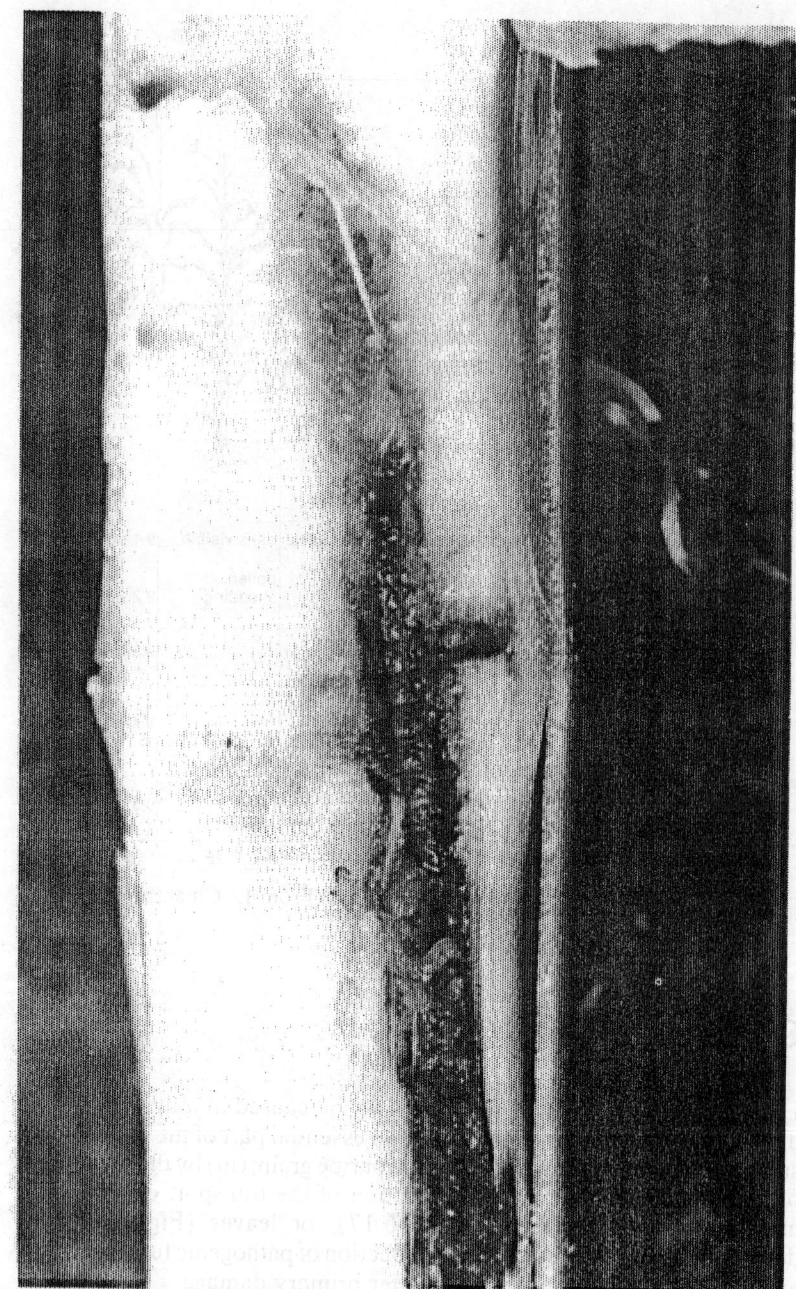

Fig. 6.17 Larva of the stemborer, *Busseola fusca* in a maize stalk. (P.T. Walker.)

Fig. 6.18 Damage to maize leaves by beetles of the genus *Adoretus* in Thailand. (P.T. Walker.)

There are few accurate estimates of crop loss due to the various pests mostly because of the extreme variability of yields caused by climate, crop variety, the type of agriculture and time of growth. In Kenya, a survey in 1963 estimated the average annual losses in maize presented in Table 6.6.

Table 6.6 Annual losses in maize due to pests in Kenya in 1963

Pest	% loss	Loss (tonnes)	Equivalent area (ha)
Borers	18	244 300	211 294
Armyworms	9	105 900	86 768
White grubs	3	13 100	11 898
Cutworms	3	12 600	10 801
Nematocerus	5	3 600	2 104

The percentage loss is calculated as the loss incurred if pests are not controlled, averaged over various provinces and weighted by area of crop. The equivalent area, also called the 'Ordish area', is the area which would be available if losses were prevented.

A recent assessment of losses due to *Busseola* borer in Kenya showed, by relating yields to infestation rates in insecticide trials, that about 1.2 per cent reduction in yield per 1 per cent plants attacked, occurred over 3 years. On small farmers' fields near Kitale, Kenya, an average of 14 per cent loss was found. In Nigeria losses of 0.9 per cent per 1 per cent plants attacked have been recorded. Losses due to *Spodoptera exempta* armyworms have been estimated in Kenya by artificial damage trials. Losses of 8–100 per cent are possible, depending on the percentage foliage area lost and time of attack. In the 1979/80 season, a loss of about 14 900 tonnes of maize was estimated in Tanzania.

Management of maize pests

Although maize is grown widely in the tropics and there has been considerable research on the control of its key pests, no deliberate attempt has been made to develop a specific well co-ordinated integrated management programme for maize pests in any particular locality. Several separate methods for the control of maize pests are available in different maize growing regions. Active attempts to apply integrated pest management (IPM) principles to maize growing are being encouraged, ranging from perhaps a single measure such as parasite release or crop resistance to the integration of several measures, as in India, Nicaragua and other countries. The commercial use of IPM is now developed in the USA, France and elsewhere, and some reviews of single or integrated methods have appeared. The guidelines for integrated control of maize pests published by the Food and Agriculture Organization of the United Nations (FAO) are largely based on the North American experience.

The remainder of this contribution will highlight different approaches to the management of maize pests in various parts of the tropics. This will thus provide a collection of components which hopefully could be combined harmoniously in a well co-ordinated integrated management system for the maize pests in an area.

Components in the integrated management of maize pests

Cultural practices

To reduce pest attack by manipulation of cultural practices needs a knowledge of the pest life-cycle and of the population of the pest at different stages. Once these are known, trials can be carried out on the effects of changes in the time and method of sowing, cultivation, fertilization, duration of crop maturation, time and method of harvest, and crop rotation.

For example, the life-cycle of the maize stemborer, *Busseola fusca*, was examined in the Kitale area of Kenya over several years by growing rows of maize every 2 weeks through the season, irrigating if necessary. It was known that the pest passes the dry season as diapausing larvae in dry maize stalks. If more than about 10 mm of rain falls (see Fig. 6.19 point *R*), diapause ends, larvae pupate, adults emerge and eggs are laid about 30 days after *R*; and damage to plants appears about 45 days after *R*. If crops at the most susceptible stage, that is 20–80 cm high, are growing during the egg-laying period, they can suffer maximum attack and damage. If grown earlier, plants may be too big during egg laying; if later they may miss oviposition and be too big when the second-generation eggs are laid, about 100 days after *R*. Thus if the time of egg laying is found by examination of plants or by forecasting from the rainfall, fewer insecticide applications will be needed to control this pest. Unfortunately the date of sowing is not easily changed because of the need to sow early for maximum yield, but it is nevertheless possible to predict the time of maximum attack. Change of sowing date is advocated in the Philippines for reducing the attack of maize by white grubs, *Atherigona*, thrips, and *Heliothis*.

The depth of sowing can affect cutworm attack and its control with insecticides. Fertilizer increases the nitrogen content of the crop and

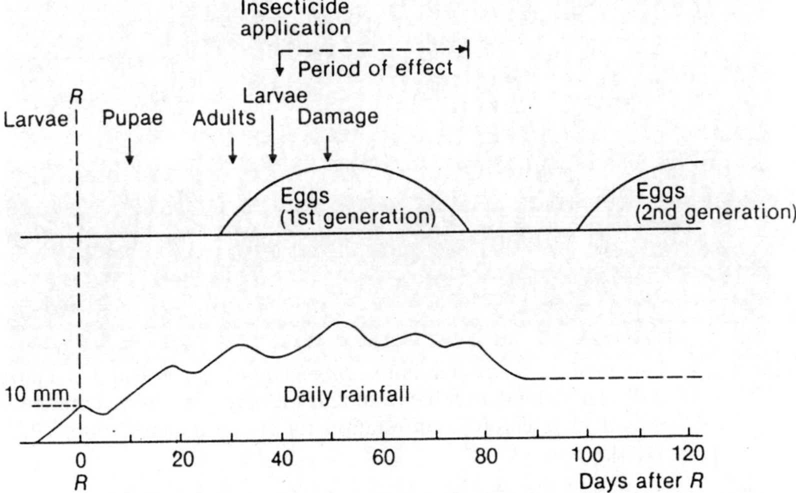

Fig. 6.19 Schematic representation of some events in the life-cycle of the maize borer, *Busseola fusca*, at Kitale, Kenya in relation to rainfall and insecticidal applications. R denotes the date by which the critical 10 mm of rain has fallen, and which results in ending larval diapause.

attack by aphids, but in Nigeria, fertilizer application reduces losses due to stemborers, due to greater plant vigour. High fertility and plant density usually favour pest and disease attack in temperate climates, but in the tropics the effects are often linked with the rainfall and the time and rate of growth, and the direct effect on pests is hidden.

Early harvesting, artificial drying and bulk storage of grain are valuable in reducing weevil damage in Kenya, where drying facilities have been built. Bulk handling has also been used to the same effect in the West Indies and in the USA.

Irrigation influences pest attack by, for example, increasing mite attack on maize in Texas. Irrigation may increase pest attack by providing additional crops and extending the life-cycle of the pests.

Fig. 6.20 Larvae of the maize stemborer, *Busseola fusca*, over-winter in maize stalks left behind after harvesting. If they are not removed the stalks provide a reservoir of infestation for the next maize crop. (P.T. Walker.)

Destruction of crop residues is the obvious way to break the pest life-cycle; this is required by law in some countries, but it is difficult to enforce (Fig. 6.20). In Nigeria felling maize stalks caused the death of borers. Burying stems deeper than 50 cm, or partial burning of sorghum stalks to kill stemborer larvae and pupae, has been suggested. Another possibility is the use of trap crops which can then be destroyed. Crop residues laid along contours for soil conservation can result in massive infestations of borers. Crop residues and weeds also encourage cutworm attack of the following crop, but the removal of green material reduces later attack.

A long interval between cultivation and sowing prevents egg laying by cutworm moths, and the response of larvae to light has been found to be important in knowing the best depth at which to sow. Weeds were important in the relation between pest, predator and weed populations in maize crops in Colombia. Cultivation exposes soil pests to predation by birds, but this may not be possible for conservation reasons. The trend to zero cultivation, no-till, or direct drilling may have increased *Chilo* borer attack in South Africa, but reduced *Elasmopalpus* borer in the USA. Disturbance of the soil affects *Heliothis* emergence. An uncultivated 10 m wide strip round a field is recommended in Central Africa to prevent leafhopper vectors introducing maize streak virus. The previous crop can have a great effect on the pest population of a later crop. Interplanting crops can affect the pattern and level of pest attack. In the Philippines, maize grown with groundnuts was less attacked by borer, and different types of maize and beans cropping affected pest attack in Central America. These mixed crops were less attacked by *Diabrotica* and *Spodoptera* than monocultures, and *Spodoptera* attack on maize was 88 per cent less when beans were sown 20–40 days earlier than the maize.

Use of host plant resistance

This is probably the most promising way of reducing pest attack, either by resistance to an initial attack, by antibiosis once plants are attacked, or by tolerance to attack. Resistance of maize to stemborer and leaf aphid is important in the USA where the location of genes on chromosomes has been studied. The International Working Group on *Ostrinia* has been selecting composites with borer resistance. Some resistance to *Busseola fusca* has been found at the International Institute of Tropical Agriculture (IITA) in Nigeria, as well as to *Sesamia calamistis*, *Buphonella* ground-beetles and streak virus. At the International Maize Improvement Centre (CIMMYT – *Centro Internacional de Mejoramiento de Maiz y Trigo)* in Mexico, several lines have been investigated for resistance to individual pathogens, tested against as many pests as possible, the selections contributing to the broad-based composites which are being used to incorporate resistance to African and Asian borers. Some resistance to *Spodoptera, Heliothis* and *Rhopalosiphum* has been found and these lines are being tested in an international programme. Maize resistance to *Sesamia nonagrioides* has been found in France, and to *Chilo* in India. Promising resistance has been found by the International Centre of Insect Physiology and Ecology in Kenya.

Host plant resistance to maize pests has been explained on the basis of anatomical and chemical factors in the maize plant. For example,

Heliothis resistance is related to the high amino acid content of the maize silks of resistant varieties. Anatomical features, such as a hard stem, loose leaf sheaths or hairyness have been found to be partly responsible for some levels of resistance.

Use of natural enemies

Parasites and predators

Parasites such as *Apanteles, Telenomus* and *Trichogramma* have proved very valuable in the control of *Diatraea, Ostrinia* and *Sesamia* borers, and some are available commercially in some countries. Other examples of biological control of maize pests which are being investigated are the microbial control of grasshoppers, parasites and predators of *Eldana*, and parasites and a fungus on *Mythimna*. Spiders serve as important means of control in India and the Far East, and in Egypt the parasites and predators of maize pests, for example *Orius* and *Chrysopa* on *Heliothis*, have been studied and used for maize pest control.

In East Africa and Madagascar, for example, the parasites and predators of maize pests have been surveyed and reviewed. The main parasites, mostly from stemborers, are two ichneumonids, three braconids, a chalcid, a eulophid, a scelionid egg parasite, and two tachinids. There are several parasites of *Spodoptera* armyworms, *Heliothis*, *Agrotis* cutworms, and aphids. Three scoliid wasp parasites of soil chafer grubs have been introduced from Mauritius. Apart from the egg parasites, however, many borer parasites attack too late to prevent crop damage, and are often too few to reduce later pest generations. Existing parasites are often either unsuccessful or of limited distribution. The Commonwealth Institute of Biological Control (CIBC) station in East Africa plans to introduce five successful parasites from Asia and the West Indies, and to redistribute East African species.

Pathogens

In East Africa natural infections by *Cordyceps* fungus disease of borers, and bacteria and virus diseases have been reported. *Bacillus thuringiensis* can be effective against borers and armyworms, although this bacterium had little effect on *Busseola* in field trials in Kenya. Virus diseases are well known in *Spodoptera exempta* armyworms and commercial formulations of virus are available for *Heliothis* control. Viral control of cutworms has been found effective in Europe.

Fungal and bacterial diseases of soil beetle larvae are sometimes very effective, and Japanese beetle milky disease has been introduced for the control of *Cochliotis* beetle larvae in Tanzania. In China, *Beauveria* applied to stalks infested with *Ostrinia* reduced the next year's moth population by 75 per cent and killed 82 per cent after application to first generation larvae. *Nomuraea* fungus is an important pathogen of *Spodoptera* in Australia.

Biological control is often compatible with chemical control in maize, because chemicals are applied usually to the top of the plant, although systemic insecticides might kill parasites living in a host. Parasites of pyralid borers, which have exposed eggs, may be repelled or killed by insecticides, and here granular insecticides are the least dangerous. The toxicity of new insecticides to parasites should be tested, before they are used.

Use of pheromones

Although the use of pheromones to reduce attack has had some success with some moth pests of fruit and cotton, with *Sitotroga* moths in maize in France, and with *Spodoptera littoralis* in Israel, their main use has been for assessing the timing and size of populations with a view to more effective and efficient insecticidal control. The big advantage over light-trapping is selectivity since only one target species is attracted to the pheromone trap. Pheromone traps are used in the forecasting system for *Spodoptera exempta* in East Africa, for assessing cutworm moth populations in Europe, and are being experimented with for *Busseola* in East and Central Africa. A cheap, plastic funnel trap may be used (Fig. 4.8) with a plastic container or bag for collection, or the pheromone capsule can be placed in the middle of a sticky board. Pheromone trap catches should be co-ordinated with light-trap catches, with field attack, with the timing of control measures and with the final crop yield.

Physiological and genetic measures

The value of changing stemborer life-cycles has been examined in the USA, where northern and southern races of *Ostrinia* have different degrees of diapause. Crossing the races and selection can reduce diapause and lead to moth emergence in the winter when they are killed. Adult stemborer emergence in the dry season in the tropics would have the same effect.

Growth and ecdysis hormones which delay or speed up ecdysis or diapause have similar effects, but it is difficult to apply the chemicals to the insect at the right time. The insect growth regulator (IGR) Dimilin, is promising against *Ostrinia*.

Sterility induced by chemicals or irradiation has been tested on a pilot-scale against *Heliothis* corn earworm, *Diatraea*, *Ostrinia* and *Chilo* borers. A field trial on *Heliothis* eradication in the Virgin Isles was promising but unsuccessful for various reasons.

Chemicals in maize which attract, repel or stimulate the feeding of pests are known, and although chemical anti-feedants have shown promise, they are of little practical application at present.

Improved use of pesticides

If pesticides are used to control maize pests, they can be used more effectively, in smaller quantities, and with less danger to the environment than they are commonly used.

Granular pesticides are environmentally acceptable because they contaminate less of the environment, fall where required and do not drift, and moreover because the release of pesticide can be controlled, repeated treatments may not be needed. They are less dangerous to humans, animals and wild life as well as beneficial insects. They have proved very effective against borers and in many countries granular alternatives to DDT have been tested against stemborers. Carbofuran and endosulfan are most effective. As little as 0.3 kg endosulfan/ha can control borers for 8 weeks after application as granules (Fig. 6.21), and no residues on the crop after harvest have been found. Residual systemic granular insecticides such as carbofuran give extended protection after application below the seed, with an added side-dressing later if necessary. Many plant-feeding

Fig. 6.21 The application of endosulfan granules to individual maize plants for control of stemborers. (P.T. Walker.)

pests as well as nematodes are controlled, with little effect on their parasites and predators.

Granular pesticides have advantages over other formulations against soil pests. The growth of plant roots and the ability to place granules accurately allow safe action for a long period. Seed dressing is also a very efficient way to use pesticide and is very selective in its action on pests and beneficial insects. Pesticide efficiency can be improved by better timing, based on pest surveillance and forecasting, and by application only at predetermined economic pest action thresholds. It is necessary to know how long the pest and the crop are at stages when control is easy, the duration of insecticidal effect, and the amount of attack which can economically be allowed to develop. Trials on *Busseola* in Kenya indicate that control of larvae is only necessary between 10 and 40 days after egg-laying starts. Endosulfan granules will control larvae for more than 40 days, so only one application is necessary soon after the start of egg-laying. Less than about 5 per cent of plant attack does not cause economic yield losses, so by scouting for egg-laying, or forecasting it from rainfall, and only using insecticide if there is more than 5 per cent attack, insecticide applications can be reduced from 3 kg DDT/ha (3 applications) to a single application of 0.3 kg endosulfan/ha.

In a 7 state maize integrated pest management programme in the USA, based on scouting, action thresholds and other methods, it was found that almost half the insecticide used on maize in Iowa was unnecessary, and in Illinois over $US8000 a year was saved on soil insecticides used on maize after groundnuts.

Design of an integrated pest management programme for maize

A successful IPM programme for maize pests can be developed only when the necessary biological, ecological and behavioural data on the target pest species are available, and the required staff and physical facilities are provided. The information presented in this contribution should form the basis for the design of IPM programmes in maize according to the following scheme.

An integrated programme may be designed as below.

(a) Identify the main pests and diseases, the key pests, occasional and minor pests, and start a programme to assess their economic importance.
(b) Establish pest assessment methods, and monitor the pests in relation to the crop and the climate.
(c) List possible control factors and assess their feasibility.
(d) Consider their impact on the environment.

(e) Establish contact with those working on similar problems in the country and outside.
(f) Start trials on the introduction of the factors singly or in combination to find their effect on the pests and on the economic yield.
(g) Start long-term studies on promising factors.
(h) Draft a theoretical model of the effects of the factors, singly and then in combination, which can be tested against reality and altered accordingly.
(i) Train extension staff and farmers, with demonstrations, on the recommended programme.

Management of pests of subsistence crops

C.P.F. de Lima

In the tropics, legumes, pulses and cereals are the major subsistence crops. The pests of each group of crops together with the relevant pest management strategies are considered below.

Legumes and pulses

These represent an extremely wide variety of grams, lentils, peas, beans and groundnuts, all very attractive to insects which attack virtually every part of the plants. The incidence of several important pests can be lessened by appropriate cultural measures since these crops grow rather quickly, within 3–4 months. The major pests and their management strategies are outlined as follows.

Ophiomyia phaseoli (Diptera, Agromyzidae)

The bean fly occurs in some parts of West Africa including Mali, Nigeria and Senegal, parts of Ethiopia, Egypt and Sudan and every country in South and East Africa. The fly is common in south-east Asia, northern Australia and the Pacific Islands.

The bean fly attacks several species of legumes but prefers *Phaseolus*. The small (2 mm) black female flies insert their eggs into the young leaves. The larvae mine their way to the base of the stem where they pupate. Plants damaged by *Ophiomyia* exhibit a characteristic thickening of the base. Several longitudinal cracks develop along the stem and the leaves turn yellow. In areas where successive crops are grown in a season, plants that have escaped early infestation may be attacked at a later stage and show a characteristic swelling of the nodes where the fly has pupated. Experiments in Kenya show that up to 30 per cent of the crop is lost in untreated bean fields. In trials at the National Agricultural Laboratories in Kenya, using continuous sowings (and no insecticidal sprays) it was found that up to 100 per cent of the crop was damaged.

Under subsistence conditions in Kenya, effective control of this pest has been obtained with minimum use of insecticides under the following management system:

(a) Seed dressing with 40 per cent aldrin (0.5 kg insecticide/100 kg of seeds);
(b) Early planting, that is as soon as the rains have appeared and the soil has sufficient moisture;
(c) Farmers in an area plant at the same time;
(d) Successive or overlapping crops are avoided;
(e) After harvest, all crop residues are removed and destroyed by burning.

In commercial plantings with several successive and overlapping crops bean flies cannot be controlled without the following insecticidal spray schedule:

(a) When seedlings are 10 days old begin the first spray with one of the following organophosphate insecticides: diazinon (600 g.a.i./ha), fenitrothion (500 g.a.i./ha) or fenthion (500 g.a.i./ha).
(b) Repeat with 2–4 sprays at fortnightly intervals using diazinon (900 g.a.i./ha), fenitrothion (750 g.a.i./ha) or fenthion (750 g.a.i./ha).

Lepidopterous pests

Caterpillars of both the spotted borer, *Maruca testulalis* (Pyralidae), and the American bollworm, *Heliothis armigera* (Noctuidae), are pests that bore into the flower buds of legumes and hollow them out. They attack green pods and consume the seeds. Losses of 15–25 per cent occur on unsprayed crops. In heavily infested crops well over 50 per cent of the crop may be lost.

Maruca testulalis larvae are usually reddish to yellowish with dark spots. The pest is common in Central and South America as far south as Uruguay, in Africa south of the Sahara and in south-east Asia and eastern Australia.

Heliothis armigera larvae are usually green to black in colour with pale longitudinal bands on the sides. Other colour variations exhibited by the larvae include orange and reddish brown. The pest is common in Africa, the Mediterranean countries and parts of southern Europe, the Middle-East, south-east Asia, New Zealand and parts of north-east Australia.

In areas with a bimodal rainfall pattern, planting of legumes may follow cotton, maize or sorghum whereas in other areas legumes are interplanted with these main crops in a mixed cropping farming system. When the crop is interplanted with cotton this results in more severe damage to the legumes from *Heliothis armigera* which migrates to cotton only when the legume pods become dry. Less infestation from lepidopterous borers on legumes occurs when the crop is planted in a subsequent season (for example the short rains in East Africa), as there is a sufficient interval during the rains to provide a 'closed' season.

A management system practised for *H. armigera* involves a combination of chemical and physical methods:

(a) All crop residues are destroyed by deep ploughing;
(b) A closed season of 2 months is observed;
(c) Chemical application once before flowering and once after pod formation with one of the following synthetic pyrethroid insecticides – permethrin (100 g.a.i./ha), fenvalerate (100 g.a.i./ha), cypermethrin (50 g.a.i./ha), deltamethrin (= decamethrin) (15 g.a.i./ha).

Aphid pests

Both *Aphis fabae* (black bean aphid) and *Aphis craccivora* (groundnut aphids) are black aphids (although *A. craccivora* has a shiny appearance), and are important pests of legumes and pulses. They are widely distributed in the tropics and found on every continent, except that *A. fabae* does not appear to be present in Australia.

Effective control is difficult without the use of chemicals. The following management practices have been in use in tropical Africa:

(a) Early planting to give the plants time to establish themselves before the aphid infestation begins;
(b) Closer spacing in groundnuts to reduce aphid populations and incidence of the aphid-transmitted rosette disease (aphids appear less willing to settle on a closely spaced crop and more aphids are usually found on an open crop);
(c) Destruction of volunteer plants (cultivated crops that have seeded naturally and have not been deliberately sown) and weeds by ploughing before the rains arrive;
(d) At aphid infestation levels in excess of 10 per cent, 1–2 sprays of the organophosphate insecticides dimethoate (600 g.a.i./ha) or diazinon (600 g.a.i./ha) applied at 2 weekly intervals;
(e) Use of resistant crop varieties selected from breeding programmes.

Cereals

In many tropical countries the staple food for the people is based on one of the following cereals, maize, wheat, rice, sorghum and millet. These crops are grown often in a single main rainfall season although in areas like East

Africa with two rainy seasons two crops of early maturing varieties are sown. Maize, sorghum and millet may be grown next to cotton, and pests like *Heliothis armigera* and *Taylorilygus vosseleri* tend to multiply on the cereals before migrating to cotton. In several countries rice is cultivated under irrigation and where sugarcane is also grown near by, *Sesamia* is a problem on both crops.

Lepidopterous stemborers

Stemborers constitute a major pest of cereals in the tropics, and include *Heliothis armigera* (American bollworm), *Busseola fusca* (maize stalkborer), *Sesamia calamistis* (pink stalkborer), *Sesamia inferens* (purple stemborer), *Maliarpha separatella* (white rice borer), *Chilo partellus* (spotted stalkborer) and *Chilo polychrysa* (dark-headed rice borer).

H. armigera is common in Africa, Asia and parts of Australia and *B. fusca* occurs in virtually every African country south of the Sahara. *S. calamistis* occurs almost everywhere in Africa while *S. inferens* is common in south-east Asia. *M. separatella* is found in south-east Africa including Madagascar, as well as in all the West African countries south of the sahara. *C. partellus* occurs in East Africa and south-east Asia, while *C. polychrysa* is present only in south-east Asia.

The eggs of such stemborers are usually laid on the leaf sheath of the plant and the larvae bore into the stem often causing death of the leaves. In early stages of growth the crop is able to compensate for this loss, for example, by putting out more tillers in rice. Under good agronomic conditions maize compensates for loss of stand by increased cob weight in surviving plants. In maize a 1 per cent infestation results in approximately 1 per cent yield loss whereas in rice there is no apparent loss with a 10 per cent infestation, but at 20 per cent infestation a 7 per cent yield loss is experienced.

Infestation can be reduced considerably by an integration of the following management practices:

(a) Farm hygiene including burning or burying all crop residues after harvest to kill dormant stages of the pest – for example in Sierra Leone larvae of *Maliarpha* remain dormant in stubble for 20 weeks;
(b) Having a short sowing period, especially for rice, to reduce the period during which the crop is susceptible to attack;
(c) Encouraging farmers in an area to plant their crops at the same time;
(d) Practising a closed season of at least 8 weeks;
(e) Under irrigated conditions continuous cropping of rice must be avoided;
(f) Stalkborers often have alternative weed hosts, for example *Scirpus grossus* which harbours larvae of *Chilo polychrysa*, and such weeds should be destroyed.
(g) Chemical control in cereals grown for subsistence is accomplished by the use of granular formulations of insecticides. Systemic insecticides

such as carbofuran 5 per cent granules may be applied in the root zone for uptake by the plant. Non-systemic insecticides like 5 per cent diazinon or endosulfan granules may be applied in the funnel of the plant. Rice requires one treatment in the nursery plot and one after transplanting. Other cereals, such as maize and sorghum, require one insecticidal application when the plant is 50 cm tall and a second treatment after 3 weeks.

Aphid pests

Rhopalosiphum maidis (maize aphid) and *Schizaphis graminum* (wheat aphid) are fairly widespread in the tropics in Africa and Asia, in Europe, North and South America and Australia. *R. maidis* is most widely distributed in Africa and occurs in virtually every country on the continent. *R. maidis* is dark green with black cornicles and a caudal process, while *S. graminum* is pale green with a dark green stripe on the back. The damage caused by these pests is serious when it affects young plants and losses in yield of up to 25 per cent have been recorded in barley when the number of aphids per tiller exceeds 20. These aphids are also vectors of several virus diseases including mosaic virus on sugarcane, wheat mosaic virus and cereal yellow dwarf of cereals.

Satisfactory control of these pests has been achieved by strict observance of the following management programme:

(a) Burning of crop stubble after harvest;

(b) Early planting of the crop to enable the crop to withstand aphid attack;

(c) Fertilization of the soil to encourage vigorous crop growth;

(d) Late-sown crops and crops grown in marginal areas with poor fertilization may suffer heavy infestation and attack on the inflorescence and insecticidal sprays of dimethoate or diazinon (600 g.a.i./ha) at 10 days intervals should be applied to supplement the control of the pest by natural agents.

Control of stored products pests

M.I. Ezueh

In the developing countries, substantial losses are sustained annually on various stored commodities due to damage by insects and other organisms during the post-harvest handling processes. Quantitative assessment of these losses has been difficult because of the extremely variable and complex traditions which influence storage practices and the lack of suitable 'loss-survey' techniques. However, assessments made on farms, in traders' stores and central storage depots from selected areas have indicated serious depreciations in both quantity and quality of various stored products due to inadequate pest control. Although comparative statistics are available for planning purposes caution must be exercised in extrapolating loss estimates from one situation to another as the concepts of damage and loss tend to vary with prevailing local cultural, economic and social factors.

World storage losses for all grains have been estimated as about 10 per cent of the annual production, which, in quantitative terms is over 100 million tonnes. In the developing world an average annual loss of 20 per cent is generally cited for planning purposes. Although this figure may be a conservative estimate, in terms of total grain production in Africa, Asia and Latin America this comes to about 77 million tonnes. The greatest proportion of storage losses occurs at the subsistence farmer's level where average losses may exceed 30 per cent. This is because of the inherent weaknesses of the traditional post-harvest operations such as poor threshing, cleaning, drying and storing techniques. Although gross national storage loss estimates are not given in these countries, specific figures are available for certain commodities. For example, about 15 per cent of maize grains harvested in Ghana are lost annually to the maize weevil, *Sitophilus zeamais*. In Nigeria, about 5.0 per cent of stored cowpea is lost to bruchid beetles at the farm level, increasing to about 30 per cent in the traders' stalls. In Uganda, maize stored in traditional stores at a 12.5 per cent relative humidity for six months may loose 8–9 per cent of its weight due to attacks of grain weevils. Average estimates of grain losses in Asia are in the order of 16.5 per cent.

Storage losses should not be visualized only in terms of weight loss, because damaged seeds have often lost their viability, and this reduces available planting materials needed to sustain production. Post-harvest losses may also result in loss of quality of the stored commodities, and this may be reflected in changes in their texture, taste, appearance and overall acceptability in the market. Infested food items may have reduced nutritional values. The most unfortunate aspect of storage losses, however, it that they occur at a time in the production systems when compensatory mechanisms are no longer operative.

Sources of primary infestation

The design and operation of a sound storage system requires a fundamental knowledge of how stored goods become infested by pests, and the following main sources have been universally recognized.

Field-to-store infestation

For many cereals and pulses, the main storage pests infest the crops at about ripening time. Such incipient field attacks, which in many instances are generally low, develop substantially during the months of storage under favourable environments. For example, attack by bruchid beetles, *Callosobruchus* spp. on cowpeas begins in the field at about pod-drying time and damage of between 2–11 per cent may occur at this stage. This later develops to 20 per cent in the stores. Infestation of sorghum heads by weevils, *Sitophilus oryzae*, and the grain moth, *Sitotroga cereallela*, usually starts in the fields 53 days before the crop is harvested.

Poor hygiene of storage structures

Another major source of infestation is the practice of using the same stores and bins year after year without proper hygiene. Many storage pests tend to hibernate in secluded places like cracks and holes in the walls, or in wooden structures in the stores. They emerge from these places when new stored products are introduced and resume feeding, multiply and damage or contaminate the commodity. This form of infestation by residual pest populations is common during storage on the farms as well as in large central storage depots where facilities may be too large for adequate maintenance.

Cross-infestation

Storage of new produce near old granaries or other forms of storage gives rise to 'cross-infestation', that is pests which are established already in the old stores migrate to the new stores if they are sufficiently near. This is particularly troublesome in large central depots like warehouses, sheds and silos, where in addition to new pests in the stores, infestations also build up in litter which is inevitably scattered around them during loading, off-loading and packaging processes. A peculiar situation arises when field crops are planted near stores or warehouses. For example, the greatest attack of maize by *Sitophilus zeamais* is found in fields near maize stores. A gradient of infestation is thus established from the edge of the field, (near the maize store) towards the centre, indicating a progressive dispersal from the main focus of infestation.

Physical factors influencing storage losses

The magnitude of storage losses in any given situation is due to the combined effects of physical and production processes, the storage climate and biological factors. The physical factors are those which directly affect the nature and appearance of the stored commodity during the post-harvest processing. The storage climate provides the favourable conditions for the interplay of the other factors. Temperature and relative humidity during storage greatly influence the storability of the commodity as well as the survival of the organisms which cause damage losses. Rėlative humidities of 78–85 per cent are optimal for rapid development of storage pests, provided the temperature conditions are also conducive. Although insect problems are greater in regions of relatively high humidity, temperature is the predominant factor that influences insect propagation. A temperature range of 27–32 °C is the most favourable for the multiplication of stored product pests. It is estimated that at temperatures of about 32 °C the rate of insect development can achieve a monthly compound increase of 50 times a month.

In undetected infestations, heat is generated usually through the biological activities of the infesting organisms giving rise to situations technically known as 'heat-spots'. When heat-spots develop in stored commodities they serve as good indicators of pest infestation. If they are ignored pest populations rapidly build up and cause extensive damage to the stored products.

In addition to the above physical factors another important determinant of damage by storage pests is the moisture content of the commodity being stored. The growth or development of most storage pests is adversely affected by very dry conditions. For example, if the moisture content in cereals such as maize and millet, is below 11.5 per cent pest development is retarded, and if the moisture content is less than 8 per cent breeding may cease completely. Seeds or grains with more than 12 per cent moisture content are therefore highly susceptible to attack and damage by stored product insects. In this regard the inherent properties of some of the local varieties of maize, such as hard endosperm and low moisture content, are attributes which reduce their susceptibility to insect attack.

Management of storage pests

General principles

Measures directed towards control of storage pests involve the manipulation of the storage environment to make it less favourable to the insects. Such measures include the design of durable storage structures and their management in accordance with the following procedures which prevent or reduce infestations:

(a) Storage structures should be constructed with protective materials that will keep out insects. If cribs are used for instance, they should be mounted on concrete platforms to prevent attack by termites.

(b) Storage of clean, unbroken grains should be a major objective because broken grains are easily attacked by insects and thus encourage pest infestation. Moreover, broken grains increasingly absorb moisture from the atmosphere thereby encouraging the growth of micro-organisms. Using improved methods of threshing, shelling and cleaning produces whole grain and prevents this source of infestation.
(c) Grains and other products must be sufficiently dry to prevent rapid deterioration during transportation and storage. Moist produce is more prone to attack than well dried produce.
(d) Grains not fully ripened contain a higher proportion of moisture and will not keep as long as mature grains because the enzyme systems are still active. There must be prompt harvesting of crops to reduce field pests being carried to storage bins. Moreover, if the crop is left in the field longer than necessary, alternate drying and wetting due to variable weather conditions leads to cracking of grains which thus become predisposed to insect or fungal attack.
(e) Stores must be kept clean. Floors, corners, crevices and old bins must be swept clean before use. Old sacks or similar containers must also be cleaned thoroughly before new produce are introduced. Residual contact insecticides should be used to kill any remaining pests before new grain is introduced.
(f) Neither litter nor waste grains, feed particles nor discarded food should be allowed to accumulate around stores, since they are favourable sites for pests to develop.
(g) When hermetic storage is practised, the containers must be adequately air-tight to prevent both physical deterioration of the stored produce, and to arrest further development of insects that may have been accidentally introduced.
(h) Old and new stocks of produce should not be mixed as this leads to contamination of the new one.
(i) There should be regular inspection of newly stored produce to detect early infestations.

Traditional storage and pest management practices

Traditional methods of storing harvests among the developing countries have been evolved to fit local requirements. Small-scale or subsistence storage is carried out in sealed clay pots, gourds and baskets which are kept within the domestic environment. Larger quantities meant for sale or consumption during famine periods are stored in larger containers called cribs made of mud, stones, and plant materials. These structures are often raised off the ground on wooden or stone platforms and the tops are protected from rain by thatched roofs (Fig. 6.22). The traditional grain storage systems generally lack adequate drying facilities and protection against insect attack. However, certain simple control methods have

Fig. 6.22 Farmer's storage containers made of local mud and thatch for storing their locally grown grain. (Courtesy M.W. Service.)

evolved with traditional farming systems which manipulate the storage environment in such a way as to disrupt pest build-up. For example, grains are mixed with sand, limestone or ash to provide a physical obstacle to the movement of insects through the grain; this tends to limit oviposition. Physical contact with these inert particles also abrades the wax coating of the insect's cuticle which results in rapid loss of body moisture and death due to desiccation.

In northern Nigeria a modification of this principle is practised by peasant farmers who mix the larger grains of sorghum or maize with fine-seeded millet. This technique has proved effective against moth and beetle pests because it limits their access by filling up the inter-granular spaces. Both local herbs and smoke from small fires have also been used as insect repellents and fumigants to deter the establishment of pests. In some areas, particularly in the drier regions, underground pits are dug and after grain has been introduced they are hermetically sealed, thus limiting oxygen supply which minimizes pest infestations. The method, however, is not suitable for large-scale storage of grain.

Modern storage pest management

Improvements which have been made in the management of storage pests in developing countries are based largely on suitable adaptions of the traditional techniques. The aim is to provide cheap and rural storage technology that can be adopted readily by the peasant farming population. Some examples of such improved schemes are now described for the storage of cowpeas and maize.

Storage of cowpeas

A practical and easily attainable method for storing cowpeas for small-scale purposes was developed in Nigeria. It is based on the principles of hermetic storage integrated with the use of protective materials which act as barriers against insects; no insecticides are used. The materials consist of a plastic bag (50 X 150 cm) sealed at one end, a cotton (baft) cloth liner (50 X 96 cm) and two pieces of string. The plastic bag is perfectly sealed at one end and folded from the top like a cuff until the open end is level with the sealed end. A cotton liner is then placed inside the plastic bag and its top folded over the plastic bag, then shelled cowpeas are poured in the cotton bag until it is nearly full (capacity of 45–50 kg of seed). The cotton bag (i.e. liner) is then folded over the cowpeas and tied with string. There should be no seeds between the cotton bag liner and the outer plastic bag. The cuff of the plastic bag is now drawn up, twisted tightly above the tied cotton bag and tied. The surplus plastic is bent on itself and the two halves tied together. The bag is now ready for storage. Bags so prepared must also be stored away from rats and contact avoided with any rough edges which might tear the bags and render the method ineffective. Each plastic bag is used only once and then discarded, since repeated use introduces certain elements which weaken the system.

Storage of maize

Basically two models of cribs have been developed to store maize and these are described here and shown in Figs. 6.23 and 6.24.

The improved crib is constructed mainly from locally available materials such as bamboo sticks, raphia sticks, string and various roofing materials. In one form, the roofing is made of dried grass stalks or similar materials and the sides are protected with metal chicken wire (Fig. 6.23). In the other model as a further improvement, particularly for large storage, corrugated iron sheets are used (Fig. 6.24) instead of grass roofing. The significant feature of these improved cribs is the provision for increased ventilation which aids drying of the stored maize cobs. The leg supports

Fig. 6.23 An example of a small version of an improved storage crib for maize. Note the inverted metal cones on the legs to prevent rodents climbing up. (Courtesy of the National Accelerated Food Production Project, Rice-Maize Centre, National Cereals Research Institute, Ibadan, Nigeria.)

Fig. 6.24 A large capacity improved maize crib with a corrugated iron roof. (Courtesy M.W. Service.)

are made termite-proof and rot-resistant by painting with preservatives such as cuprinol-cresosote or waste engine-oil. Metal conical rodent-guards are provided and fitted as shown in Fig. 6.23. The length of the crib is orientated to face the prevailing winds. With this set-up ripe maize even with 35 per cent moisture can be dried down to about 13 per cent moisture content within 2–3 months, subject to certain physical specifications. For instance, in humid climates the width of the crib should be only about 60 cm and hold less grain to ensure faster drying. In drier regions a width of about 150 cm is suitable. Insect control is achieved by using a safe contact-type insecticide, such as Actellic 50 emulsion concentrate (primiphos methyl) applied at the rate of 15 ppm. This is achieved by mixing 30 ml of the formulation (commercial product) in 200 ml of water and spraying the insecticide with a flit-gun type of hand pump. Fifteen strokes of the pump will discharge enough insecticide to treat 20 kg of maize cobs. Chemical treatment is done when the crib is loaded with the cobs in order to ensure uniform application.

Construction of the crib from locally available materials and minimal use of a safe pesticide are usually within the competence and management resources of an average farmer in developing countries. This system h[illegible] substantially reduced storage losses in many areas where it is being practised. In several West African countries such as Ghana, Nigeria and Gambia, and in some East African countries like Zambia, Malawi and

Kenya, after the maize cobs stored by the above method have properly dried they are shelled. The grain is then stored in impervious containers (44 gallon capacity drums) for long periods without losses due to moulds, insects and rodent pests.

Large-scale storage

Storage problems encountered by produce dealers and milling industries are usually of greater magnitude because of the much larger quantities of commodities involved. In many African countries stored products are packed generally in cotton or jute bags which are stacked in warehouses. Under such conditions of storage, a good dunnage (a pile of mats or straw sheets on which the produce is placed) is essential to prevent deterioration of the lowest bags caused by water vapour absorption from the floor. Warehouses must be properly constructed of adequate masonry having smooth cement rendering both on the interior and exterior surfaces. The floors and walls should be free of cracks and water-proof. The position of storage bags should be rotated at intervals. In addition to maintaining good storage hygiene, insect control can be further ensured by use of insecticides. For example, applying insecticides as 'sandwich' treatments between the bags coupled with spreading barrier dusts, such as HCH, around the bases of stacks. In cases of pest outbreaks, rapid control must be achieved by fumigation, with chemicals like methyl bromide or aluminium phosphate, of stacks of bags placed under gas-proof sheets such as polythene.

Recommended reading

Coffee

Abasa, R.O. (1976) 'A review of the biological control of coffee insect pests in Kenya'. *East African Agricultural and Forestry Journal*, **40**, 292–9.

Abasa, R.O. and Mulinge, S.K. (1972) 'Effect of the fungicide "Du-Ter" on the giant coffee looper *Ascotis selanaria reciprocaria* Walk.' *Turrialba,* **22**, 99–100.

Anon. (1975) 'Pests of coffee and their control'. *Coffee Research Foundation,* 103 pp.

Crowe, T.J. and Greathead, D.J. (1970) 'Parasites of *Leucoptera* spp. (Lepidoptera:Lyonetiidae) of coffee in East Africa: An Annotated List'. *East African Agricultural and Forestry Journal*, **35**, 364–71.

Le Pelley, R.H. (1968) *Pests of Coffee*, Longman, London, 590 pp.

Le Pelley, R.H. (1973) 'Coffee Pests'. *Annual Review of Entomology*, **18**, 121–42.

McNutt, D.N. (1975) 'Pests of coffee in Uganda, their status and control'. *PANS*, **21**, 9–18.

Wanjala, F.M.E. and Dooso, B.S. (1979) 'Effects of rainfall on systemic insecticides in the control of *Leucoptera meyricki* in Kenya'. *Turrialba*, **29**, 311–15.

Cotton

Bishara, I. (1934) 'The cotton worm *Prodenia litura* F. in Egypt'. *Bulletin de la Société entomologique d' Egypt*, **18**, 288–420.

Bottrell, D.G.and Adkisson, P.L. (1977) 'Cotton insect pest management'. *Annual Review of Entomology*, **22**, 451–81.

Campion, D.G., McVeigh, L.J., Murlis, J., Hall, D.R., Lester, R., Nesbitt, B.F., and Marks, G.J. (1976) 'Communication disruption of adult cotton leafworm *Spodoptera littoralis* (Boisd.) (Lepidoptera, Noctuidae) in Crete using synthetic pheromones applied by microencapsulation and dispenser techniques'. *Bulletin of Entomological Research*, 66, 335–44.

Falcon, L.A. and Daxl, R. (1973) 'Report to the Government of Niçaragua on the Integrated Control of Cotton Pests'. *Food and Agriculture Organization*, Rome, Italy, **NIC/70/002/AGPP**, 60 pp.

Falcon, L.A. and Smith, R.F. (1973) 'Guidelines for integrated control of cotton pests'. *Food and Agriculture Organization*, Rome, Italy, **AGPP**, **18**, 92 pp.

House, V.S., Ables, J.R., Jones, S.L. and Bull, D.L. (1978) 'Diflubenzuron for control of the boll weevil in unisolated cotton fields'. *Journal of Economic Entomology*, **71**, 797–800.

ICAITI (Instituto Centroamericano de Investigacion Technologia Industrial). (1977) 'An environmental and economic study of the consequences of pesticide use in Central American cotton production'. *United Nations Environmental Programme – Project Final Report, Guatemala*, 295 pp.

Moussa, M.A., Zaher, M, and Kotby, F. (1960) 'Abundance of the cotton leaf worm, *Prodenia litura* (F.) in relation to host plant. I Host plants and their effect on biology'. *Bulletin de la Société Entomologique d'Egypt*, **44**, 241–51.

Newsom, L.D. (1974) 'Pest management: history, current status and future progress'. In F.G. Maxwell and F.A. Harris (eds) *Proceedings of the Summer Institute of Biological Control of Plant Insects and Diseases*, 1–18.

Pearson, E.O. and Maxwell Darling, R.C. (1958) 'The insect pests of cotton in tropical Africa'. *Empire Cotton Group Corporation and Commonwealth Institute of Entomology*, London, 355 pp.

Ripper, W.E. and George, L. (1965) *Cotton Pests of the Sudan*, Blackwell Scientific Publications, Oxford, 345 pp.

Tunstall, J.P. and Matthews, G.A. (1972) 'Insect pests of cotton in the Old World and their control'. In *Cotton Technical Monograph* No. **3**, Ciba-Geigy Ltd Basle, Switzerland, 46–59.

Cocoa

Entwistle, P.F. (1972) *Pests of Cocoa*, Longman, London, 750 pp.

Kumar, R. (1979) 'Biological control of cocoa pests in West Africa'. *Rapport 5ième Conférence d'entomologistes du cacaoyer de l'Ouest-African, Yaoundé*, 1976, 109–16.

Legg, J.T. (1972) 'Measures to control spread of cocoa swollen shoot disease in Ghana'. *PANS*, **18**, 57–60.
Leston, D. (1970) 'Entomology of the cocoa farm'. *Annual Review of Entomology*, **15**, 273–94.
Majer, J.D. (1976) 'The influence of ants and ant manipulation in the cocoa fauna'. *Journal of Applied Entomology*, **13**, 157–67.
Owusu-Manu, E. (1976) 'Natural enemies of *Bathycoelia thalassina* (Herrich-Schaeffer) (Hemiptera:Pentatomidae), a pest of cocoa in Ghana'. *Biological Journal of the Linnaean Society*, **8**, 217–44.
Wood, G.A.R. (1975) *Cocoa*, Tropical Agricultural Series, Longman, London 292 pp.

Sugarcane

Agarwal, R.A. and Siddiqui, Z.A. (1964) 'Sugar cane pests'. *Indian Journal of Entomology,* **25**, 149–86.
Box, H.E. (1953) 'List of sugar cane insects'. *Commonwealth Institute of Entomology*, London, 101 pp.
Jepson, W.F. (1954) 'A critical review of the world literature on lepidopterous stalk borers of tropical graminaceous plants'. *Commonwealth Institute of Entomology*, London, 127 pp.
Long, W.H. and Hensley, S.D. (1972) 'Insect pests of sugar cane'. *Annual Review of Entomology*, **17**, 149–76.
Williams, J.R., Metcalfe, J.R., Mungomery, R.W. and Mathes, R. (1969) *Pests of Sugar Cane*, Elsevier, Amsterdam, 568 pp.

Rice

Anon. (1967) *The Major Insect Pests of the Rice Plant*, Proceedings of a Symposium at the International Rice Research Institute, 1964, John Hopkins Press, Baltimore, Maryland, 729 pp.
Anon. (1976) *'Pest Control in Rice'. PANS* Manual No. **3**, (2nd ed.), 295 pp.
Anon. (1979) *The Brown Planthopper: Threat to Rice Production in Asia,* International Rice Research Institute, Los Baños, Philippines, 369 pp.
Anon. (1979) *Guidelines for Integrated Control of Rice Insect Pests*, FAO Plant Protection Paper No. **14**, Rome, Italy, 115 pp.
Anon. (1980) *Rice Protection in Japan. Text for Group Training Course on Control of Rice Diseases and Insect Pests.* I *Plant Pathology*, 111 pp., II *Entomology*, 154 pp., III *Miscellaneous*, 233 pp., Japan International Center.
Breniere, J. (1976) *The Principal Insect Pests of Rice in West Africa and their Control*, West African Development Association, Monrovia, Liberia, 52 pp.
Cheaney, R.L. and Jennings, P.R. (1975) 'Field problems of rice in Latin America'. *Centro Internacional de Agricultura Tropical (CIAT),* Colombia, Series **GE-15**, 90 pp.
Pathak, M.D. (1975) *Insect Pests of Rice*, International Rice Research Institute, Los Baños, Philippines, 68 pp.

Maize

Berger, J. (1962) *Maize Production and the Manuring of Maize*, Centre d'étude de l'azote, Geneva, Switzerland, 315 pp.

Bottrell, D.G.(1979) *Guidelines for Integrated Control of Maize Pests*, FAO Plant Production and Protection Paper No. **18**, Rome, Italy 91 pp.

Brown, E.S. and Mohamed, A.K.A. (1972) 'Relation between simulated armyworm damage and crop loss in maize and sorghum'. *East Africa Agricultural and Forestry Journal*, **37**, 237–57.

Chiang, H.C. (1976) 'Assessing the value of components in a pest management system: maize insects as a model'. *FAO Plant Protection Bulletin*, **24**, 8–17.

Chiang, H.C. (1978) 'Pest management in corn'. *Annual Review of Entomology*, **23**. 101–23.

Chiarappa, L. (1971) *Crop Loss Assessment Methods: FAO Manual on the Evaluation and Prevention of Losses by Pests, Diseases and Weeds*, FAO and Commonwealth Agricultural Bureaux, 276 pp; Supplement I, 1973; Supplement II, 1977.

Cramer, H.H. (1967) *Plant Protection and World Crop Production,* Pflanzenschutz-nachrichten Bayer, **20** 524 pp.

Jepson, W.F. (1954) *A Critical Review of the World Literature on the Lepidopterous Stem Borers of Tropical Graminaceous Crops*, Commonwealth Institute of Entomology, London, 127 pp.

Leyenaar, P. and Hunter, R.B. (1977) 'The effect of stem borer damage on maize yield in the coastal savanna zone of Ghana'. *Journal of Agricultural Science*, **10**, 67–70.

Luckman, W.H. (1978) 'Insect Control in Corn: Practices and Prospects'. In E.H. Smith and D. Pimentel (eds), *Pest Control Strategies*, Academic Press, New York, 137–55.

Mohyuddin, A.I. and Greathead, D.J. (1970) 'An annotated list of parasites of graminaceous stem borers in East Africa, with a discussion of their potential for biological control'. *Entomophaga*, **15**, 241–74.

Nishida, T. (1978) 'Management of corn planthopper in Hawaii'. *FAO Plant Protection Bulletin*, **26**, 5–9.

Nye, I.W.B. (1960) *The Insect Pests of Graminaceous Crops in East Africa*, Colonial Research Publication No. **31**, HMSO, London, 48 pp.

Odiyo, P. (1977) 'A forecasting system for the African armyworm'. In R. Kumar (ed.) *Proceedings of the 3rd Scientific Meeting, International Centre of Insect Physiology and Ecology, 1976*, 60–6.

Sarup, P., Sharma, V.K., Panwar, V.P.S., Siddiqui, K.H., Karwaha, K.K. and Agarwal, K.N. (1977) 'Economic threshold of *Chilo partellus* infesting maize crop'. *Journal of Entomological Research*, **1**, 92–9.

Walker, P.T. (1960) 'Insecticide studies on the maize stalk borer, *Busseola fusca*, in East Africa'. *Bulletin of Entomological Research*, **51**, 321–51.

Walker, P.T. (1960) 'The relation between infestation by the maize stalk borer, *Busseola fusca*, and yield of maize'. *Annals of Applied Biology*, **48**, 780–6.

Walker, P.T. (1975) 'Pest control problems, pre-harvest, causing major losses in world food supplies'. *FAO Plant Protection Bulletin* **23**, 70–7.

Walker, P.T. (1981) 'The relation between infestation by lepidopterous stem borers and yield in maize'. *European Plant Protection Bulletin,* **11**, 101–6.

Walker, P.T. and Hodson, M.J. (1976) 'Developments in maize stem borer control in East Africa, including the use of insecticide granules'. *Annals of Applied Biology*, **84**, 111–14.

Subsistence crops

Ackland, J.D. (1975) *East African Crops*, Longman, London, 252 pp.
Chiarappa, L. (1971) *Crop Loss Assessment Methods: FAO Manual on the Evaluation and Prevention of Losses by Pests, Diseases and Weeds*, FAO and Commonwealth Agricultural Bureaux, 276 pp., Supplement I, 1973; Supplement II, 1977.
Hall, D.W. (1970) 'Handling and storage of food grains in tropical and subtropical areas'. *FAO Agricultural Development Paper*, No. **90**, 350 pp.
Hill, D.H. (1975) *Agricultural Insect Pests of the Tropics and their Control*, Cambridge University Press, 516 pp.
Singh, S.R. and van Emden, H.F. (1979) 'Insect pests of grain legumes'. *Annual Review of Entomology*, **24**, 255–75.

Stored products

Adams, J.M. (1976) 'A guide to the objective and reliable estimation of food losses in small-scale farmer storage'. *Tropical Stored Products Information*, **32**, 5–12.
Anon. (1978) *Post-harvest Food Losses in Developing Countries,* Board on Science and Technology for International Development Commission on International Relations National Research Council, National Academy of Sciences, Washington DC.
Christensen, C.M. (1974) *Storage of Cereal Grains and their Products*, American Association of Cereals Chemists, St. Paul, Minnesota, 549 pp.
Davies, J.C. (1970) 'Insect infestation and crop storage'. In J.D. Jameson (ed.) *Agriculture in Uganda*, Oxford University Press, 274–80.
Giles, P.H. (1964) 'The storage of cereals by farmers in northern Nigeria'. *Tropical Agriculture, Trinidad*, **41, 197–212.**
Hall, D.W. (1977) 'Grain storage'. In C-L.A. Leaky and J.B. Wills (eds) *Food Crops of the Lowland Tropics*, Oxford University Press, Oxford, 255–71.
Harris, K.L. (1972) *Methodology on Assessment of Storage Losses of Food Grains in Developing Countries,* report prepared for the staff of the International Bank for Reconstruction and Development, Washington, DC.
Nyanteng, V.K. (1972) 'The storage of food stuffs in Ghana'. Institute of Statistical Social and Economic Research, University of Ghana, Legon, Ghana, *Technical Publication Series*, No. **18**.

Schulten, G.G.M. (1975) 'Losses in stored maize in Malawi and work undertaken to prevent them'. *Bulletin of the European and Mediterranean Plant Protection Organization* 5, 113–20.

Tyagi, A.K. and Girish, G.K. (1975) 'Studies on the assessment of storage losses of food grains by insects'. *Bulletin of Grain Technology*, **13**, 84–102.

7 Management of vectors

The intentions of this chapter are to outline the methods at present being employed to reduce vector populations and disease transmission, to focus attention on the difficulties being encountered, and at the same time to outline more integrated strategies for the management of vectors. The emphasis will vary according to the vectors, for example with some species there may already be a trend to curtail insecticidal usage because environmental measures are proving practical whereas in other instances there may as yet appear to be few realistic alternatives to almost total reliance on insecticides.

Mosquito control

M.W. Service

Mosquito control may be required because local mosquitoes are vectors of malaria, filariasis or various arboviruses, or simply because they constitute an intolerable biting nuisance. Control methods can be directed at the larvae, adults or both stages; the actual methods used will depend on the biology and behaviour of the species and resources available.

The following account summarizes firstly the methods presently used or being developed to control mosquitoes, and then describes the role of integrated vector management in reducing mosquito populations.

Larval control measures

Control measures directed at mosquito larvae can be divided into the following three convenient categories: (i) physical, mechanical or environmental methods; (ii) biological methods; and (iii) chemical methods.

Physical or environmental control

The methods described here are concerned not with killing mosquitoes by chemical or biological means but with reducing their breeding places, that is 'source reduction'. A very simple form of this which is a component of 'species sanitation' consists of filling in larval habitats, thus permanently eradicating them. For example, water-filled tree-holes which are known to be larval habitats can be filled up with sand and cement, larger collections of waters such as ponds, borrow pits and marshes can be filled in with rubble, earth or refuse, or alternatively they can be drained. Understandably, however, there may be opposition to either filling in or draining ponds and borrow bits if they are used as a source of water for people or livestock. It is also impossible to drain or fill in rice fields, although it may be possible to grow rice strains that require flooding for only a limited time. Large collections of water may also be used for recreational activities or as refuges for wildlife, thus again making drainage impractical. Many species of mosquitoes breed in small and temporary habitats such as small pools, puddles, roadside ditches, water-filled vehicle tracks and cattle hoof-prints, and it is not usually feasible to either fill in or drain such numerous and scattered collections of water.

Sometimes mosquito breeding may be reduced or eliminated by modifying the habitat. For example, larvae may occur in small pools and marshy areas that have formed at the edges of streams which have a twisting course, but it may be possible to prevent the formations of such habitats by straightening the course of streams and increasing the speed of their water flow. Marshes and swamps, especially those with sloping margins allowing fluctuations in the water line, are often very attractive to mosquitoes such as *Aedes* which lay their eggs not on the water surface but on damp mud and leaf litter. Instead of draining or filling in such habitats they can be excavated and the shore line steepened to form almost vertical banks enabling them to be flooded permanently, thus eliminating exposed muddy areas (Fig. 7.1). This method is called impoundment. Although converting a marsh to a shallow pond or lake deters breeding of *Aedes* mosquitoes there is the danger that it may provide ideal habitats for other mosquitoes, such as *Anopheles* species. Another problem is that such drastic environmental changes may be unacceptable ecologically.

Physical methods to control mosquito breeding, such as drainage, filling in and impoundment, can initially be costly but they may incur little expenditure for their maintenance; they have the advantage of usually giving long-lasting control because larval habitats are destroyed permanently.

In some areas, such as Japan, China and south-east Asia, considerable reduction of mosquito breeding in rice fields seems to have been achieved by changing to 'dry rice' cultivation methods combined with early ripening

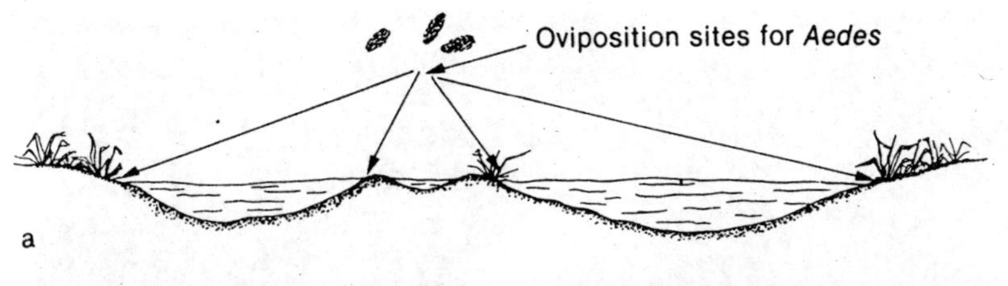

Fig. 7.1 Diagrammatic representation of how the system of impoundment reduces mosquito breeding; (a) a typical shallow accumulation of water with scattered areas of exposed mud and a muddy sloping shore line which constitute attractive oviposition sites for *Aedes* mosquitoes; (b) the same habitat after it has been excavated to leave a deeper and more permanent collection of water and steep and well-defined banks. The numbers of *Aedes* oviposition sites are substantially reduced, moreover this new habitat attracts water birds and fish, which apart from making the habitat more attractive ecologically, may well prey on the mosquito larvae.

rice varieties. Rice fields are flooded during only the first week following transplanting, and although fields are afterwards irrigated daily, water is not allowed to accumulate. This approach, however, can only be successful when applied to areas with porous sandy or loamy soils, and is not feasible during monsoon weather when fields are inundated with rain.

Mosquito-breeding in water-storage pots in villages (Fig. 7.2), or in water cisterns, reservoir tanks, and faulty soak-away pits in towns, can in theory be prevented by very simple methods. For instance, by covering the openings of the pots or tanks with mosquito netting, by repairing broken covers on soak-away pits and septic tanks, or by emptying the water from pots about every 5 days to prevent completion of the life-cycle and emergence of adult mosquitoes. In practice, however, it proves extremely difficult to persuade people to cover pots or periodically empty them, and to keep reservoirs and cisterns covered. It is sometimes thought that a piped water supply leads to a reduction in mosquitoes such as *Aedes aegypti*, which commonly breed in water-storage pots. Frequently, however, water is piped only to certain points in a village or town, and the supply often dries up, so people will still tend to store water in pots. In

Fig. 7.2 Typical clay water-storage pots being inspected for the presence of mosquitoes such as *Aedes aegypti.* (M.W. Service.)

India an important urban malaria vector, *Anopheles stephensi*, breeds in wells, but this can be prevented by fitting lids to the wells and supplying each with a hand pump to raise the water.

A novel method that has been experimentally tried in Kenya consists of placing small polystyrene balls in pit latrines to form a complete physical barrier over the water to prevent oviposition by *Culex quinquefasciatus* (= *C. fatigans*). These balls are cheap, non-toxic, virtually indestructable and have little value or attraction for people to steal them; and even if the latrines dry out they will refloat on the surface when water is present again. This is an example of a very simple method that might substantially reduce breeding of a particular mosquito species in one type of habitat.

Biological control or biocontrol

This method is sometimes referred to as naturalistic control but there is little if anything natural about the approach, because either exotic pathogens, parasites or predators have to be introduced into habitats, or their incidence increased to try to obtain effective control. Because of the growing concern over insecticidal contamination of the environment with powerful synthetic insecticides, the continuing appearance of insecticide resistance in vectors, and more recently the rising costs of insecticides, increasing attention has been focused on biological control methods.

Predators

Predatory fish have been used for mosquito control since the early 1900s. The most widely used species belong to the *Gambusia affinis* subgroup, fish endemic to North America, but which have been introduced into many tropical countries. Another common species is the guppy, *Poecilia reticulata* (= *Lebistes reticulatus*), which although not as voracious as *Gambusia* often appears to withstand greater organic pollution and is therefore better in some habitats against *Culex quinquefasciatus* (= *C. fatigans*). These fish have the advantage of being viviparous, having a high reproductive rate, and as they grow to only about 3–6 cm they are not removed and eaten by the people which is an important practical consideration! Unfortunately *Gambusia* are very aggressive fish and may kill small indigenous fish and immature stages of larger species and hence can endanger valuable local fish supplies. *Gambusia* also preys on crustacea and various aquatic insects including species that graze on phytoplankton and if their populations are substantially reduced by fish there may follow undesirable increases in algal growth. Despite these problems, other limitations and the fact that many other species of fish prey on mosquito larvae, *Gambusia affinis* and *Poecilia reticulata* remain the most useful mosquito predators in many situations. *G. affinis* is reported sometimes to give good control of mosquitoes breeding in rice fields when stocked at the rate of 250–750 fish/ha; other times, however, it fails to give effective control.

In several towns in the Indian subcontinent *Gambusia* and *Poecilia* are specially bred for introducing into wells, cisterns, water-storage tanks and other man-made collections of water, mainly to control *Anopheles stephensi*, but also to a lesser extent *Culex quinquefasciatus* (= *C. fatigans*). They survive well in many of these situations and are reported to reduce larval populations substantially, but there is no convincing documented evidence that fish, or any other mosquito predators, have ever significantly reduced transmission of a mosquito-borne disease.

Larvae of many malaria vectors, such as the *Anopheles gambiae* complex in Africa and *Anopheles culicifacies* in the Indian subcontinent, occur in very small collections of water liable to repeated desiccations. In these habitats *Gambusia* and *Poecilia* species would be unable to survive. The so called 'instant' or 'annual' fish of the genera *Nothobranchius* and *Cynolebias*, which have drought resistant eggs, are better adapted to such temporary habitats, but may nevertheless not always be able to establish breeding populations because of unsuitable environmental conditions. Furthermore, it would be impractical to introduce them into the numerous pools, puddles, vehicle tracks and hoof-prints which in the rainy season may be extremely important mosquito larval habitats. An Asian fish, *Aplocheilus latipes*, has the ability to move from large and permanent pools and ponds across damp earth to temporary small pools and puddles, and might prove useful in some situations.

Pathogens and parasites

There is an impressive number of pathogens, including viruses (iridescent and cytoplasmic polyhedrosis viruses), bacteria (*Bacillus thuringiensis*, *B. sphaericus*), fungi (*Coelomomyces, Lagenidium, Culicinomyces*), protozoa (*Nosema*, *Thelohania*), and parasites such as parasitic nematodes (*Romanomermis culicivorax*) which can kill mosquito larvae. Despite this array of potential enemies and the commercial production of *R. culicivorax* and certain strains of *B. thuringiensis* there are relatively few instances where either pathogens or parasites have given worthwhile and sustained control of tropical mosquitoes. Unfortunately many reports on reductions in larval population size have been based largely on casual observations on an *ad hoc* basis.

A common problem is that in the laboratory pathogens and parasites may show promise as control agents whereas when they are tested in the field they frequently fail to survive due to unfavourable environmental conditions, for example organic pollution, microbial flora, pH of water, oxygen deficit, strong sunlight, adverse temperatures or drought. Sometimes introduced parasitic nematodes appear to have established themselves in mosquito habitats for at least 6 years, whereas in many other instances they require repeated reintroductions. Little is as yet known about the environmental factors that affect the development, maturation and hatching of *R. culicivorax* eggs. When parasites and pathogens are unable to survive and reproduce and consequently have to be introduced into larval habitats repeatedly, this is likely to become costly as well as labour intensive.

Mortality caused by *Bacillus thuringiensis* serotype H14 (= var. *israelensis*) is due to its spores producing toxic protein crystals called endotoxins which when ingested by mosquito larvae act as a stomach poison. The bacteria together with the toxins can be formulated as a wettable powder, but as the endotoxins are rapidly denatured in the environment repeated applications are required to give good mosquito control. *B. thuringiensis* is a microbial insecticide and therefore should not strictly be regarded as a true biological control organism. Caution may be required because a few strains produce *exotoxins* that are harmful to birds and mammals, and resistance has been reported from laboratory studies, but the results are questionable. Moreover, even parasitic nematodes are not proof against the development of resistance. Certain mosquito species (*Anopheles quadrimaculatus* and *Culex quinquefasciatus* in the USA) have in the laboratory undergone genetic selection to produce populations that are either successful in avoiding being parasitized by *Diximermis peterseni* and *Romanomermis culicivorax*, or able to develop physiological resistance involving melanization and encapsulation of the nematode parasites. However, this resistance has been recognized only in the laboratory under intensive selective pressure, and it is likely to develop slowly, if at all, in the field. Some mosquitoes, such as *Culex tritaeniorhynchus* a common man-biting mosquito that breeds in rice fields and is a vector of Japanese encephalitis in Asia, are more or less refractory to parasitism by *R. culicivorax*. Because these nematodes cannot usually develop in brackish or polluted water they are unlikely to be very useful against either salt marsh mosquitoes, or *C. quinquefasciatus* (= *C. fatigans*) in stagnant habitats.

Ecological considerations

Although much interest has been generated in the feasibility of biological control of mosquito larvae, and this should be encouraged, there are at present extremely few biocontrol measures that can be used, even on a limited scale, to give effective control. Fish, such as *Gambusia*, *Poecilia* and various indigenous species, have sometimes given effective biocontrol of certain mosquito species in some habitats, but there are many situations where fish are completely ineffective.

Two promising and relatively recent biocontrol candidates are *Bacillus thuringiensis* and *Romanomermis culicivorax*. Although a commercial company started producing *Romanomermis culicivorax* they lost interest after a feasibility study indicated that their sales would be very limited, rendering the venture economically unsound. While there are no technical difficulties in mass culturing *B. thuringiensis*, there still remain some technical problems of storage in hot climates without deterioration, standardization and formulation. As yet these agents have been evaluated mostly in relatively small-scale field trials. Biocontrol methods can only begin to be considered seriously as an important component of vector management when the most promising pathogens and parasites are commercially available.

Apart from the difficulties of many predators, parasites and pathogens becoming successfully established in mosquito habitats, there are ecological reasons why it will probably prove difficult to subtantially reduce mosquito populations by biological methods. For example, mosquito larvae are not randomly or evenly dispersed in their aquatic habitats, but have a very patchy distribution. Consequently the likelihood of predators, pathogens or parasites encountering mosquito larvae will depend on their biological distributions as well as their searching abilities. Moreover, many important man-biting mosquitoes are excellent examples of what, in today's ecological jargon, are referred to as 'r-strategies'. That is, they are exploiters of the environment and are continually colonizing temporary habitats, and are characterized by a high reproduction rate, short generation-time and high potential for dispersal. They typically have populations that are resilient to disturbances and which are repeatedly booming and crashing. They often suffer large natural density-independent mortalities, such as caused by desiccation of temporary larval habitats. These characteristics tend to make it difficult to significantly reduce their population size by density-dependent methods, such as by using predators and pathogens. In contrast, catastrophic density-independent mortality as caused by insecticides or microbial agents like *Bacillus thuringiensis* can be very effective in reducing, or even eradicating, r-type pests. Not all mosquito species, however, are r-strategists. Larvae of *Anopheles stephensi*, for example, are usually found in India in low densities in wells, and the species is probably not affecting an r-strategy life-style. This probably explains why predatory fish have been reported to give good control of this urban malaria vector.

Chemical methods

Before the advent in the 1940s of the organochlorine insecticides, chemical control of mosquito larvae was mainly achieved by the application to breeding places of Paris green or kerosene (paraffin) and fuel oils. Paris green is a copper aceto-arsenite which is spread as a fine dust on the water surface and which when ingested by surface feeding mosquitoes, such as *Anopheles*, acts as a stomach poison. Oils spreading on the water surface kill culicine and anopheline mosquitoes by both release of toxic aromatic hydrocarbons and by the more stable fractions of the oils interferring with their respiration at the water surface. Although resistance has not developed against these chemicals their use was more or less discontinued when the more powerful insecticides such as DDT and HCH became available and were sprayed on larval habitats. In some areas mosquitoes became resistant, firstly to DDT then to HCH and dieldrin, necessitating a change to the organophosphate insecticides such as malathion, fenthion and temephos (Abate). Several species then developed resistance to the organophosphates, and consequently the carbamate insecticides such as carbaryl (Sevin) had to be used, all of which are more expensive than DDT.

In addition to promoting resistance these insecticides kill, not just mosquito larvae, but a broad spectrum of invertebrates and sometimes also vertebrates. In soak-away pits, septic tanks and other highly polluted waters, all of which support large larval populations of *Culex quinquefasciatus* (= *C. fatigans*), an important filariasis vector, non-specificity will not be very important as few other insects will be found in such habitats. In contrast, when these insecticides are sprayed on ponds, borrow pits, marshes and streams, a large variety of non-target organisms are likely to be killed. Of greater concern, however, is the persistence of organochlorine insecticides in the environment and their biological magnification in soil, water, plant and animal tissues. Because of this environmental pollution the World Health Organization no longer recommends that organochlorines are used as mosquito larvicides; instead the much less persistent and biodegradable organophosphate and carbamate insecticides are recommended. However, although these insecticides rapidly break down in the environment, they will nevertheless kill a wide range of invertebrates, and some are specially toxic to certain vertebrates such as fish, and others to ducks.

Fig. 7.3 A pressurized sprayer being used to spray an insecticidal emulsion of an organophosphate insecticide on town drains, a typical larval habitat of the mosquito *Culex quinquefasciatus*. (M.W. Service.)

Because neither Paris green, oils nor residual insecticides remain effective on the water surface for more than one or two days, larval habitats must be sprayed every 7–10 days in the tropics to ensure that larvae hatching from eggs are killed before they pupate and give rise to adults. This requisite for regular spraying at frequent intervals prohibits wide-scale control of mosquito larvae (see p. 275).

In some parts of the world there is renewed interest in some of the earlier chemical control methods, which although quite effective were abandoned with the arrival of the more powerful insecticides. For example, Paris green is being used in parts of India not only as a dust but as an oil emulsion. Elsewhere this chemical has been formulated as granules which either float on the water or sink, and which slowly release their toxicants and hence give control over several days to weeks. In some countries there has been a return to using oils as larvicides, but this may be hampered by the increasing price of petroleum.

To try to overcome some of the problems of using residual insecticides less orthodox chemicals including insect growth regulators (IGRs), monomolecular films of various wetting agents such as egg or soya lecithins, aliphatic amines and other novel compounds have been experimentally evaluated. There have been difficulties in getting some of these compounds to spread efficiently on the water, and many rapidly degrade in strong sunlight, at high temperatures, or in the presence of organic pollution, mud or detritus. At present none of these unconventional chemicals can be advocated for mosquito control in the tropics. However, in the future, insect growth regulators like Dimilin, which is a chitin inhibitor, and Altosid, which is a juvenile hormone mimic, might prove useful, but there remain a number of practical problems. In most situations insect growth regulators are either rapidly hydrolyzed in larval habitats, degraded due to their extreme sensitivity to ultraviolet light, or become adsorbed onto organic matter and therefore may be of little use in polluted waters. Because of their lack of persistence they remain effective for only 2–3 days and this is a problem when they act only on certain stages of mosquitoes, such as late 3rd- or early 4th-instar larvae. Slow-release formulations, as achieved by their incorporation onto sand granules, into brickettes and micro-encapsulated forms, have given greater residual activity, but further improvements are still needed. Although insect growth regulators are more specific than insecticides they may still kill a variety of aquatic invertebrates, while at the same time allowing others to increase exceptionally in population size. Insecticide-resistant strains of houseflies, flour beetles (*Tribolium castaneum*) and the *Anopheles gambiae* complex have already developed resistance to some insect growth regulators, and this serves as a warning as to what might arise if they were extensively used.

Adult control measures

Housespraying

In the 1940s control of malaria vectors in the tropics was based almost entirely on destruction of larval habitats by the application of mineral oils, Paris green, DDT, or by physical methods such as drainage and filling in. These control procedures were only feasible in towns and centres of economic importance such as mining camps, and tea, coffee and rubber estates, because the immensity of the task prohibited spraying breeding places in rural areas every 7–10 days, or alternatively draining or filling them in. With the availability of DDT, and later other organochlorine insecticides, larval control measures were gradually replaced by control directed at the adults. The method adopted was to spray DDT wettable powder on the interior walls, roofs or ceilings of houses to leave a residual deposit of about 2g/m^2. The rationale behind housespraying is based on the belief that malaria vectors rest inside houses either prior to and/or after feeding, and will therefore pick up a lethal dose of insecticide – but see p. 276. DDT is such a persistent and stable insecticide that houses need to be sprayed only twice a year to maintain an effective deposit. Not surprisingly, however, some malaria vectors, and also other indoor-resting mosquitoes, developed resistance to DDT and consequently houses had to be sprayed with other insecticides such as HCH, dieldrin, malathion, or propoxur. All these insecticides cost more than DDT and, moreover, have to be sprayed more frequently (3–4 times a year), again adding to the cost of control.

The application of residual insecticides to the interior of houses, unlike spraying larval habitats with organochlorine insecticides, results in negligible pollution of the environment. For example, it has been estimated that in areas of tropical Africa having 100 people/km^2, spraying houses with DDT results in depositing only about 10 g DDT/ha outside the houses, twice a year, a very small contamination when compared with the application of as much as 80 kg of insecticide/ha each year to control cotton pests.

Apart from resistance problems and the ever-increasing costs of insecticides, there are many other difficulties which have led to the failure or breakdown of residual house spraying to control malaria vectors. In India, for example, there were an estimated 75 million malaria cases a year in 1947, that is before residual spraying started. By 1965, however, there were only 100 185 cases and clearly malaria had almost been eradicated, but thereafter deterioration set in and by 1977 there were an estimated 7–10 million people with malaria. (Since then conscientious surveillance and control operations have apparently succeeded in reducing malaria transmission, but exact figure are lacking.) Nevertheless, this was still a great improvement on the situation before spraying started. This disastrous resurgence of malaria in India was due to a combination of factors, including the development of insecticide resistance in the vectors, over-simplification of malaria epidemiology, and to socio-economic reasons. For instance, there was a lack of trained manpower, increasing apathy towards control, reduced budgets for malaria control at a time when costs of insecticides and labour were rising, and poor entomological and parasitological surveillance.

Elsewhere malaria remains a problem. In Africa there are some 450 million people but there has never been any real reduction in malaria transmission in tropical Africa. Malaria remains the most important endemic disease, being responsible for more than a million infant deaths a year. In Central and South America malaria has been reduced in some areas, while in many others it is still rampant. The situation in the Americas is aggravated by the fact that several of the important malaria vectors do not rest inside houses and, consequently, are unaffected by housespraying. This problem of outdoor-resting mosquitoes is also encountered elsewhere, and it seems that in some areas there has been an increase in its incidence.

Fogging and ULV spraying

Fig. 7.4 A landrover with an ultra-low-volume (ULV) spraying machine discharging a fine fog of a synthetic pyrethroid insecticide which will kill mosquitoes resting amongst vegetation, as well as those inside houses. (M.W. Service.)

Alternative control strategies are obviously required to kill outdoor-resting mosquitoes, and one method consists of outdoor application of insecticidal fogs or aerosols. Greater operational efficiency is obtained as well as reduced costs, if concentrated insecticidal formulations are applied sparsely than if the same quantity of insecticide is diluted in a solvent and a large volume sprayed. In this technique, known a ultra-low-volume (ULV) spraying, fine droplets forming a fog or aerosol are blasted out from the spraying machine (Fig. 7.4). Application rates can be as little as 74 ml–1 litre/ha. The most suitable insecticides for ULV spraying have high specific gravity and low volatility and should be biodegradable such as the organophosphates – malathion, fenthion and temephos or carbamates like propoxur, and the synthetic pyrethroids such as bioresmethrin and deltamethrin (= decamethrin). Applications need to be made in calm weather when air turbulence is minimal, and this usually means in the evenings or early mornings when few thermals are rising from the ground. Aerosols can be produced by portable or vehicle-mounted spraying machines, but the terrain may be difficult and inaccessible, making it impossible to get adequate coverage from the ground. Aerial ULV spraying gives better coverage but is more costly, requires expert pilots and may cause more environmental contamination due to insecticidal drift. Whatever method or insecticide is used, there is little residual effect. Consequently although outdoor, and also sometimes indoor, resting mosquitoes are killed, the treated areas soon become reinvaded by mosquitoes flying in from unsprayed surrounding areas. Repetitive applications are needed before mosquito populations are substantially reduced. Clearly large numbers of insects other than mosquitoes are killed, consequently ULV spraying may not always be acceptable ecologically.

Both ULV spraying and the application of insecticides to larval habitats will favour the faster selection of resistance than residual housespraying, because they exert selective pressure on both sexes of mosquitoes. Similarly, using the same insecticide for controlling larvae and adults will again enhance the appearance of resistance. The selection of resistant populations has sometimes undoubtedly been accelerated, or even initiated, by agricultural pesticides. For example, repetitive spraying against rice and cotton pests on irrigation schemes has several times resulted in local populations of mosquitoes developing resistance (see p. 35), despite the fact that insecticides may not have been directed against mosquitoes.

Genetic control

The successful eradication of the screw-worm fly (*Cochliomyia hominivorax*) from certain areas of the southern USA by repetitive releases of sterilized male flies gave encouragement to the idea that genetic methods could be used to control or eradicate mosquito vectors. However, there are innumerable technical difficulties as well as ecological reasons why this approach will prove difficult against mosquitoes, many of which are r-strategists (see p. 271). At present genetic control methods cannot be used to control mosquitoes.

Integrated mosquito management

The belief that mosquitoes, whether they constitute a biting nuisance or are vectors, can be controlled by a single method needs to be abandoned in favour of a more flexible system that incorporates the best of all available techniques. This type of holistic approach has in the past been integrated control, but it is now becoming increasingly common to talk of integrated management of pests and vectors. Management should aim at suppressing mosquito populations to a level where they are no longer a serious problem, this being a more realistic goal than eradication. Effective mosquito management will often involve the co-operation and co-ordination of people with different disciplines. For instance, management of mosquitoes breeding on crop irrigation schemes may entail combining the expertise of medical and agricultural entomologists, agronomists, water engineers and, possibly, economists and sociologists.

The philosophy of mosquito management should not be to ban insecticides but to minimize their use in order to reduce environmental contamination and postpone as far as possible the spread of resistance. But ecological damage is not caused just by insecticides. For instance, draining marshes or filling in ponds may also destroy important aquatic habitats, whereas careful management might allow such habitats to be retained but their attractiveness to mosquitoes reduced. The actual control measures adopted will depend on local conditions, the biology of the mosquitoes, resources available and sometimes the life-style of the local people. Mosquito nuisance in urban areas, for instance, might be reduced by a combination of simple and relatively inexpensive measures, such as introducing predatory fish into wells and water reservoirs, the application of oils or Paris green to polluted ditches and streams, repair of defective covers on soak-away pits and septic tanks, and educating people that unwanted water should not be allowed to accumulate in discarded pots and other containers. About 2 million malaria cases in the Indian subcontinent are transmitted by urban mosquito vectors and this incidence could be reduced substantially by such simple control measures.

In many areas of Asia and South America, malaria is relatively unstable and only about a 5-fold reduction in vector population size may be required to arrest malaria transmission. In other situations, as in many African countries where malaria is holoendemic and very stable and where 1 in 20 or 100 of the vectors biting man may be infected with malaria parasites, there must be an exceedingly high reduction (possibly 1000-fold) in vector densities before there can be any appreciable decrease in malaria transmission. Such large reductions in vector population size will be difficult to obtain without insecticides playing a dominant role. It is recognized therefore that there are situations where residual housespraying accompanied perhaps by spraying larval habitats with oils, Paris green or organophosphate insecticides, may be necessary. Stocking rice fields with predatory fish has occasionally reduced the numbers of insecticidal applications required to maintain acceptable low levels of pest mosquitoes, but difficulties are frequently encountered which do not make the method universally appropriate.

An integrated management programme should not be considered static. There must be continual efforts to improve it and modify the management strategies to take into consideration changes in the ecology of the mosquitoes and local conditions. Vector management will normally require more detailed knowledge of the ecology and population dynamics of vectors, such as mosquitoes, than insecticidal control measures. In environmental approaches to vector management the major difficulties will not usually be the implementation of control measures, but their continual operation and maintenance. It has to be recognized that there is no general panacea for mosquito control and that in some situations little can be done to alleviate mosquito breeding. It is often difficult, for instance, to substantially reduce mosquito breeding in rice fields, consequently the best procedure might be to provide irrigation workers with mosquito nets and screened houses, and also to provide them with anti-malarial drugs and an efficient health service. There will be situations where it is considered uneconomical to reduce mosquito populations to a tolerable level or to reduce disease transmission, and others where the benefits gained would not offset environmental damage.

In emergencies, such as epidemic situations, it has to be realized that insecticidal methods, especially aerial ULV spraying, will often be the most effective in quickly reducing mosquito populations and preventing the spread of vector-borne diseases. Nevertheless, although such practices may produce spectacular reductions in vector densities, the benefits are usually short-lived, they therefore must be regarded as valuable stop-gap measures and not the answer to the problem.

Conclusions

Man has repeatedly failed to recognize the great resilience of most vectors and pests, their ability to adapt and modify their behaviour to ensure their survival, and their opportunistic nature that allows them to exploit changing or new ecological situations. A lesson that must be learnt is that the delicate ecological balance that normally prevents potential vectors and pests from causing a serious threat to mankind is very easily upset by man's actions, for example irrigation schemes, hydro-electric projects, urbanization. Man not only seems to have caused most of his pest and vector problems but often fervently resists changing his behaviour to redress the situation. Common sense tells us that man should strive to work with nature rather than against it, but this demands a better understanding of the ecological mechanisms regulating field populations of vectors and pests.

Control of simuliid blackflies

J.F. Walsh

There are over 1000 species of simuliid blackflies and the females of most of these take the blood of mammals or birds. Many species cause considerable nuisance by attacking man and his domestic animals. In northern parts of North America and Europe they may occur in such vast numbers as to become intolerable, disrupting forestry, causing serious losses to livestock farmers and annoyance to tourists. In tropical Africa several blackflies are vectors of the filarial parasite, *Onchocerca volvulus*, the causative organism of human onchocerciasis or river blindness (Fig. 1.8). In some areas of Central America, *Simulium ochraceum* occurs both as a biting pest and as a vector of *O. volvulus*.

Blackflies are characterized by breeding in fast flowing rivers and streams, and the ability to disperse long distances.

The larval stages are aquatic sessile organisms obtaining their food (bacteria and unicellular algae) by filtering particulate matter from the water. Owing to their preference for fast flowing waters, breeding sites tend to be clearly defined and of restricted distribution. In contrast the adults are widely dispersed and most species which have been studied in detail can fly more than 20 km. In the case of some members of the *S. damnosum* complex, the most important group of *O. volvulus* vectors in Africa, there is good evidence that some individuals disperse more than 300 km. Adults, however, are rarely encountered away from their breeding sites, except when biting man or his domestic animals. There is a lack of detailed knowledge of the behaviour and ecology of the adults, for example where they rest between blood-feeding and oviposition. As a result, most attempts at control are directed against the larval stages. Occasionally, however, aerial spraying and ground-fogging with insecticides have been used to protect small communities, such as logging and mining camps and military installations in northern Canada, from adult biting populations.

Larval control measures

Insecticidal control

The control of blackflies started at the beginning of the modern insecticide era. In 1944 DDT proved to be very effective against the onchocerciasis vector, *Simulium ochraceum*, in Guatemala, and until relatively recently was the favoured insecticide for *Simulium* control. This insecticide was responsible for such notable successes as the eradication of *S. neavei* from Kenya and the elimination of *S. damnosum* from Zaire at Kinshasa. DDT is now little used; most control schemes use emulsifiable concentrates of either the organophosphate insecticide temephos (Abate), or the organochlorine methoxychlor. The selection of suitable insecticides takes into account their environmental effects. All insecticides now considered for blackfly control must be biodegradable, have low fish and mammalian toxicity, and not have too drastic an effect on the non-target aquatic invertebrates. Fortunately blackfly larvae tend to be markedly susceptible to insecticides and at dosage rates currently employed many of the invertebrate organisms characteristic of *Simulium* breeding sites are able to persist despite regular larviciding operations. However, at environmentally acceptable treatment rates these insecticides have no effect on the eggs or pupae of blackflies, so the sole target is the larvae.

The amount of insecticide used to control *Simulium* larvae is dependent on the volume of water being discharged by the larval habitats. This discharge rate is measured in m^3/sec. The amount of insecticide used is calculated to give a dosage rate of between 1 part of insecticide to 20 million parts of water, and 1 part of insecticide to 1 million parts of water, maintained for 10–30 minutes from a fixed site. In the WHO Onchocerciasis Control Programme (OCP) in West Africa, temephos is applied at rates of between 0.05 ppm and 0.1 ppm for 10 minutes and is carried downstream in turbid waters for about 45 km. In the treatment of very small streams, such as those in which the South American, *S. ochraceum*, habitually breeds, effective dosages will probably have to be higher.

The application of blackfly larvicides is usually a relatively simple matter, most of the work being done by the natural flow of the river. The insecticide may be introduced into the water in several ways. It can be dripped through a hole in the bottom of a 25 litre can of insecticide from a conveniently sited bridge or some other point for a 30 minute period (Fig. 7.5), or the entire dose can be dropped from an aircraft in about one second (Fig. 7.6). The method chosen will depend on the scale of the operation, the size of the rivers to be treated, and the resources available. In most operations the insecticide is applied not from an aeroplane but from ground-based spraying equipment. Providing the treatment point is some distance from where the larvae occur, and a properly formulated insecticide is adequately spread across the width of the river, it will be passing the larval sites for up to 30 minutes after travelling a distance of

Fig. 7.5 Insecticidal emulsion being dropped from an old 4 gallon kerosene tin into a *Simulium damnosum* complex breeding place. The insecticide is carried downstream. (Courtesy of R.W. Crosskey.)

about 2 km, even if it is applied very rapidly. However, where treatment has to be made directly above a breeding site, owing to lack of access upstream or poor flow, it is desirable to slow down the application rate of the insecticide. This will ensure a reasonably prolonged contact with the blackfly larvae and allow the toxic ingredient to take effect, either by ingestion or contact.

Non-insecticidal control measures

Of value in a few control situations is the physical destruction of *Simulium* breeding sites. Many of these are formed by man-made structures such as dam outlets, bridges and causeways. Such structures may play an important part in onchocerciasis transmission, not only by providing local breeding places but also by attracting migrant females, some of which may contain infective worms, to areas commonly visited by man. Many such structures fall into disrepair and cease to be functional. They could be removed without any loss of amenity and assist in reducing breeding.

Insect growth regulators (IGRs) have also been tested against blackflies. Their effects are usually confined to a very precise stage, often only a

Fig. 7.6 Aerial application in the OCP area of West Africa of temephos. The helicopter is dropping a measured quantity of this organophosphate insecticide into a breeding site. (Courtesy of the World Health Organization.)

single larval instar, in the life-cycle. So their utility against fast-developing species with overlapping generations must be in doubt. Control has been demonstrated in field trials but the level has never been sufficiently high to substantially reduce populations. More developments and trials are needed, slow release formulations may give better control. Insect growth regulators may have a future role in integrated programmes against pest species.

Of great interest is the possible use of *Bacillus thuringiensis* var. *israelensis* (= serotype H14). This spore-producing bacterium should be considered more as an insecticide of biological origin rather than as a biocontrol agent, because it does not perpetuate itself or persist in the habitat but has to be repeatedly sprayed. It kills blackfly larvae (and larvae of other medically important insects such as mosquitoes) by the release of endotoxins which act as a stomach poison when ingested by the larvae. Formulations of *B. thuringiensis* have been tested in several regions including West Africa. Stability at tropical temperatures appears good, but formulations suitable for large-scale field use await development.

Blackflies are attacked by a wide range of fungi, microsporidia and mermithidae though generally little is known about their natural importance or even about their life-histories. Progress in evaluating the potential of biocontrol agents is severely limited by the lack of self-perpetuating laboratory colonies of blackflies. Field trials on mermithids affecting *Simulium damnosum* complex did not offer much hope that they could be used as very successful control agents. Although biocontrol of blackflies is desirable and its possibilities have generated considerable interest, there does not appear to be any prospect of the early operational use of biological control agents.

Examples of control measures

Although larvicides are most likely to be chosen for the control of blackflies, the actual strategy will depend on the biology of the particular blackfly species and whether it is just a biting nuisance or a vector of river blindness.

To illustrate the question of appropriate strategies, three simplified examples from different zoo-geographical regions are considered. These are: (i) *S. damnosum* complex in the Afrotropical region; (ii) *S. ochraceum* in the Neotropical region; and (iii) *S. venustum*, and *S. arcticum* in the Holarctic region.

The onchocerciasis control programme (OCP) in West Africa

The *S. damnosum* complex consists of some 25 sibling species distributed widely throughout tropical Africa. It also occurs in subtropical areas to the south and in a small focus in the Yemen. About 10 different species are found in West Africa where the commoner of those occuring in the savannah areas are very efficient vectors of onchocerciasis. Although occasionally the vector may be present in such large numbers as to cause a severe biting nuisance, this is not often the case and many areas of onchocerciasis exist where maximum biting rates are only about 200 bites/man/day. However, at such densities considerable transmission may be occurring, with 10–15 per cent of the older flies carrying infective stages of the filarial parasite (*Onchocerca volvulus*). In these areas the

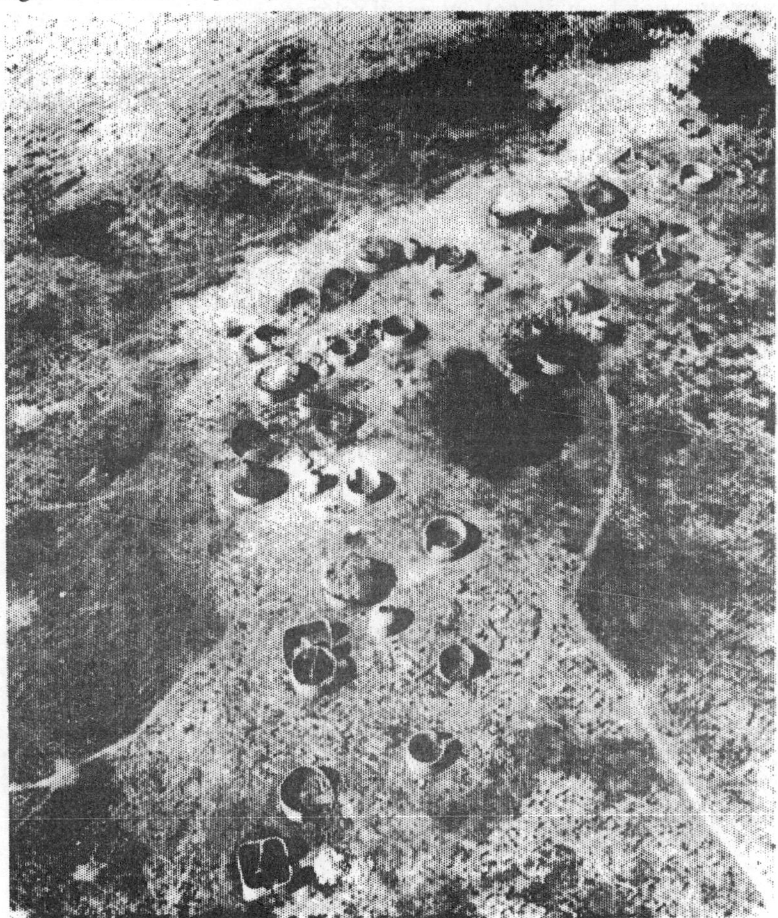

Fig. 7.7 The incidence of river blindness (onchocerciasis) is sometimes so high that people leave their villages to escape the disease. This photograph shows the village of Brimi, Chad, which has been abandoned: of a population of 400 only 9 remained, all are blind. (Courtesy of the World Health Organization.)

flies may prefer biting man to other hosts and in some communities almost 100 per cent of the human population suffer from onchocerciasis.

In West African savannah areas, the *S. damnosum* complex may pass through about 16 generations in one year, with a larval development time of about 8–12 days. Breeding is confined to fairly fast flowing rivers and streams, seldom less than 4 m wide. However, adults of *S. damnosum* are capable of extremely long range movements. Thus, we have a disease of major socio-economic importance carried by a vector breeding in restricted areas which can be readily detected and treated with insecticides. However, the vector has a very rapid development time and is capable of recolonization from distant untreated sources.

Owing to the severity of the disease in the Volta River basin (Fig. 7.7) a control programme, the OCP, was initiated in 1974. Although the full magnitude of the vector's dispersal was not known, it was nevertheless appreciated that this was considerable and could result in regular reinvasions of the control zone from adjacent untreated areas. To keep reinvasion to a minimum the operational area was made as large as was politically, economically, and technically feasible at the time. Initially this covered an area of 654 000 km^2 in the Volta River basin of West Africa taking in parts of 7 countries – Ghana, Mali, Niger, Upper Volta, Ivory Coast, Togo and Benin, but this was later expanded to 764 000 km^2 (Fig. 7.8).

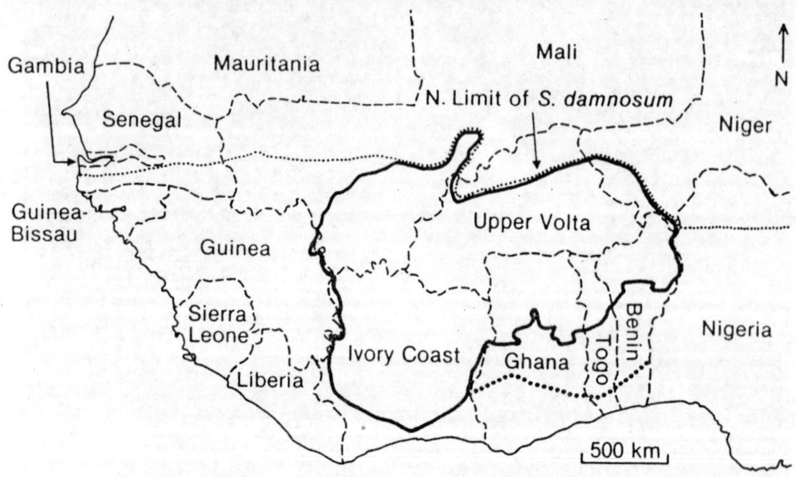

Fig. 7.8 West Africa showing the area (————) of the Onchocerciasis Control Programme (OCP) in relation to the seven participating countries (Benin, Togo, Ghana, Upper Volta, Mali, Ivory Coast and Niger), and the southern limit (............) of the extension area. The northern limit of the distribution of the *Simulium damnosum* complex (..........) is also shown. (Redrawn and modified from original prepared by the World Health Organization.)

It was realized that permanent elimination of the vector was not possible. The programme was therefore designed to last between 15 and 20 years during which time transmission of the disease was to be interrupted, if possible, throughout the area. This time-scale was dictated by the fact that the parasite can live in its human host for between 11 and 17 years. The basic aim of the scheme was to prevent transmission for as long as the human population in the area were infected with onchocerciasis. When this goal was achieved the vector could be allowed to repopulate the disease-free area.

To attain this objective it was assumed that all breeding sites throughout the control area might have to be treated weekly with temephos. It was not thought that any other type of control measures such as environmental changes to the blackfly breeding places, biological control of the larvae, or control aimed at the adults could result in the requisite high level of control. The only way to obtain the necessary coverage of breeding sites over such a large area was considered to be by insecticidal spraying from fixed-winged aircraft and helicopters. Even so the control zone had to be restricted to savannah country where breeding sites could be readily located and treated from the air.

Although the objective of eradicating a serious scourge of mankind was generally considered to be meritorious, the possibility of causing serious environmental damage could not be ignored. Many of the more productive blackfly breeding sites were located on fairly large rivers which supported important indigenous fisheries, vital to the economy of local communities as well as being major suppliers of scarce protein. Also the scheme, although offering the prospect of major benefits, was inevitably very expensive, requiring large amounts of support from external sources. Many of the potential donor countries had active environmental lobbies, antagonistic to any use of insecticide, regardless of potential benefits. In consequence WHO, the executing agency for the OCP, undertook careful insecticidal screening to discover the insecticide that would do least environmental damage. Furthermore, WHO instigated a scheme for regular aquatic monitoring by independent workers to check on any changes in riverine ecology that might nevertheless be caused by the operation. This environmental caution has created a precedent which will probably have to be followed in future multi-national pest control activities reliant on funding from many donor governments.

Control of blackflies in East and Central Africa

To contrast with the above exceptionally large-scale operation, the possibility of controlling onchocerciasis vectors in East and Central Africa is briefly considered. In Uganda and Tanzania both onchocerciasis and its vectors are restricted to discrete foci. Vectors include, in addition to the *S. damnosum* complex, species of the *S. neavei* group, the larvae and pupae of which are phoretic on crabs, that is they are attached not to weeds or stones but to the bodies of live aquatic crabs. It is very likely that several of the disease foci could be eliminated by carefully planned, small-scale, cheap, ground-based control operations. An important point is to ascertain which foci are normally isolated throughout the year and not subject to invasions of adult blackflies from elsewhere. If it is clear that a small focus is isolated then a short intensive vector eradication programme with larvicides applied over about 4 months should prove cost effective. Even where isolation is less than complete 'intermittent eradication', such as achieved with the *S. damnosum* complex on the Victoria Nile (Uganda) may be possible and economically worthwhile.

It should be remembered that insecticides applied against the larvae were mainly responsible for the eradication of *S. neavei* from Kenya in the 1950s.

In most of Africa vectors of onchocerciasis seem to be very efficient and it therefore requires greater reductions in fly numbers to eliminate or even substantially reduce onchocerciasis transmission. At present this is unlikely to be achieved by any means other than with larvicides. Nevertheless, environmental contamination should be minimized in these small-scale activities by applying insecticides at very precise dosage rates and at intervals determined for each individual river. In these circumstances the effects on non-target organisms should be minimal.

Control of *S. ochraceum* in Central America

In Central America *S. ochraceum* is the main vector of onchocerciasis. It breeds in many of the numerous very small streams flowing through heavily forested, precipitous mountain country. In such terrain the use of aircraft is precluded. The steams preferred by *S. ochraceum* are so small (optimum width 0.3–1.5 m) that insecticides will not carry far, so innumerable dosing points would be necessary for complete larvicidal coverage. Moreover, as larval development may take only about 15 days, and breeding occurs throughout the year, larvicides would have to be applied many times each year. Carrying out regular and efficient ground-based treatments in such an area would present insuperable logistic problems, so it is not surprising that entomologists have usually considered that control of onchocerciasis in Central America must be by medical rather than entomological methods. Hitherto no control operation has been mounted, though a Japanese team now plan to do so.

However, one aspect of *S. ochraceum* biology may provide a weak link which can be exploited. In the Americas onchocerciasis is confined to relatively small foci where *S. ochraceum* biting densities are extremely high, whereas the vector is found more widely in lesser numbers. This fact and recent transmission studies strongly suggest that *S. ochraceum* is an inefficient vector of *O. volvulus*. Moreover, other potential vectors, such as *S. metallicum*, that may be present are not likely to be able to maintain the disease in the human population in the absence of *S. ochraceum* acting as the primary vector. Possibly therefore even 'inefficient' control measures might sufficiently reduce the *S. ochraceum* population to cause a cessation of transmission, as well as a great reduction in nuisance biting. Thus it may be feasible to carry out a ground-based larviciding campaign, which either does not treat all larval habitats in each spraying-cycle, or in which spraying coverage is more complete, but the numbers of treatments are fewer.

Now, if it were found that relatively small reductions in the population size of *S. ochraceum* could break the transmission cycle or at least greatly reduce transmission then less efficient and drastic control methods might be appropriate. Thus the use of self-perpetuating biological control agents, such as parasitic nematodes which usually cause considerably less larval mortality than insecticides might prove feasible. An integrated control approach combining biodegradable insecticides with other appropriate control measures might be effective and at the same time minimize environmental damage.

Control of blackfly pests in North America

In North America *S. venustum* is a serious biting pest of man and his animals. Larvae occur in vast numbers of small rivers and streams. Despite there being o ly two or three generations a year which means very few insecticidal treatments are required, ground-based control poses tremendous logistic problems. Consequently aircraft are used, as they can cover larger areas having dense networks of streams needing to be sprayed. The insecticide is applied in a spray swathe 200 m wide, and not topically as in West Africa. This technique causes some adult mortality but a direct attempt to control adults is not recommended because *S. venustum* can disperse an average of about 10 km from its emergence site and may travel much farther. The blackfly season, although causing an intense nuisance, is often very short, so personal protection by wearing protective clothing and using repellents may play a role which it does not have in tropical areas.

S. arcticum is another pest species in Canada, attacking mainly cattle. There is good evidence for wind-assisted movements over distances of 220 km so that any attempt to control the adults would be very difficult. However, larvae are restricted to very large rivers where only two generations a year are produced, and they are therefore relatively easily killed by conventional applications of larvicides from boats or bridges.

Environmental issues arouse considerable emotion and controversy in North America, and therefore many useful ecological studies have been made during the control of *S. arcticum* in the Saskatchewan and Athabasca rivers of Canada. In the USA much of the control of blackflies breeding in small rivers is designed to protect sports fishermen. Thus any action resulting in decreased fish production would be self-defeating. The pressure to find non-insecticidal methods of control is consequently considerable.

Selecting control strategies

The different situations discussed above demand variations in the basic control tactics. With efficient tropical vectors there is little latitude in the type of control that can be used, and at present the only solution is to use larvicides. Whether they should be sprayed by ground-based equipment or by aircraft depends on the nature of the terrain, the scale of the operation, and the biology of the blackfly vector and epidemiology of onchocerciasis. Where the blackfly is either an inefficient or non-vector there is opportunity for more flexibility and innovation in planning control measures.

It is desirable that in future there should be a combination of environmental control methods, biological control agents and the use of more selective insecticides leading to an integrated vector management programme. However, such an approach is much more demanding of biological knowledge and skilled manpower than the more simple campaign of larviciding.

Control of tsetse flies

D.A.T. Baldry

Tsetse flies (*Glossina* spp.) have attained notoriety because they transmit pathogenic trypanosomes which in man give rise to human trypanosomiasis (sleeping sickness) and in livestock cause animal trypanosomiasis (nagana). There are 21 species of *Glossina* and their distribution extends over approximately 10 million km^2 of Africa between latitudes 15° N and 22° S. They occupy a wide variety of vegetation types and transmit 6 different trypanosomes of socio-economic importance: *Trypanosoma brucei gambiense* to man in western Africa; *T.b. rhodesiense* to man in eastern Africa; and *T. congolense, T. vivax, T.b. brucei* and *T. simiae* to livestock.

Bovine trypanosomiasis is in many areas the major veterinary problem of livestock in Africa, and it is commonly the principal constraint to rural development. It impedes economic development which in many countries depends heavily on the agricultural sector for economic growth.

The current integrated approach to trypanosomiasis control

The methods which have developed during the last quarter century for combating trypanosomiasis fall into the following three main groups:

(a) The use of trypanocidal drugs for both therapeutic and prophylactic purposes;

(b) The rearing of trypano-tolerant breeds of cattle in areas where susceptible breeds are unable to endure or survive local trypanosome challenges;

(c) The control or eradication of the tsetse vector.

Trypanocidal drugs

Drugs are available which are higly active against the trypanosome species which affect man and livestock. However, the use of trypanocidal drugs is not simple, especially in the treatment of human trypanosomiasis, because some drugs in routine use, for instance melarsoprol and nitrofurazone are basically dangerous to administer and can thus only be used under strict medical supervision. Fortunately this does not apply to pentamidine, the most important prophylactic drug for Gambian sleeping sickness. Another major problem encountered in the treatment of trypanosomes in both man and livestock is the development of drug resistance. Despite these and other various limitations the therapeutic and preventive administration of trypanocidal drugs still constitutes the backbone of the animal trypanosomiasis control effort.

Trypano-tolerant cattle breeds

It has long been recognized that some West African cattle breeds, for example N'dama and Muturu, have low susceptibilities to trypanosomiasis. However, the degree of trypano-tolerance is relative to the magnitude of the trypanosome challenge to which they are exposed.

Introductory remarks on the concepts of tsetse management and integration

Before examining in detail various concepts and aspects of tsetse control, and of eradication in some cases, it is pertinent to consider what exactly is meant by the three terms 'vector management', 'integrated control', and 'integrated vector management' which are often used synonomously with regard to tsetse control, but which in strict terms do not mean exactly the same thing.

In tsetse (i.e. vector) management the primary objective is to manage a tsetse population, to reduce it or its collective trypanosome challenge to a level below which trypanosome transmission and its resultant socio-economic impediment are negligible; that is levels below which the disease situation can readily and economically be mastered by normal medical and veterinary services.

Integrated tsetse control means the combined use of appropriate forms of tsetse suppression, both chemical and biological, for effective, and economical reductions of tsetse populations and trypanosome challenges. Ideally the individual control techniques which are being integrated, especially those involving chemicals, should be as selective as possible, and less hazardous to the environment.

Thus in considering on the one hand tsetse management and economics, and then integrated control, it is not difficult to see how the two together constitute *integrated tsetse management*, which is what we are really concerned with here. The two concepts are distinct, but in collective context they present a concept of effective and economic tsetse control which is environmentally acceptable, and within which there should not be complete dependence upon any one type of control method, be it chemical, biological or genetic.

Integrated tsetse management makes good technical and environmental sense, but it must also make good economic sense through increased economical development, whether by improved human health or by increased animal production and health. Although in recent years there have existed good possibilities for well balanced and integrated tsetse control in some African countries, they have not always been realized because of lack of funds, equipment and appropriately trained staff. Consequently, a second best approach has often been adopted which has been less technically and environmentally ideal, but feasible within the local technical and budgeting infrastructures.

The concepts of integrated tsetse management, is by no means a new one, as it was advocated many years ago. But interest was later directed to insecticidal control to the almost virtual exclusion of all other techniques due to the availability in the late 1940s of the synthetic and powerful organochlorine insecticides.

The current tsetse control approach

Here the term 'control' is implied as meaning an attack on a tsetse population which either reduces it to an acceptable low level or results in total eradication. Far too often eradication campaigns fail because of inadequate protection of the treatment zone required to keep it free from tsetse reinvasion. Thus, many eradication campaigns become control campaigns requiring recurrent control effort and funding.

In situations where tsetse control is an integral part of land-use development – usually in relation to animal trypanosomiasis control – the control effort usually has two components. One, the attack component, has as its objective the control of tsetse flies in a zone delineated for development. The other, the maintenance or consolidation component, has as its objective the control of tsetse flies in a barrier zone externally adjacent to the development zone to prevent, or at least limit, the movement of flies from nearby untreated areas into the development zone. Both components of the overall tsetse control campaign may involve the same or similar tsetse control techniques, but it is more usual for them to be different. For example, a well-maintained bush-clearance barrier of appropriate dimensions is usually the most effective method of protecting a development area which has been sprayed, but within which there remains vegetation suitable for reoccupation by the fly after the effects of the insecticide have worn off.

In most instances tsetse control operations are initiated only after long and careful planning (which should take account of economics and the environmental impact) because the aim is to achieve medium to long-term control of endemic or enzootic situations and to assist rural development.

Tsetse control is a very complex subject, and it is not possible to describe all aspects here. Because of its complexities and the belief that generalities can be misleading and often difficult to interpret, attention is focused on the control of only one tsetse species, *G. morsitans*, which exists as three subspecies : (i) *G.m. submorsitans*, in western Africa; (ii) *G.m. centralis* in the savannah woodlands of eastern Africa which surround the central African rain forest belt; and (iii) *G.m. morsitans* occupying the savannah woodlands adjacent to the East African coast.

In the following paragraphs, accounts are given of the control of *G.m. submorsitans* in Nigeria to illustrate a largely unintegrated approach, and of *G.m centralis* and *G.m morsitans* in Zambia to illustrate a largely integrated management programme.

Control of *G.m. submorsitans* in northern Nigeria

With the advent of the persistent organochlorine insecticides, relatively high dosages of DDT, and to a less extent dieldrin, were ground-sprayed from the late 1940s to the early 1960s to all accessible vegetation within the *G.m. submorsitans* habitats. The drastic effects of this type of spraying on non-target organisms are too well known to be enumerated here.

As more information became available on the localized dry season distribution of the flies and of their preferred resting sites, there emerged the concepts of discriminative and selective spraying. Insecticide applications became discriminative in that they were applied only to those parts of the habitat which were essential for dry season survival. Applications became selective in that they were orientated to special resting-site structures and to the height at which flies preferred to rest (usually up to heights of 3.5 m). Refinement of the combined discriminative/selective techniques led to the standard ground-spraying technique as used today, in which only about 12 per cent of the overall area is sprayed. Relatively narrow insecticide swathes are applied to riverine forests, woodland/grassland ecotones, hill-slope ecotones, the edges of roads, paths and cattle tracks, and across extensive areas of uniform woodland.

Throughout the 1970s much attention was focused on helicopter spraying techniques with a view to: (i) increasing operational speed; (ii) further refining the discrimination principle (selectivity does not apply to aerial spraying techniques); and (iii) evaluating alternative insecticides to DDT and dieldrin. The standard helicopter spraying technique which emerged from such research involved a single application of endosulfan along 30 m wide swathes which gave the same type and percentage discrimination as was obtained during ground-spraying. Endosulfan, at the swathe dosages employed (up to 900 g/ha) is, however, toxic to fish. Consequently, the potential of some of the newer insecticides, is being evaluated, especially that of the synthetic pyrethroids which although less persistent than DDT and endosulfan, may prove to be economically and environmentally acceptable substitutes in both ground and aerial spraying operations.

In many places the *G.m. submorsitans* fly belts are sufficiently small and isolated that they can be treated without having to conduct maintenance control operations. Where the latter, however, are required, combinations of bush clearance and chemical control methods are employed.

The excellent tsetse control achieved in northern Nigeria is a good example of control which is almost totally dependent upon the use of highly persistent insecticides. However, every attempt is being made to reduce insecticide dosage rates by increased discrimination and selection, and to find acceptable substitutes for the relatively toxic compounds which are in current use and which have given the Nigerian authorities such outstanding successes in recent years.

Control of *G.m. centralis* and *G.m. morsitans* in Zambia

Tsetse habitats

In Zambia, as in many other parts of eastern Africa, *G. morsitans* is most frequently associated with extensive areas of 'miombo' woodland, in which the commonest trees are species of *Julbernardia* and *Brachystegia*. In many cases the monotony of the 'miombo' is broken only by reticulated 'dambo' grasslands ('mbuga' in East Africa). These are flood-plain grasslands, of low agricultural potential, in which there is no defined drainage channel. Riverine forest of the type so typical of West Africa, rarely occurs, although the dambos may contain thickets (often associated with termite mounds) and/or isolated trees or clumps of trees. During the wet season *G. morsitans* is widely distributed through the miombo woodland but in the dry season it becomes much more concentrated around dambos which provide the best dry season grazing for game and livestock. Flies become particularly conspicuous in dambo/woodland ecotones, and provide an excellent example of a 'fringe effect'. These ecotones contain fauna and flora of both the dambo grasslands and miombo woodlands, but because dambo and miombo are such discrete communities, the fringing ecotone is very narrow; often so narrow as to be almost indistinguishable.

The 'attack' component of tsetse control

The early use of insecticides for tsetse control in Zambia was very similar to the approach reported for northern Nigeria. However, it was also recognised that there were other tsetse control strategies which could be advantageously used in combination with insecticidal techniques. Consequently, for a number of years there has been a largely integrated approach to tsetse control and at the present time the Zambian situation, probably provides the best example of a relatively advanced system of integrated tsetse control.

In Zambia three 'attack' methods are employed during the dry season: ground-spraying; aerial spraying; and bush clearance.

Discriminative ground-spraying is the most commonly employed technique. In most instances DDT wettable powder is applied by twin sprayers mounted on the rear of a tractor, but in hilly and/or thicket situations it is applied by means of pressurized knapsack sprayers. The basic strategy is to apply the insecticides across 20 m wide swathes, with particular attention being given to the dambo/miombo ecotones. Supplementary spray swathes are applied at 300 m intervals through uniform woodlands, and along communication lines. Spot spraying is directed at thickets and woodland clumps within the dambo grasslands. The level of discrimination is about 12–18 per cent.

In areas where the terrain is flat and its vegetation uniform, twin-engined fixed-wing aircraft are used to make 5 sequential aerosol applications of endosulfan. In escarpment and other hilly situations, ruthless manual bush-clearing techniques replace insecticidal methods.

Although very few environmental studies have been made of tsetse control operations in Zambia, it would appear that discriminative ground-spraying is the most acceptable technique. The impact on non-target arthropods inhabiting dambo/miombo ecotones is admittedly probably very high during spraying and shortly thereafter, but the potential for repopulation should be good, since only 12–18 per cent of the total area is sprayed. Contamination of dambo dry-season grazing and of streams is generally regarded as minimal.

Discriminative ground-spraying is also an important 'consolidative' or 'maintenance' technique, but it is not usually employed alone, more often it is one component of an integrated strategy designed to stop tsetses moving into the control area, and to prevent livestock having contact with tsetses outside the control area.

Consolidative or maintenance component of control

By employing one or more of the control techniques described above during the attack components, the overall control policy aims at controlling tsetse flies in an area which can be stocked with livestock. At the same time the control areas need protection (or consolidation) from fly reinvasions and this is done by so called 'holding lines'. The basic holding line system is illustrated schematically in Fig. 7.9 and consists of the following methods.

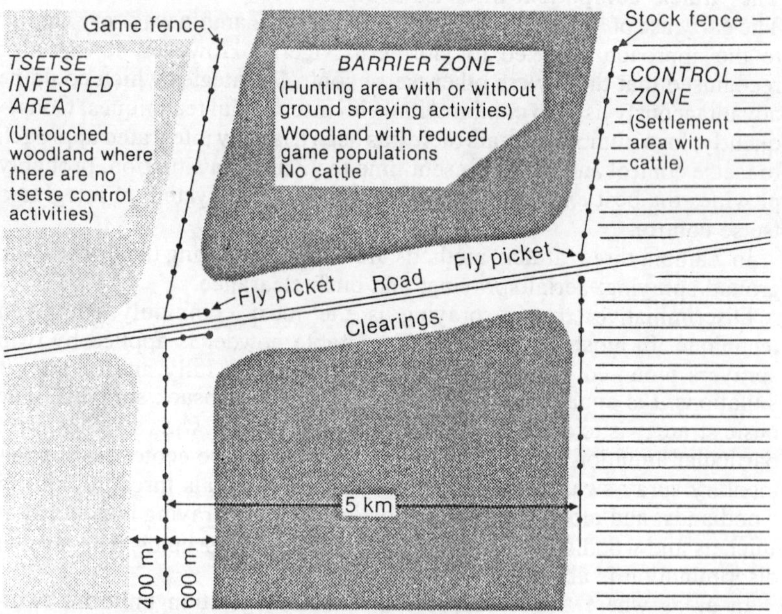

Fig. 7.9 Schematic representation illustrating the holding line concept of tsetse control as practised in Zambia.

Stock fences These are perimeter fences which either surround a settlement area or protect that part of the perimeter where there is a trypanosomiasis threat from outside. The primary objective is to restrict the movement of cattle and reduce their chances of contacting infective flies. They also help prevent some game animals entering the settlement area.

Although well maintained stock fences stop cattle moving into dangerous areas, they do not *per se* prevent trypanosomiasis transmission, since flies may already be present in the settlement area, or they may fly into it across the fence.

Game fences Game fences are constructed 5 km outside the stock fences and usually parallel to them. Their purpose is to reduce the numbers of game animals that would otherwise have access to the intervening 5 km wide barrier zone (Fig. 7.9). This action reduces the food supply of flies living in the barrier zone.

Even strongly constructed game fences fail to prevent the passage of some small animals such as warthog (*Phacochoerus aethiopicus*) and bushpig (*Potamochoerus porcus*) which can burrow under them, or strong jumpers like the kudu (*Tragelaphus* spp.). Moreover, where game fences cross gullies and broken ground many game species (e.g. bushbuck, *Tragelaphus scriptus;* duiker, *Cephalophus* and *Sylvicapra* spp.; warthogs and bushpigs) find their way across.

Normally all woodland vegetation is cleared for 600 m on the barrier side of the game fence and for 400 m on the other side. This allows easy access for ground-spraying teams (if required), and provides good visibility for hunters who may patrol the fence for game elimination purposes.

Barrier zones Barrier zones which are bounded by game and stock fences may themselves be subjected to tsetse control activities. The commonest strategy involves ground application of insecticides; less commonly, bush clearance activities are introduced.

However, in most cases game elimination is practised. This aims at reducing fly density by eliminating or reducing the preferred hosts of the barrier tsetse population. Frequently, this approach is not very effective, especially when game densities are low. Some species, e.g. warthog and duiker, are very difficult to eliminate because of their concealed behaviour.

Fly picket Fly pickets are most commonly sited at places where communication lines cross stock and games fences. Their purpose is to reduce the numbers of flies which would otherwise be carried by pedestrians, cyclists and vehicles from tsetse-infested areas into barrier zones, and settlement areas. Fly picket orderlies check traffic passing through their pickets and may catch any flies which can be detected, and apply insecticides to inaccessible parts of vehicles.

To further reduce the movement of tsetse flies along communication lines it is customary to clear woodland vegetation to a width of 1–2 km adjacent to those parts of roads which cross barrier zones (Fig. 7.9).

Although the above account is a good example of an integrated approach to *G. morsitans* control in Zambia, totally satisfactory results have not always been achieved. This is partly because some flies are always able to cross the barriers and partly because the farming practices of people living in the settlement areas are not always sufficiently intensive or extensive to render the areas unsuitable for repopulation by *G. morsitans*.

Some final points for consideration

It is anticipated that the measures currently used in trypanosomiasis control will continue for several years. However, it must be emphasized that an immunization procedure would eventually be the method of choice for trypanosomiasis control. With regard to animal trypanosomiasis control much detailed and long-term research is being pursued, but regrettably for the time being immunization is only a research objective. Nevertheless, the objective could become reality in the not too distant future, in which case the whole concept of trypanosomiasis and tsetse control would need to be drastically altered.

Recently much attention has been devoted to various novel techniques that seem to be less hazardous than insecticides to the environment and might more readily be combined into integrated control strategies. In the forefront are techniques which depend on the release of sterile male tsetse flies or the use of insect growth regulators and pathogenic viruses. Synthetic pheromones combined with insecticides may also become important. However, until such alternative methods have been adequately tested under field conditions and been subjected to detailed cost/benefit analyses, it is more appropriate to consider some more realistic alternative methods which can be introduced now, or in the very near future. For example, efficient trapping devices, such as the biconical trap (see p. 104), which can be used alone or in combination with insecticides and be so efficient as to reduce tsetse populations. In parts of West Africa this principle, using insecticide-impregnated screens to attract and then kill tsetses, is now giving excellent results.

There is also considerable scope for devising better and cheaper ways of applying insecticides as well as introducing new compounds and formulations. However, the need to replace persistent organochlorine insecticides is not so great in relation to tsetse control as it is in relation to some other disease vectors and crop pests. Fortunately tsetse flies have not so far developed insecticide resistance. Furthermore, because of their slow reproductive rate a single insecticidal application may under optimum condition achieve very effective control, if not eradication. Even if control measures have to be repeated, the inter-spray interval is usually sufficiently extended to give ample time for the recovery of those non-target organisms which may have been adversely affected by the first spraying. (A great many arthropod, fish and bird species are highly mobile with high potentials for population re-establishment and multiplication). Environmentally this should be reassuring, but, in many cases it is largely irrelevant, particularly when tsetse control is considered in relation to animal trypanosomiasis control and livestock production. Cattle are grazers and eat grasses not trees, shrubs and creepers which constitute a tsetse habitat. Similarly, arable farmers require open ground on which to cultivate their crops. Consequently, man's activities in clearing land after a savannah tsetse control operation, either for the purpose of pasture improvement and/or for cultivation, have a much greater and longer lasting impact on the environment than the original tsetse control operation, which made the land available to him and his livestock.

Control of triatomine vectors of Chagas' disease

O.P. Forattini

Insecticidal control

Theoretically the best and most permanent measure for controlling triatomine vectors is to rehouse people living in dilapidated mud and thatch dwellings (Fig. 7.10). These new houses will be unattractive to the vectors because they lack cracks and crevices in the walls and roof or ceilings in which the vectors can shelter. Nevertheless, as yet such an expensive rehousing programme is not practical, except on a very minor scale, and other control measures therefore have to be used to control the bugs. The most widespread method is based on residual insecticides, which are sprayed both inside houses and in a variety of outbuildings and man-made animal shelters etc. The more commonly used insecticides are the organochlorines, such as HCH in Brazil and Argentina, and dieldrin in Venezuela.

The aim is to leave a minimum surface deposit of 0.5 g/m^2 of HCH gamma-isomer (lindane) or 1 g/m^2 of dieldrin on the interior surfaces of buildings. HCH gives high kills of bugs immediately after spraying but remains effective for only about 30 days; in contrast the residual activity of dieldrin extends over several months, but it is not such a powerful insecticide. An advantage of using HCH is that it has a much lower toxicity for man and other vertebrates than dieldrin, the toxicity of which imposes restrictions on its use. Mainly owing to the probability that the bugs will develop insecticide resistance to HCH and dieldrin, organophosphate and carbamate insecticides are often used, of which the more common are malathion, fenthion and propoxur.

In addition to spraying the interior surfaces of houses with water-dispersible powders of residual insecticides, other techniques including fumigation, fogging, especially with ultra-low-volume (ULV) dosages, and the use of slow-release insecticides may be undertaken. These alternative methods are usually used with malathion, dichlorvos (DDVP), propoxur or other suitable organophosphate and carbamate compounds. Some health services combine a number of these methods, such as

Fig. 7.10 Typical thatch-roofed and mud-walled delapidated houses in South America which provide numerous cracks and crevices in which triatomine vectors of Chagas' disease rest. Ideally such dwellings should be replaced by houses of more modern construction. (O.P. Forattini.)

applying two sprays of HCH separated by 2–3 months and an annual ULV fogging with malathion, and sometimes houses may in addition be fumigated with dichlorvos. Without doubt insecticidal control of triatominae represents the most practical approach that can at present be adopted. For highly domiciliated bugs like *Triatoma infestans* total elimination of house infestations may be achieved, but afterwards there must be repeated house inspections to detect any reinfestations. This may arise due to several reasons, including passive reinvasion by bugs being brought into houses with furniture, or by active reinfestations originating from outdoor natural foci of the insects, as sometimes occurs with *Rhodnius prolixus* and *Panstrongylus megistus*.

Chemical control, however, is not without its problems. Some of the common insecticides used, especially dieldrin, but to a lesser extent HCH, are relatively toxic to man and have to be handled with caution if accidental poisonings are to be avoided. Moreover, insecticides are not cheap and the residual effects of the most persistent last only 2–3 months, thus necessitating several applications a year, and this increases both labour and insecticidal costs. There is no clear evidence that any of the insecticides used kill triatomine eggs. In Venezuela insecticide resistance to dieldrin and HCH has been recorded in both *R. prolixus* and *Triatoma maculata*, but the epidemiological importance of this has not been completely established. More investigations are needed on the dangers of insecticidal resistance to determine how important this is likely to be in control operations. Finally, another problem is that environmental pollution may arise from widespread insecticidal applications, especially if organochlorine insecticides are used.

Because of these problems – increasing costs of insecticides, evolution of insecticide resistance, and environmental contamination – and the gradual realization that insecticides alone are unlikely in the long-term to be the solution to controlling Chagas' disease, more attention is slowly being paid to employing insecticides in an integrated approach. Some of the other components that might be incorporated into a pest management programme are outlined briefly below.

Non-insecticidal control methods

Alternative methods include biological control with natural enemies like wasps (*Telenomus*) which parasitize triatominae eggs. There are also several natural pathogens and predators of triatomine bugs, but although they may be assisting in natural population regulation they have not as yet been exploited very much as control agents. One difficulty is that many natural enemies are unlikely to prove easy to mass rear in the large numbers needed for releasing into the field in biological control programmes. Another problem is that many predators are very susceptible to insecticides and are greatly reduced in numbers if houses have been sprayed in malaria and Chagas' disease control programmes. For these and other reasons, it seems that predators do not, unfortunately, have much potential for reducing bug populations.

The most likely biological control candidates are pathogenic bacteria or viruses that could be sprayed on the insects, but such microbial insecticides would need to be extremely safe to man if they were to be used in houses. Other possible approaches include the use of insect growth regulators (IGRs), dehydrating substances such as silica gel to cause rapid desiccation of the insects, or even possibly genetic control with sterilized male bugs. Methyl bromide is a non-residual chemical that has sometimes been used to kill agricultural pests, and has occasionally been employed to fumigate houses to destroy triatominae. Its toxicity to vertebrates makes it useful also in killing rats and other pests, but at the same time makes it dangerous to use in houses unless special precautions are taken. As yet none of these alternative control measures has been used to control Chagas' vectors except on a very localized scale.

Planning a control or management programme

An efficient programme for triatominae control should incorporate at least three stages: preparation; attack; and surveillance. The preparation stage involves geographical reconnaissance; epidemiological studies on Chagas' transmission; biological, behavioural and ecological investigations on the vectors; and most importantly, community education to explain the strategies and objectives of the control programme. This last aspect is essential if the full co-operation of the people is to be obtained.

The attack stage starts with the first widespread spraying of all dwellings in the area. The interval between this initial spraying and respraying varies greatly, but if HCH is used it is generally recommended to be 6 months. After this second treatment the attack stage enters the second phase and houses are searched for bugs about every 6 months and only those harbouring bugs are sprayed again. Thus, this part of the attack stage comprises more selective spraying.

During both the initial attack phase and the subsequent selective spraying the insecticide is carefully applied not only to the walls, ceilings and roofs of houses, outbuildings and animal shelters etc., but also to furniture and other objects. In addition to this, various outdoor refuges, rodent holes, and cracks and crevices that might harbour the bugs, must be sprayed with higher dosages of insecticides. It is essential that spraying operations are efficiently and conscientiously performed and this requires considerable training of the personnel engaged.

Surveillance is the final stage of the control campaign and perhaps the most difficult to sustain because theoretically there are no more triatominae resting indoors, but the potential still exists for houses to become

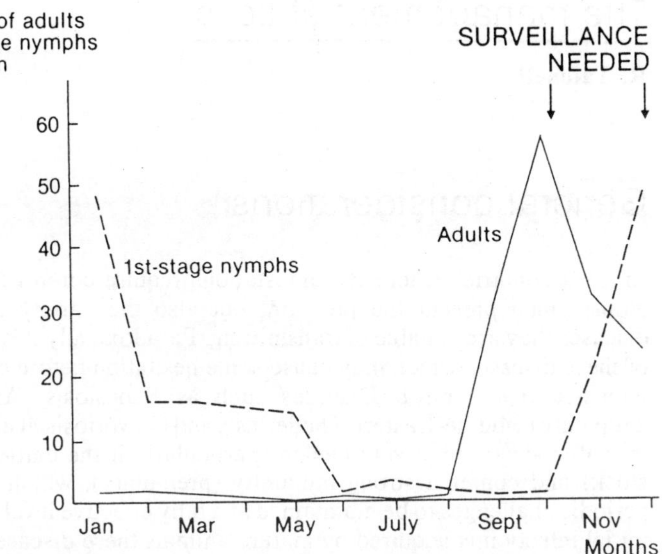

Fig. 7.11 Monthly percentages of adults and first-stage nymphs of natural populations of *Panstrongylus megistus* colonizing artificial shelters, such as fowl houses. The sudden drop in numbers of adults during September is a result of dispersal and the reinvasion of other shelters, such as houses. This exodus of bugs warns that surveillance at this time of year is very important if control measures are to be maintained and evaluated. (Redrawn and modified from O.P. Forattini *et al.*, 1979.)

reinfested. The surveillance period also provides excellent opportunities for important studies to be undertaken on the biology of both domiciliated and sylvatic populations of bugs, such as their natural development times, duration of their life-cycle, population build-up (Fig. 7.11), dispersal, and the factors leading to reinvasion of houses with bugs. Such information on the behaviour and population dynamics of the triatominae is useful both in planning better control tactics and in establishing more appropriate sampling methods in surveillance programmes.

In conclusion, control of the vectors of Chagas' disease is in theory not too difficult, at least for the more highly domiciliated species, because if insecticides are correctly applied they are very efficient in killing the bugs. Nevertheless, there are frequently administrative difficulties in ensuring that spraying operations are properly carried out. It is also essential to undertake surveillance after the control programme to discover whether permanent control has been achieved or whether after a few years reinfestation occurs. In other words, as there is little likelihood of improving dwellings on a large scale, post-control surveillance is indispensible. More reliable surveillance procedures are needed to detect the early stages of triatomine infestations, so that action can be taken quickly to prevent both the reinfestation of sprayed dwellings and new infestations of new houses.

The management of ticks

R. Tatchell

General considerations

In most countries where tick infestations require control it is not ticks alone which present the problem, but also the variety of tick-borne diseases they are capable of transmitting. Paradoxically it is the presence of these diseases which may cause some hesitation before embarking on intensive tick control. Diseases such as Babesiosis, Anaplasmosis, Tropical or Middle-Eastern Theileriasis and Cowdriosis, if acquired early in calfhood, are relatively benign (particularly if the cattle are of local stock) and confer a strong immunity (premunity), which may require periodic challenges to be maintained at a fully effective level. If, however, initial infection is acquired by mature animals these diseases cause high mortalities. Under these circumstances an individual or government should analyse very carefully the benefits that are being sought from tick control and the hazards which may develop as a consequence. Thus, if the benefits are largely 'cosmetic' with the removal of light tick burdens (10–20 engorging females) and if the concomitant results will be the production of a disease-susceptible population, then tick control should not be considered. In contrast, when highly productive and valuable exotic cattle are exposed in East and Central Africa to *Theileria parva* infections (East Coast Fever), intensive tick control is essential because even calves have an unacceptably high mortality rate from this disease. The level of control required to protect cattle in hyperendemic *T. parva* areas is by far the highest in the world as no other tick-borne disease kills virtually 100 per cent of susceptible stock. Even local small East African Zebu cattle, once mature, are fully susceptible and will die from *T. parva* administered by a single tick. During times of vector abundance (throughout the year in Kenya and Uganda and lowland areas elsewhere, and during the rains in highland areas of Tanzania and Central Africa) valuable stock must be dipped in acaricides twice weekly. Well-fenced farms with a long history of excellent tick control need to dip only once weekly.

Tick control at this intensity is very effective but very expensive. It has an additional virtue in that any acaricide-resistant individual ticks present in a natural population are unlikely to have any opportunity to mate and have progeny. With less intensive regimens selection of acaricide-resistant strains is a great danger. Where immunization or treatment against tick-borne diseases is available it is preferable to reduce the number of acaricide treatments to the minimum necessary to prevent economic losses from 'tick worry'. In this way the cost of inadvisable dipping is avoided and resistant ticks are swamped by an excess of susceptibles. Dipping for cosmetic purposes alone is expensive and merely tends to increase the selection pressure needed to encourage resistance.

As implied above, tick management in the past has been mainly a matter of chemical control but in recent years successful alternative measures have become available. Increased knowledge of tick biology in some instances has allowed more effective timing of acaricide applications and also allowed a desirable reduction in their frequency. In addition, management of the tick environment on the host (selection of resistant animals and breeds) and off the host (habitat modification and pasture spelling) have been advantageous in many instances.

The incidence of *Ornithodoros moubata* in African dwellings is believed to have been reduced largely as a result of improved living standards.

Environmental control

Pasture spelling

This involves excluding suitable hosts from areas of pasture and thus controls tick populations by denying the free-living stages any opportunity to feed. The method demands that the free-living stage is short-lived and that uncontrolled wild hosts are not present. In practice this means only larvae of *Boophilus* spp. which are 1-host ticks, because nymphs and adults of 2- and 3-host ixodid ticks survive too long (6–12 months).

In Australia *B. microplus* larvae die rapidly in the summer and removing cattle from grazing paddocks for 8–10 weeks results in greatly reduced tick numbers. There are management problems because good fencing and complete herding together of the cattle are essential. In addition ungrazed pasture may deteriorate rapidly under some conditions and could pose a serious long-term problem. It is therefore necessary to design a grazing and spelling system that minimizes tick damage within the constraints imposed by the need to maintain good pastures. In regions where cold winter temperatures inhibit oviposition and larval hatching, it is possible to have a single spelling system. In this instance the cattle are held in small paddock for the 10 week period needed to allow larvae to hatch during the 'spring rise', that is the spring-time mass hatching of eggs laid earlier which are induced to hatch by rising temperatures. The cattle are then free to graze over the remainder of the property. Alternatively, a 2 paddock rotational system could be used but with the additional penalty of increased management costs for handling and fencing. Because of these unfortunate constraints of pasture deterioration and increased costs, most farmers only indulge in opportunistic spelling rather than a fully planned regimen.

Habitat modification

Modification of the habitat has been used successfully to alter the microclimate needed for the development of free-living stages. In Europe *Ixodes ricinus* relies on poorly drained pastures with a dense growth of grass for shelter. Drainage and pasture improvement combined with heavy grazing can cause the disappearance of the tick. Similar pasture improvement and heavy grazing has had beneficial effects on the numbers of *Rhipicephalus appendiculatus* and *Haemaphysalis longicornis* in Africa and Japan, respectively.

In North America *Amblyomma americanum* has been well controlled by clearing bushes and trees providing shelter for ovipositing females and developing larvae. Once again it is unfortunate that the costs of these operations are high and they will only be carried out as part of a general programme of improvement and not solely for tick control.

Host resistance

Host resistance, whereby host factors prevent feeding and maturation of more than small numbers of ticks, has only been fully appreciated for a short time. Resistance by cattle to *Boophilus microplus* is acquired only after exposure to ticks. New cattle lacking previous exposure to ticks are apparently highly susceptible, but even with these cattle the yield is only 20 per cent; as many as 80 per cent of infesting larvae are unable to feed or develop successfully through to the engorged female in this 1-host life-cycle. Exposed cattle fairly rapidly reach their individual level of resistance which is generally high in *Bos indicus* breeds (0–1 per cent yield). In Australia the past decade has seen the demonstration of successful tick control by resistant hosts pass from being a scientific experiment to acceptance and full-scale use by practical farmers. Although *Bos taurus* cattle (European breeds) have a small proportion of resistant individuals and this characteristic is heritable, the majority are highly susceptible. This means that the duration of any breeding programme to enhance tick resistance within a *B. taurus* herd will be too long to be practicable. Fortunately, the frequency distribution of resistant individuals in *B. indicus* cattle (Zebu breeds) is skewed in the opposite way, so that the vast majority are resistant and again this factor is highly heritable. In addition, the gains in live-weight and other performances of *B. taurus*/*B. indicus* crosses show a useful measure of hybrid vigour. These crosses and animals with a greater *B. indicus* content are also able to thrive under harsher conditions (climate and nutritional) than *B. taurus*. These latter factors have been decisive in encouraging the adoption of *B. indicus* and its crosses in Australia. As in tick control by both habitat modification and pasture spelling, the benefits of tick control are often incidental to the other benefits arising from husbandry changes. In northern Brazil the value of

Zebu and Zebu cross cattle in aiding tick control has been well appreciated for many years.

To take full advantage of control by host resistance, an accurate assessment of the resistant status of animals in a herd is essential. This can be obtained by careful counts (see p. 116) of natural ticks infestations that have been allowed to increase to reasonable levels, or by standard artificial infestations. In the latter case counts are made on days 20, 21 and 22 after infestation with *Boophilus* spp. In this way animals in the herd can be ranked and highly susceptible animals culled. If culling is impracticable it is beneficial to graze susceptible animals separately where they can be dipped regularly and the ticks they produce are not affecting the high level of control exercised by the resistant animals.

The timing of the counts is important because resistance is only fully expressed when hosts are in good condition. Under stress of any kind, resistance can be almost completely lost and at these times differences in resistance between hosts may largely disappear. These stress factors can be environmental (cold, starvation, etc.) or physiological (senility, pregnancy, lactation, etc.) or the result of concurrent infections. (*Babesia bovis* infected calves acquire resistance to *Boophilus microplus* more slowly than uninfected controls). At these times acaricidal treatments may be necessary.

Chemical control

Acaricide resistance

Chemical control of ticks has had a relatively long history, starting with the use of arsenic in the early years of this century. There has been continued general progress in the effectiveness of chemicals in killing ticks, their stability in use and safety. The great problem has been the ability of ticks to respond to this chemical selection pressure by normal evolutionary processes to produce resistant strains.

It should be noted that the commonly held belief that poor control management increases the risk of acaricide-resistance development does not agree with current computer analyses. The latter support the view that high selection pressures following efficient use of chemicals increases the risk of selection of resistant strains. It may well be that variable control efficiency under poor management in which high selection pressure is followed by poor control may be a significant factor. Certainly continual poor control, as was seen in one country using an unstable HCH formulation in dips to combat *Rhipicephalus appendiculatus*, did not result in resistance to this tick, although resistance in *Boophilus decoloratus* and *B. microplus* did emerge. Organochlorine insecticides are largely

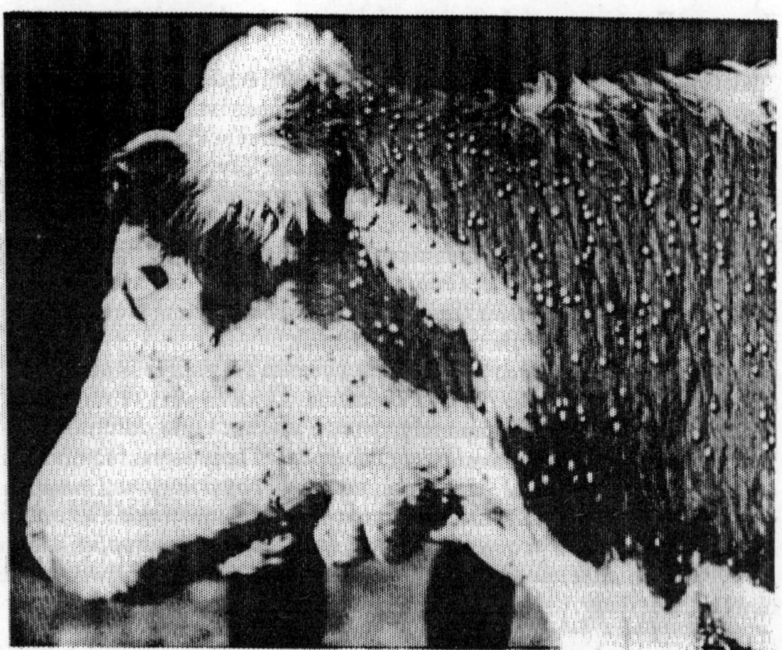

Fig. 7.12 Heavy infestation of the tick, *Boophilus microplus*, on a Hereford cow from a herd that had been treated with the organophosphate compound chlorpyrifos (Dursban) every 10–15 days for more than four months. (Courtesy of the Food and Agriculture Organization of the United Nations.)

outmoded because of residue problems and the high resistance factors in all countries where their use has been long established. The organophorous acaricides are still effective but in some areas (Australia, South America) resistance has developed progressively to most acaricides in some localities within a few years. Country-wide, the distribution of resistance may be variable but the result is to reduce the usefulness of the chemical (Fig. 7.12).

Over a 10 year period in Australia at least 8 acaricide-resistant strains of *B. microplus* evolved with increasingly effective protective mechanisms and the process continues. New chemical groups are continually being developed but this occurs as a result of the search for new insecticides which represents the main financial incentive for the large chemical companies. At one time it was feared that the cost of developing and registering new acaricides was increasing so rapidly that these financial considerations would limit the interest of chemical companies in the relatively small acaricide market which was so prone to resistance problems. This does not appear to have been the case and a useful reserve of very active chemical groups such as the formamidines, cyclic amidines, synthetic pyrethroids etc. is available.

Fig. 7.13 Cattle being lead through an acaricidal dip. Note complete coverage. (Courtesy of the Food and Agriculture Organization of the United Nations.)

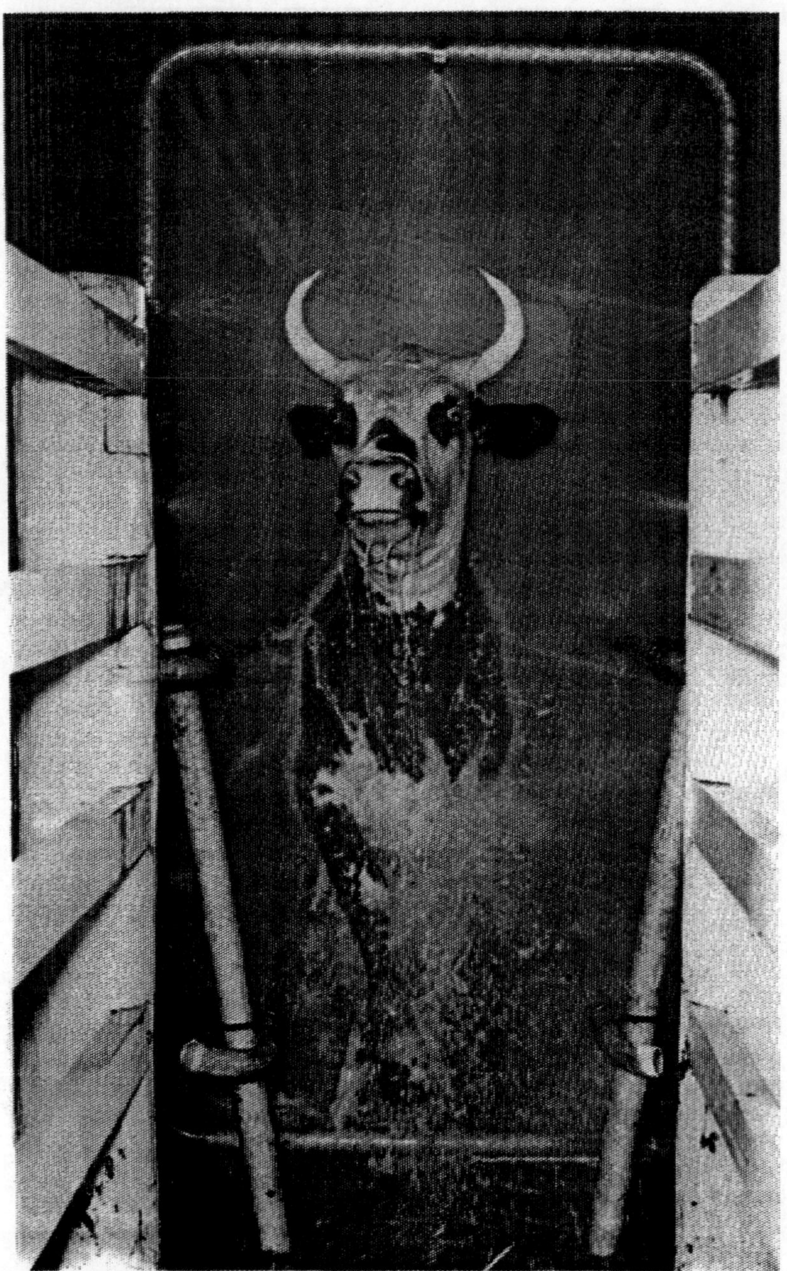

Fig. 7.14 An animal being dosed with an acaricide by the spray-race technique. (Courtesy of the Food and Agriculture Organization of the United Nations.)

However, there can be no complacency as cross-resistance from previously selected strains can pose a threat to even new chemical groups. Thus, Australian DDT-resistant ticks in the field show significant cross-resistance to new highly active and stable synthetic pyrethroids. Artificial selection of this strain by permethrin led to a rapid increase in resistance.

Dipping and spraying techniques

The application of acaricides by dipping or spraying cattle remains the normal method of controlling ticks on livestock. There is no doubt that dipping is much more reliable as mechanical failures cannot occur (Fig. 7.13). Mechanical failures in hand-pump sprayers, motorized sprayers and spray-races are a continual hazard to the farmer who has produced a disease-susceptible herd by good tick control. In the developing countries the use of spray-races (Fig. 7.14) cannot be recommended. Far too often they have been installed and after a period of successful operation have broken down or become prone to failure, with the result that tick-borne disease epizootics have caused heavy economic losses. Admittedly spray-races do have the advantage of allowing more rapid treatment of large herds and the acaricide is always freshly made up. However, adding extra acaricide ('boosting') to overcome preferential loss of the active chemical ('stripping') from the spray fluid becomes necessary. With modern stable dip-chemicals the advantages over a dip tank are only slight and cannot compensate for the only-too-obvious failure rate.

Hand-spraying (Fig. 7.15) although usually regarded as ineffective, can be a useful method of tick control for the conscientious small farmer. Its risks lie in the need to supervise closely the application of the chemicals. Too frequently an operator waves the spray over an animal's back without working the jet closely over the body in a routine sequence to ensure complete wetting. The method is expensive in terms of acaricide usage as recovery of fluid dripping off the animal is usually not possible. The advantage is that a small herd of up to 10 head can be treated without having to walk them through tick-infested countryside to a public dip. Treatment of greater numbers of cattle increase the risk of ineffective application because the effort needed becomes too great. Agricultural hand-sprayers are not suitable without modification as the nozzle produces too fine a spray; complete wetting is essential.

As with most aspects of animal husbandry it is essential to keep good records of acaricide treatments. For the successful operation of a dipping tank a record book is essential. A sample page heading should be as follows: Date; Dip level before dipping; Water Added; Acaricide added; Number of head of cattle dipped; Dip level after dipping; Remarks. Using this form, dip supervisors can pinpoint operational faults and correct them. Good training and good record keeping are prerequisites for a successful dipping operation.

Fig. 7.15 An animal being sprayed with an acaricide by hand in Burma. Application is only as thorough as the operator. Here special attention is paid to the axillae. (Courtesy of the Food and Agriculture Organization of the United Nations.)

To obtain meaningful dip level readings it is obviously essential that a dip be calibrated before being put into use. Surprisingly, this step is often omitted and 'standard' markings or dip sticks used. To measure dip levels it is better to use a dip stick resting by a cross-piece on the side wall of the dip. Markings on dip walls rapidly become obscured by filth and dip sticks resting on the dip bottom tend to rest on accumulating soil and other debris. The dip stick should be marked with a hack-saw blade at 1000 litre intervals up to the Lowest Dip Level and thereafter at 100-litre intervals to the Highest Dip Level. The capacity of the dip should be durably marked on the outside dip wall as well as entered in the record book. These points may appear trivial and too obvious to need mention but practice shows that they are neglected too often.

Successful dip operation demands an effective concentration of active chemical. The effectiveness depends on correct charging and replenishment and also on the degree of contamination of the dip fluid. Cattle bring with them soil, hair, urine and faeces which accumulate in the dip and alter its chemical stability. The seriousness of the problem posed by a badly fouled dip varies according to the dipping regimen. Thus in Australia where dipping can be reduced, by the use of tick resistant cattle, to 2–3 dips a year it is simple to arrange for an analysis of the dip contents and then replenish as needed before each dipping. Under these circumstances dips can last for years without emptying, cleaning and recharging. The concentration of active chemical can be relatively high in Australia because of the 3 week dipping interval and this gives an additional safety margin. In East Africa where weekly dipping is the rule dips become foul rapidly. Chemical concentrations are relatively low because of host toxicity problems with such frequent dipping and the safety margin for killing ticks is thus reduced. In these circumstances degradation can be a problem and frequent changing of the dip contents is advisable; for example, every 6 months or after every 50 000 head dipped.

Alternative control methods

New chemical control methods have been effective in trials but their use is still limited. Thus, plastic ear tags or bands impregnated with an acaricide such as dichlorvos, have given good results against *Amblyomma maculatum* and *A. americanum*. These ticks are seasonal and the 3 month duration of effectiveness of the tag covers the period of adult activity. Control is necessary mainly because of the lesions produced by large number of ticks and these attract screw-worm flies (*Cochliomyia* spp.) which can themselves cause economic loss to the cattle industry. In the absence of serious disease transmission a few ticks can be tolerated. The situation, however, does not apply to *Rhipicephalus appendiculatus* (vector of *Theileria parva*), which also has a strong predilection to attach on the ears, beause a single infected tick can transmit a fatal attack of East Coast Fever.

Good results have also been shown by slow-release devices such as intra-ruminal boluses (Famphur) against *A. americanum* and *A. maculatum*. *Boophilus annulatus* and *B. microplus* were also completely eliminated by the target release of 7 mg/kg/day. The boluses have a life of 50–60 days but obviously problems arise in connection with the regulations concerning acceptable residues. There are also hopes that a new class of anti-helminthics and acaricides (Avermectins) derived from *Streptomyces avermitilis* will prove satisfactory in field trials. The dose rate is low, in the order of only μg/kg.

More novel immunological control measures are also being considered at a number of centres but so far field trials have not been attempted although the preliminary results are encouraging. The more normal immunological approach of attempting to stimulate hosts to reach their characteristic level of immunity (i.e. host resistance) has limitations. Thus, the innate capacity of the host to respond effectively to tick antigens may be limited, particularly with non-adapted breeds. Ticks and tick-borne disease have a definite capacity for immuno-suppression as do a wide range of stress factors. All these points result in an increase of host suspectibility and a loss of control over tick numbers.

The new approach rests on the fact that host immunoglobulins pass freely across the gut epithelium into the haemolymph and are thus distributed to all organs. More or less crude antigens from tick guts, ovaries and other tissues have been used to immunize animals. Success has been achieved in that obvious impairment of tick metabolic processes has been noted. In both *Dermacentor andersoni* and *Amblyomma maculatum* feeding on such immunized hosts has led to only partial engorgement and either no egg production or no viable offspring.

Conclusions on control methods

We can see therefore that although tick control at the moment rests on well-tried if not frankly old-fashioned control techniques, there is every hope that some exciting new approaches will be successful. Success should be measured in terms of reduction in economic cost of treatments or reduction in the risk of acaricide-resistance developing rather than in terms of the dream of eradication. There are areas where eradication is no doubt feasible but elsewhere economic factors should remain the basic for planning a control strategy.

As mentioned earlier tick control in East Africa means the unremitting use of acaricidal treatments on a weekly or twice weekly basis. Elsewhere in the world relatively small numbers of ticks can often be tolerated in order to increase the relative contribution of the control strategy to the overall profit of the livestock operation. Any reduction in acaricidal treatments which does not result in too great damage, reduces costs and hence increases profits. In the past control decisions have been based on the judgement of experienced owners and there has been little possibility of evaluating all the factors involved in these empirical decisions. The factors that need consideration include: (i) tick damage; (ii) management costs and actions, for example mustering (= herding); (iii) pasture spelling and dipping; (iv) biological factors affecting tick population dynamics, for example climate; (v) tick population size; (vi) susceptibility/resistance status of herds; and (vii) the costs associated with the need to minimize the risk of acaricide resistance.

By constructing a tick population computer model it is possible to optimize control regimens for a given season or for a given farm. It is also possible to assess the risks, costs and benefits associated with different schedules. This has been done for south-east Queensland in Australia and its extension to other parts of Queensland is possible. Similar models have proved very useful in the rational control of fascioliasis in Europe, but it will, admittedly, be more difficult to implement such computer technology in many tropical areas.

Computer modelling of this kind can only be useful according to the quality and amount of information available. To this end extensive research programmes are required to obtain data on tick population dynamics, climate, tick-borne disease transmission and, most importantly reliable estimates of the economic losses and gains.

It is clear that chemicals must continue to be the main line for tick control in the foreseeable future but more rational control strategies must support the chemicals to prolong their effective life by minimizing the risk of resistance developing. Computer modelling, use of resistant hosts, artificial stimulation of immunity, and disease control of killing infections such as East Coast Fever, all have a role and should be integrated into a control system which will benefit the livestock owner by reducing costs, and the chemical industry by prolonging the trouble-free life of their chemicals.

Prevention and control of scrub-typhus

M. Nadchatram

In 1948 the newly discovered antibiotic chloramphenicol was shown to be very effective in curing patients infected with scrub-typhus (*Rickettsia tsutsugamushi*). Within the next few years the efficiency of a number of broad-spectrum antibiotics in treating scrub-typhus was clearly demonstrated and it appeared that rickettsial diseases, such as scrub-typhus, would no longer be a problem. Unfortunately this was not so, because the reservoir of the disease remained and people still became infected, though mortality was drastically reduced. In rural peninsular Malaysia the occurence of the disease is regarded as common as malaria.

Transmission of scrub-typhus

Scrub-typhus is a vector-borne disease and a zoonotic disease. The vectors are small mites belonging to the genus and subgenus *Leptotrombidium* (see p. 117), and the transmission of the causative agent, *Rickettsia tsutsugamushi*, is as complicated as is the life-cycle of the mite. Since only the larval stages in the mite's life-cycle are parasitic and as they take a single blood-meal, then the only method by which the disease can be spread is through trans-stadial and transovarian transmission (p. 25). Thus, through transovarian transmission the next generation larvae acquire the infection. *Leptotrombidium* larvae accumulate in small as well as large foci, termed 'mite islands'. The disease is directly related to the distribution of these mite islands in an endemic area, and the scattered distribution of the disease is termed as 'typhus islands'. Hosts such as field rodents and ground-dwelling birds disperse very little and this behaviour maintains or even increases the formation of mite islands. Other hosts, such as migratory birds and larger mammals which disperse further, carry the mite larvae to more distant localities and are responsible for establishing new areas of infection.

General control principles

Because the vectors, termed chiggers, are scattered over vast areas where scrub-typhus is endemic, it is not possible to conduct very effective control programmes. Chemicals can be applied to the ground which substantially reduce the chigger populations for 2 years or more, but their toxicity to man and wildlife has restricted their application on a large scale. Repellents used directly on the body or impregnated in clothing help prevent attacks by chiggers, but they are of limited use because of their offensive odour and other side effects.

Cultural control methods, including removal of ground vegetation to destroy the ecological balance of the environment and killing the animal hosts, may rid areas of chiggers. But burning fields of the grass *Imperata cylindrica* which harbour *Leptotrombidium fletcheri* eliminates the vector only temporarily. This is because the grass's underground rhizomes are unharmed and new growths of grass soon appear and the area once again becomes attractive to rodents and chiggers. Contrary to certain beliefs, the removal of rats does not always result in reducing the incidence of scrub-typhus, because killing natural hosts such as rats may increase the numbers of hungry larvae attacking man.

The most practical control methods are based on acaricides which kill the mites and chemical repellents for personal protection; these methods are outlined below.

Insecticidal control

Residual insecticides such as dieldrin, aldrin or HCH are the most suitable for area control of chiggers. Dieldrin applied at a rate of 28 kg/ha by high-pressure sprayers or 'Swing-fogs' over 2 years can reduce populations of *L. fletcheri* and *L. deliense* between 91–97 per cent. Although effective, such large application rates should give serious concern for environmental contamination. DDT has been used with little success against chiggers. Organophosphate and carbamate insecticides have the advantage that they are less persistent and therefore less harmful to man, domestic animals and wildlife. One of the better known short-acting compounds is malathion. it produces a rapid reduction in the chigger numbers after a single application, but respraying is necessary after 2 weeks for really effective control. However, the most effective and lasting control is achieved if chemicals can be used that have a residual persistence of several months.

Interest has recently increased in the use of systemic acaricides for the control of chiggers and other avian and mammalian ectoparasites. These are toxic compounds which are mixed with food which is then used as baits for rodents. After eating the bait the rodents are not killed but the acaricide passes in minutes amounts into their blood and other body tissues, with the result that ectoparasites feeding on the rodents are killed. Preliminary trials in Malaysia have used dimethoate, an organophosphate compound, as a systemic acaricide to control chiggers in oil palm estates and in 'lalang' fields of the grass *Imperata cylindrica*. Good control was achieved within 3 months following application of a rodent bait consisting of 0.2 per cent dimethoate with a 1:1 mixture. Control of chiggers with systemic acaricides would be effective mainly in certain areas such as parks, plantations and permanent field-work sites.

Repellents

The most convenient and practical means of preventing infection with scrub-typhus is through personal protection by using repellents which deter the parasitic mites from climbing onto a person. Repellents can be applied directly to parts of the body likely to be exposed to chigger bites such as the legs and waist, or impregnated into outer garments of clothing. The three compounds that have been widely used are DMP (dimethyl-phthalate), DBP (dibutyl-phthalate) and BB (benzyl benzoate). All are irritant to the eyes and to tender parts of the body, but this is least noticeable with DBP; whereas benzyl benzoate is sufficiently irritant to limit its use to non-sensitive individuals. Wearing impregnated clothing can give 90 per cent or more protection against the vectors. DMP should be reapplied after each laundering, but DBP remains effective for 5–7 washings and is therefore the most useful repellent. Mixtures of various toluamides, such as diethyltoluamide (DEET), have also proved to be effective. Several brands of insect and mite repellents in aerosol dispensers are readily available in shops and are ready to use.

Environmental methods

Another method used to control chiggers is to clear the ground of vegetative cover by felling and burning trees and using bulldozers to remove low-lying scrub vegetation. Alternatively, herbicides can be sprayed to destroy ground vegetation and possibly more sturdy growth, such as bushes. These measures completely change the habitat and make it unsuitable both for rodents and chiggers; but although pesticide pollution is avoided the drastic destruction of vegetation may in itself not be ecologically acceptable. In any case, such operations are really only feasible on a very limited scale and under speical circumstances such as when a school, clinic, or camp is being constructed. It is essential that such cleared land is maintained free of vegetation otherwise regrowth will occur and rodents will soon reinvade the area and mite populations become re-established.

Control of snail intermediate hosts of schistosomiasis

A. Amin

Schistosomiasis (bilharzia) is one of the more important parasitic diseases of man in the tropics and subtropics. It is caused by infection with parasitic blood flukes (*Schistosoma* spp.) called schistosomes which have certain aquatic and semi-aquatic snails as intermediate hosts.

Snail control is one of the more rapid and effective means available for reducing transmission of schistosomiasis, but its efficiency is enhanced if combined with other methods of control such as chemotherapy, provision of a clean water supply and sanitary facilities. In other words an integrated approach is needed. Different methods of snail control have been tried in various countries with varying success. Most of the approaches can be grouped under the following headings.

Molluscicidal control

Many chemicals have been screened for their molluscicidal activity. The habitats in which chemicals have to be used are variable and several formulations and application methods are needed to meet these conditions. Because of their inhibitive cost it is recommended that molluscicidal applications are restricted to places where schistosomiasis transmission is actually a problem. In other words there should be focal or spot treatment as opposed to wide-scale mollusciciding. This approach requires a thorough knowledge of the distribution of the snails and their seasonality, and the incidence of disease transmission.

One of the earlier and formerly widely used chemicals in snail control is copper sulphate. It is less toxic to fish than most other molluscicides but is ineffective at high (alkaline) pH and is not particularly ovicidal; moreover it is rapidly adsorbed by organic matter and soil. Copper sulphate is effective at a concentration of 30 ppm, but is now rarely used except in the Sudan and Egypt, having been replaced by newer compounds as described below.

Frescon (N-tritylmorpholine)

This is the most active available molluscicide. Large scale trials in several countries have demonstrated its efficiency in controlling snails when sprayed at dosages as low as 0.01–0.1 ppm for 10–30 days; but at these application rates snail eggs are not killed. A disadvantage of Frescon is that it breaks down quickly and loses its activity in acid water, and in turbid water its activity is also reduced. Frescon is available as an emulsifiable concentrate containing 16.5 per cent active ingredient.

Bayluscide (Niclosamide)

Bayluscide is at present the molluscicide being widely used in many countries. It kills snail eggs as well as snails, and is relatively non-toxic to man and therefore safe to use, and moreover remains chemically stable during long-term storage. The chemical is available as a wettable powder containing 70 per cent active ingredient and as an emulsifiable concentrate containing 25 per cent active ingredient. The recommended dosage rates range are 1–2 ppm for about 10 hours.

Bayluscide and Frescon can be applied by a drip-feed technique. This involves applying the molluscicide to the head waters of irrigation canals, small rivers, streams, etc. so that it will be carried by the flow to penetrate the main and subsidiary water courses (c.f. *Simulium* control p. 281). This very simple dispensing equipment is usually sited at narrow or turbulent points along the course of flowing water to ensue complete mixing of the chemical with the water. For stagnant or slowly flowing waters, molluscicides can be sprayed from a sprayer mounted on the back of the operator if relatively small accumulations of water, such as small ponds or drains, are to be treated (Fig. 7.16). If, however, much larger areas like lakes and swamps need spraying then more sophisticated pressurized equipment is probably needed. If long distances have to be covered, tractor-mounted sprayers may be necessary. Aerial applications of Frescon and Bayluscide have been used in the Sudan with great success.

Other molluscicides, including sodium pentachlorophenate (NaPCP) and Yurimin, have also been used with varying degrees of success, as have compounds such as paraquat and Acrolein (Aqualin) which are basically herbicides, but are also molluscicidal.

Fig. 7.16 Spraying the molluscicide Bayluscide (Niclosamide) from a pressurized sprayer on water in an irrigation canal supplying cotton fields in the Sudan. (M.A. Amin.)

Plant molluscicides

A number of plants, especially their fruit, contain chemicals that are poisonous to aquatic snails. For example, in the massive Gezira irrigation scheme in the Sudan the fruit of the plant *Balanites aegyptiaca* was formerly placed in the water to kill snails, while in south-east Asia residues of the fruit of *Thea (Camellia) oleaso* and *Schima avgenta* were found to be effective against *Oncomelania* snails. A plant locally known as

'Damsissa' is now in use in upper Egypt. The only plant molluscicide available in large quantities comes from Endod, *Phytolacca dodecandra*, a common creeper in parts of Africa, of which the active ingredient is a saponin. This compound is used in the Sudan and Ethiopia.

Biological control

Although the possibilities of biological control of snails have been studied in several laboratories as yet there has been little success in practical field control programmes.

Several vertebrates such as fishes, frogs, turtles and birds are natural predators of aquatic and semi-aquatic snails, but most work on assessing the usefullness of predators in schistosomiasis control has been on fishes. In the Sudan the mud fish (*Protopterus aethiopicus*) seems to be an efficient natural predator of snails. In Western Sudan snails were extremely rare in rain water pools containing this fish compared to other habitats where no fish were found. Field studies in Kenya over 16 years showed the value of the snail-eating fish *Astatoreochromis alluaudi* in dams in destroying *Biomphalaria* species. Several other animals prey on snails or their food. *Tilapia zillii* is a weed-eating fish that has been of some use in East Africa in destroying aquatic weeds and thus making habitats less suitable for snails. The Chinese grass carp and the South American snail, *Marisa cornuarietis*, are voracious feeders on vegetation which constitutes the food of schistosomiasis snails and by acting as competitors they may reduce available food sufficiently to cause a reduction in snail populations. In Puerto Rico after introducing *M. cornuarietis* in certain habitats, it has eradicated *B. glabrata* from them. It competed with *B. glabrata* for food and at the same time acted as an incidental predator of snail eggs and newly hatched minute snails. *Marisa* snails have virtually eradicated *B. glabrata* from the Viegues Islands near Puerto Rico. This competitor has been used routinely in Puerto Rico in an integrated snail control programme, which includes molluscicides and environmental sanitation comprising land drainage and keeping canals and streams free of weeds and silt.

Experiments have also been conducted on other potential competitors of medically important snails, such as snails of the genera *Tarebia*, *Helisoma* and *Pomacea* with promising results. For example, field studies in both Puerto Rico and Egypt showed that species of *Helisoma* and *Biomphalaria* do not co-exist for long. It seems that a harmful secretion produced by *Helisoma* causes a reduction of the *Biomphalaria* populations.

Although predators and competitors appear to have considerable potential for the control of medically important snails they have as yet not been widely used, probably because they are not so easy to use as molluscicides. While it is envisaged that biological control will at most comprise a relatively minor role in an integrated schistosomiasis control programme, there is nevertheless need for carefully designed field trials to evaluate its full potential.

Sanitation and environmental control

Schistosomiasis transmission can be broken by people avoiding contact with water likely to contain infective snails. In practice, this obvious solution is difficult to implement because it is very difficult to prevent people bathing in dangerous waters. Moreover, it is impossible to persuade people in hot countries habitually coming into contact with infected waters, such as workers in rice fields, to wear protective clothing. The spread of schistosomiasis could be curtailed by the provision of a piped water supply, improved sanitation and hygiene education and persuading people to use latrines, or at least avoid urinating or defecating in water.

Another approach is to change the environment to make it unsuitable for snail breeding, for example by draining or filling in marshy areas, alternately flooding and drying irrigation canals. In the Philippines engineering measures and improved agricultural practices have led to the removal of weeds from irrigation ditches, and the filling in of ponds and in consequence schistosomiasis transmission has been reduced. Similarly in China, land reclamation and proper management of water levels in irrigation schemes together with various ecological methods have helped reduce transmission in some areas.

Bulinus and *Biomphalaria* snails survive and breed only if the water flow is less than 0.3 m/sec, so if irrigation canals are lined with concrete and maintained free of vegetation and silt, water flow can be increased so as to prevent colonization. This simple environmental approach, of course, cannot be used to prevent breeding in dams, ponds, lakes and reservoirs.

The eradication of *Bulinus* and *Biomphalaria* snails from a habitat is extremely difficult because they are hermaphrodite, consequently a single survivor can give rise to a new population. In contrast, in south-east Asia *Oncomelania* species, the intermediate hosts of *Schistosoma japonicum*, are dioecious (i.e. the sexes are separate) and also have a slower reproductive rate. Thus they should be easier to eradicate, but they are amphibious and have a protective operculum covering their openings and are therefore more difficult to kill with molluscicides.

One of the major difficulties with environmental control methods is that they are often difficult and costly to implement, take a relatively long time to become effective, and their upkeep is usually expensive and requires considerable effort. Canals and ditches blocked with weeds and silt are commonly encountered on irrigation schemes and show that even relatively simple environmental measures are not being maintained.

In many situations neither preventive measures nor environmental or biological control methods have proved very successful in reducing schistosomiasis transmission. Chemotherapy and molluscicides therefore form the basis for schistosomiasis control in many areas.

Recommended reading

General

Anon. (1979) 'Safe use of pesticides'. Third Report of the WHO Expert Committee on Vector Biology and Control, *World Health Organization, Technical Report Service*, No. **634**, 44 pp.

Brown, A.W.A. (1978) *Ecology of Pesticides*, John Wiley, New York, 525 pp.

Clements, A.N. (1979) 'Integrated control and pest management'. *SPAN*, **22**, 23–5.

World Health Organization (1976) 'Resistance of vectors and reservoirs of disease to pesticides'. *World Health Organization, Technical Report Series*, No. **585**, 88 pp.

World Health Organization (1978) 'Chemistry and specifications for pesticides'. *World Health Organization, Technical Report Series*, No. **620**, 36 pp.

World Health Organization (1980) 'Environmental management for vector control'. *World Health Organization, Technical Report Series*, No. **649**, 63 pp.

Mosquitoes

Anon. (1973) *Mosquito Control. Perspectives for Developing Countries,* National Academy of Science, Washington DC, 62 pp.

Service, M.W. (1980) *A Guide to Medical Entomology*, Macmillan, London, 226 pp.

World Health Organization (1973) *Manual on Larval Control Operations in Malaria Control Programmes*, World Health Organization, Geneva, Offset Publ. No. **1**, 199 pp.

World Health Organization (1974) *Equipment for Vector Control*, World Health Organization, Geneva, 179 pp.

Blackflies

Crosskey, R.W. (1973) 'Simuliidae (Black-flies, German Kriebelmucken'). In K.G.V. Smith (ed.) *Insects and Other Arthropods of Medical Importance*, British Museum (Natural History), London, pp. 109–53.

Garms, R., Walsh, J.F. and Davies, J.B. (1979) 'Studies on the reinvasion of the Onchocerciasis Control Programme in the Volta River Basin by *Simulium damnosum* s.l. with emphasis on the south-western areas'. *Tropenmedizin und Parasitologie*, **30**, 345–62.

Jamnback, H. (1973) 'Recent developments in control of blackflies'. *Annual Review of Entomology*, **18**, 281–304.

Laird, M. (1981) *Blackflies. The Future for Biological Methods in Integrated Control,* Academic Press, London, 399 pp.

McMahon, J.P., Highton, R.B. and Goiny, H. (1958) 'The eradication of *Simulium neavei* from Kenya'. *Bulletin of the World Health Organization*, **19**. 75–107.

Walsh, J.F., Davies, J.B. and Le Berre, R. (1979) 'Entomological aspects of the first five years of the Onchocerciasis Control Programme in the Volta River Basin'. *Tropenmedizin und Parasitologie*, **30**, 328–44.

World Health Organization (1974) 'Research and control of onchocerciasis in the western hemisphere'. *Pan American Health Organization, Scientific Publication*, No. **298**, 154 pp.

Tsetse flies

Anon. (1977) 'Insecticides and application equipment for tsetse control'. *FAO Animal Production and Health Paper*, No. **3**, FAO, Rome, 72 pp.

Anon. (1979) 'The African trypanosomiasis'. *Report of a Joint WHO. Expert Committee and FAO Expert Consultation*, FAO, Rome, 96 pp.

Anon. (1980) 'Republic of Zambia Ministry of Lands and Agriculture Annual Report for the Department of Veterinary and Tsetse Control Services for the year 1977'. Government Printers, Lusaka, 43 pp.

Davies, H. (1977) *Tsetse Flies in Nigeria*, Oxford University Press, Ibadan, 340 pp.

Koeman, J.H. and Takken, W. (1977) 'The environmental impact of tsetse control operations – a report on present knowledge'. *FAO Animal Production and Health Paper*, No. **7**, FAO, Rome, 44 pp.

Laird, M. (ed.) (1977) *Tsetse. The Future for Biological Methods in Integrated Control*, International Development Research Centre, **077e**, Ottawa, 220 pp.

Mulligan, H.W. (ed.) (1970) *The African Trypanosomiasis*, Allen and Unwin, London, 850 pp.

Swynnerton, C.F.M. (1936)The tsetse flies of East Africa. A first study of their ecology, with a view to their control'. *Transactions of the Royal Entomological Society of London,* **84**, 1–579.

Ver Boom, W.C. (1965) 'The use of aerial photographs for vegetation surveys in relation with tsetse control, and grassland surveys in Zambia'. *Report International Training Centre for Aerial Survey*, Delft, Series B, No. **28**, 1–20.

Triatomine bugs

Forattini, O.P. Santos, J.L.F., Ferreira, O.A., Rocha E. Silva, E.O. da and Rabello, E.X. (1979) 'Aspectos ecólogicos da tripanossomiase Americana. XVI - Dispersão e ciclos anuais de colônias de *Triatoma sordida* e de *Panstrongylus megistus* espontaneamente desenvolvidas em ecotopos artificiais'. *Revista de Saúde Pública*, Sao Paulo, **13**, 299–313.

Rocha E. Silva, E.O. da, (1979) 'Profilaxia'. In Z. Brener, and Z.A. Andrade (eds) *Trypanosoma cruzi e Doenca de Chagas.* Editora Guanabara Koogan, Rio de Janeiro, 425–49.

Ticks

Food and Agriculture Organization (in press) *Practical Field Manual for the Control of Ticks and Tick-borne Diseases*, FAO, Rome.

Scrub-typhus mites

Audy, J.R. (1949) 'Practical notes on scrub-typhus in the field'. *Journal Royal Army Medical Corps*, **93**, 273–88.

Dohany, A.L. Huxsoll, D.L., Saunders, J.P. and Phang, O.W. (1980) 'The efficacy of dimethoate as a synthetic acaricide for the control of chiggers (Acarina:Trombiculidae)'. *Journal of Medical Entomology*, **17**, 30–4.

Harrison, J.L. (1956) 'The effect of withdrawal of the host on populations of trombiculid mites'. *Bulletin Raffles Museum*, Singapore, **28**, 112–9.

Harrison, J.L. (1956) 'The effect of grassfires on populations of trombiculid mites'. *Bulletin Raffles Museum*, Singapore, **28**, 102–11.

Kulkarni, S.M. (1977) 'Laboratory evaluation of some repellents against larval trombiculid mites'. *Journal of Medical Entomology*, **14**, 64–70.

Traub, R. and Wisseman, C.L. (1968) 'Ecological considerations in scrub-typhus. 3 Methods of area control'. *Bulletin World Health Organization*, **39**, 231–7.

Snails

Ansari, N. (ed) (1973) *Epidemiology and Control of Schistosomiasis (Bilharziasis)*, S. Karger, Basel, 752 pp.

Jordan, P. and Webbe, G. (1969) *Human Schistosomiasis*, William Heinemann Medical Books, London, 212 pp.

Lemma, A. (1970) 'Laboratory and field evaluation of the molluscicidal properties of *Phytolacca dodecandra*'. *Bulletin of the World Health Organization*, **42**, 597–612.

McCullough, F.S., Gayral, Ph., Duncan, J. and Christie, J.D. (1980) 'Molluscicides in schistosomiasis control'. *Bulletin of the World Health Organization*, **58**, 681–9.

Miller, M.J. (ed) (1972) *Symposium on the Future of Schistosomiasis Control*, Tulane University, New Orleans, 135 pp.

World Health Organization (1980) 'Epidemiology and control of schistosomiasis'. *World Health Organization, Technical Report Series*, No. **643**, 63 pp.

Wright, W.H. (1972) 'A consideration of the economic impact of schistosomiasis'. *Bulletin of the World Health Organization*, **47**, 559–66.

8 Relationships between development programmes and health

Starting some 10 000 years ago when man discovered the benefits of farming and animal husbandry and began to give up his nomadic existence, he has increasingly altered the environment to suit his needs. Initially this involved relatively minor alterations, such as clearing patches of land for crops, a closer association with animals, and a communal life-style leading to pockets of relatively high population density. These changes resulted in the acquisition of new diseases and parasites, several of which originated from animals and developed firstly as zoonoses and then became modified so that they infected only man. With time, environmental changes became more drastic culminating in urbanization and deforestation of large areas for cultivation, but the rates of these changes varied; generally they took place more slowly in tropical countries. More recently, however, large and ambitious development programmes, such as the construction of large artificial lakes and dams and massive irrigation schemes for crop production, have been rapidly spreading in the tropics. These far-reaching ecological changes have frequently altered disease patterns by either intensifying exposure to existing pathogens and parasites or by introducing new diseases and vectors.

Some of the health problems and reprocussions created by development programmes are discussed in this chapter.

Urbanization

Although in many developing countries towns are expanding and becoming overcrowded, more than 70 per cent of the population is still rural and 'lives off the land'. Population density varies from about 5 people/km^2 in Zaire to as many as 390/km^2 in Java, and 960/km^2 in Bangladesh. In many tropical countries about 40–50 per cent of the population is under 15 years old and only 3–5 per cent over 65, whereas in developed countries these figures are about 25 per cent and 10 per cent respectively. This difference is due to high mortality rates in early life in tropical countries, so that as many as about half of the people may die before reaching 20 years. In absolute terms the rural population is increasing the most, but the *rate* of increase is usually greatest in towns. In most African countries the population is growing at the rate of 2–3 per cent a year, that is they are doubling their population every 27 years, but this is an uneven expansion. In urban areas, which contain about 21 per cent of the people, the population is more likely to double every 15 years, and in semi-urban shanty towns every 4–6 years! In India the urban populations comprises about a quarter of the country's total population of 630 million (1979 census).

The escalating number of urban people is due partly to high birth rates, but mainly to immigration from rural areas where lack of economic opportunities drive people to seek their fortune in the towns. In India 64 per cent of the people living in the tremendously overcrowded cities of Bombay and Delhi are migrants. The population of Karachi, Pakistan, increased enormously from 387 000 in 1941 to about 3.8 million in 1968 mainly due to immigration, and this capital city is still expanding! Water is piped to the city from a source about 160 km away across semi-desert. Because of constantly leaking pipes, effluents from sewage systems and poor sanitation, numerous collections of water form which are colonized by *Anopheles stephensi*, an important urban vector of malaria. In the late 1960s an estimated 20 per cent of the population in Karachi had malaria. In Kinshasa, Zaire, the urban population has also exploded, in this instance from 126 000 people in 1948 to 1.3 million in 1975. Formerly malaria was under good control and the larval habitats of the *An. gambiae* complex were not allowed to persist but after 1960, control operations deteriorated and during the continual expansion of the city, water was allowed to accumulate all over the city. This led to greatly increased populations of the *An. gambiae* complex and in 1973 it was estimated that in certain parts of Kinshasa some 25 per cent of the population had malaria.

Not only is there migration within countries to the cities but also between countries. Large numbers of refugees from Ethiopia, Vietnam and Kampuchea, for example, have poured into the cities of neighbouring countries which are already grossly over-populated. The exponential growth of urban and semi-urban populations causes enormous health and social problems, including malnutrition and increases in communicable diseases such as respiratory infections, particularly tuberculosis, gastro-enteritis, measles and infectious hepatitis. In urban areas of some South American countries, diarrhoeal diseases are responsible for more than 90 per cent of the deaths of children under 5 years old.

Lack of adequate sewage disposal systems in densely crowded urban areas results in the accumulation of waters polluted with human and animal excreta, piles of decaying refuse (Fig. 8.1), and defective latrines and septic tanks, all of which provide ideal habitats for *Culex quinquefasciatus*

Fig. 8.1 A ditch containing water polluted with decaying vegetable debris, animal faeces and other rubbish, providing ideal conditions for larvae of the mosquito *Culex quinquefasciatus*, an important urban vector of bancroftian filariasis. (M.W. Service.)

(= *C. fatigans*) the principal urban vector of bancroftian filariasis. Populations of this mosquito have increased enormously in many towns, and have become established in others from which it was previously unknown. There has consequently been dramatic increases in the urban transmission of bancroftian filariasis. The cities of Bangalore and Hyderabad in India for example, were free of filariasis transmission until the early 1960s, when greatly increased human populations accompanied by inadequate sanitation and sewage facilities, resulted in an invasion of *C. quinquefasciatus* and filariasis then became a problem. Similarly, in both East and West Africa urbanization has caused this vector to become the dominant man-biting mosquito in many towns, and whereas formerly bancroftian filariasis was mainly a rural disease transmitted by the *Anopheles gambiae* complex, it is now becoming an important urban disease. Increased population densities in towns has also resulted in greater communication between them and neighbouring villages, and this had led to filariasis spreading centrifugally from urban to rural areas.

In the absence of an adequately piped water supply, domestic water is stored in pots, overhead tanks and cisterns in the middle of towns; in addition rain accumulates in the litter of assorted tin cans, tyres (Fig. 8.2)

Fig. 8.2 A tyre dump in India. Rain water accumulates in the tyres and they become important larval habitats of the mosquito, *Aedes aegypti*. (Courtesy of the Director, National Institute of Virology, Pune, India.)

and other discarded receptacles so commonly found in the slums of urban and semi-urban communities. All these collections are ideal habitats for *Aedes aegypti*, the main yellow fever vector in Africa and the tropical Americas, and of dengue over much of the tropics. Increased *Ae. aegypti* populations are most probably responsible for the increased numbers of dengue outbreaks that are occurring in Central and South America and in the Caribbean, and for the spread in Asia of a fatal form of dengue called haemorrahagic dengue, which was first recognized in 1953 and is associated mainly with urban populations. In many areas of the Indian subcontinent, water-storage tanks, cisterns and wells in towns also provide the only larval habitats of *Anopheles stephensi*, the most important urban vector of malaria. In Calcutta, Bombay, and other over-populated Indian cities, about half the population may be sleeping out of doors during the hot dry season, thus increasing their exposure to mosquito bites and chances of becoming infected with malaria, filariasis or dengue.

Another problem sometimes encountered is the increase of urban transmitted schistosomiasis, brought about by snails establishing themselves in drains and borrow pits, but this disease is nevertheless more common in rural areas.

Population predictions are notoriously inaccurate and are frequently adjusted to suit the particular theories of the proposer, but it seems reasonable to expect that by the year 2000 there will be some 6200 million people in the world and 80 per cent of us will live in the developing countries, and many of us will likely still be living in poverty. People will continue to flock to the towns – Mexico City will probably become the world's largest capital, with an estimated 31 million people in the year 2000. Because of the inability of towns to cope with the great influx of migrant people it is difficult to envisage that in the foreseeable future there will be any worthwhile decrease of vector-borne disease associated with urbanization; a gloomy but realistic prediction.

Roads, mining and construction camps

The 'opening up of Africa' and resultant increased travel of people between villages and different areas, most likely contributed to the devastating epidemics of sleeping sickness that suddenly flared up at the beginning of this century. More recently (1967–70) epidemics of trypanosomiasis have occurred in Ethiopia and population movements may result in the disease invading the Blue Nile and Omo Valleys.

The construction of roads frequently creates roadside excavations which fill up with rain water and constitute suitable aquatic habitats for various mosquito species and snail intermediate hosts of schistosomiasis. Roads across the Trans-Amazonian basin, the Sahara, and other vast areas are also bridging natural barriers and facilitating the spread of certain diseases among rural populations which become more exposed to infections from outside. When roads are cutting through sparsely inhabited or uninhabited terrain, semi-permanent road camps have to be established to house the construction workers and their families and this itinerant population may be exposed to a variety of diseases. For example, workers constructing roads through the jungles of South America may be bitten by mosquitoes infected with yellow fever derived from the jungle monkey populations. Similarly, in South America leishmaniasis is a zoonotic disease transmitted among forest rodents and marsupials by minute phlebotomid sandflies, but man is likely to get bitten by infected sandflies when he works or lives in the jungle. Another South American zoonotic disease is Chagas' (see p. 106) which is transmitted by blood-sucking triatomine bugs, many of which normally feed on forest rodents and live in animal burrows, but which are likely to invade houses and bite man when he enters the area.

Construction of the Trans-Amazonian highway which runs east to west across previously uninhabited jungle of Brazil started in 1970, and a government plan for colonization of cleared areas alongside the road commenced in 1971. By the end of 1974 there were some 28 000 colonists and their families living in new settlements. Medical surveys of those attending hospitals or clinics showed that arbovirus infection rates had risen from 2.6 per cent in early 1973 to 18 per cent by the beginning of 1974. After 6 months of colonization a 'new' illness termed 'Haemorrhagic Syndrome of Atamira' (HSA) appeared, and three people died. No infective organisms were involved but patients suffered from localized or disseminated cutaneous and mucosal bleeding, thought to be caused by the enormous numbers of bites they received from simuliid blackflies. In South America leishmaniasis is regarded as an occupational disease of forest workers engaged in clearing forests for road building, timber, agricultural or mining purposes, and although no human cases were recognized up to 1974, the abundance of animal reservoirs in the area and large numbers of sandfly vectors make it likely that this zoonotic disease will eventually infect the colonists. Similarly, schistosomiasis has as yet not been actively transmitted, although there are snail hosts breeding in the area and 310 imported cases of the disease were recorded during 1971–74. Malaria, however, is the worst threat, and the malaria incidence appears to have increased slightly from 1971–74 to reach a prevalence rate of about 20 per cent.

The exploitation of mineral deposits in mainly uninhabited areas may also expose workers to vector-borne infections. In fact, many of the workers extracting iron ore from the Brazilian jungle for one large commercial company contracted leishmaniasis; records show that for every kilometre of road constructed into the interior there was a new case of leishmaniasis. In India coal mining and other mining projects, especially those involving open cast mining, have created numerous larval habitats for *Anopheles* vectors and as a consequence there have often been outbreaks of malaria among the workers and their families. Even small-scale operations such as gem collection has sometimes caused disease transmission, for example, local epidemics of malaria in Sri Lanka.

Economic development during the 1950s involving forest clearance and increased numbers of cattle grazing at the edge of forests in Karnalaka (formerly Mysore) State in India was probably one of the main reasons for the upsurge of tick populations and the sudden appearance, in 1957, in man of a new zoonotic tick-borne virus disease called Kyasanur Forest Disease. In recent years this disease has appeared elsewhere in India, due it is thought to ecological changes in habitat structure brought about by man's activities. Similarly, in Asia destruction of scrub and forests for extended cultivation has drastically altered the ecology of large areas. This has allowed enormous increases of certain rodents and their parasitic leptotrombiculid mites, several of which transmit the rickettsial disease, scrub-typhus, from rodent reservoirs to man. Scrub clearance has therefore increased the risk of rural people becoming infected with scrub-typhus.

Irrigation of crops

About 30 per cent of the world's land surface is classified as arid and about 14 per cent of the world's population, i.e. some 600 million people, strive to make a living in this inhospitable environment. Large-scale irrigation schemes are a natural product of agricultural development policies aiming to improve people's nutrition and prosperity. In 1975 the total world hecterage under irrigation was 223 million, representing about 13 per cent of available arable land, but this is expected to rise to 273 million hectares by 1990. A variety of crops may be irrigated such as cotton, sugar, vegetables, citrus fruits and date palms, but the most commonly irrigated crop is rice which results in by far the greatest accumulation of standing water for vector mosquitoes and snail hosts of schistosomiasis (Fig. 8.3). Outbreaks of malaria due to rice cultivation, giving rise to the name 'irrigation malaria', have been known for nearly a hundred years, but while it may not be a new problem it is nevertheless an increasingly important one.

In Africa extension of rice cultivation has led to greatly increased populations of the *Anopheles gambiae* complex. About 65 per cent of the

Fig. 8.3 The proliferation of irrigation schemes for rice, and for other crops, has in many parts of the world led to large increases in the incidence of mosquito vectors of malaria, filariasis and a variety of arboviruses, as well as snail intermediate hosts of schistosomiasis. (M.W. Service.)

man-biting mosquitoes in a village on the Ahero irrigation project in western Kenya belonged to the *An. gambiae* complex, and some 70 times more were biting man than in a nearby village not affected by irrigation. In contrast in the irrigated village *Mansonia* mosquitoes formed only 30 per cent of the mosquitoes biting man whereas in the non-irrigated village they comprised 98 per cent of the man-biting mosquitoes. In many other areas, ranging from South America to south-east Asia, irrigation has provided a prolific source of mosquitoes which not only transmit malaria but filariasis and various arboviruses. In Tanzania and Uganda large populations of *An. gambiae* breeding in rice fields have resulted in a much higher infection rate of bancroftian filariasis in people living in villages near the irrigation fields than in those living in villages further away and where vector populations are small.

In the Indian subcontinent through to south-east Asia and Japan, larvae of *Culex tritaeniorhynchus*, an important vector of Japanese encephalitis, are found extensively in rice fields. Not surprisingly therefore there has been a marked increase in many areas in the transmission of this disease. In Sarawak, for instance vector populations in the paddy fields reach a peak at the end of each year and this is accompanied by increased infections of Japanese encephalitis in pigs which act as important amplifying hosts. Up to about 1960 the only important malaria vectors in the semi-desert like area of the Kunduz Valley in northern Afghanistan was *Anopheles superpictus* which bred in small seepages of clean water in the foothills. Irrigation then changed the ecology of the area. Not only were large-scale agricultural schemes implemented such as the planting of rice, cotton and melons, but there were increases in populations of both man and cattle, and in addition to water in rice fields, standing water accumulated in seepages, irrigation ditches and numerous pools. These new larval habitats favoured the increase of two important malaria vectors. *An. pulcherrimus* and *An. hyrcanus*, and as a consequence malaria transmission quadrupled in some villages. In Venezuela populations of malaria vectors such as *An. pseudopunctipennis* and *An. albimanus* increased markedly each year during about May and June when the rice fields are flooded.

It is worth remembering that the sudden outbreaks of malaria following dam constructions in 1921 and 1927 in the Tennessee Valley of the USA were controlled by environmental manipulation. Water levels in the man-made lakes were regularly fluctuated and all weed was removed from the edges, actions which were detrimental to the breeding of the local *Anopheles* vectors.

In addition to providing mosquito larval habitats, irrigation raises the relative humidity and this may increase the survival rate of adult mosquitoes resting among vegetation, and thus enhance their probability of transmitting infections.

Irrigation not only causes increased mosquito populations but may be accompanied by more widespread ecological changes. For example, there is usually a substantial increase in aquatic birds, such as herons, storks, ibises and egrets, several of which are important reservoirs of arboviruses and therefore increase the likelihood of viral transmission among people working or living near irrigated crops. In the Wada Halfa area adjoining Lake Nubia there have been such serious explosive outbreaks of non-biting chironomid flies of the genus *Tanytarsus* that local people developed severe allergies to these insects, and as a consequence an asthma clinic had to be established. Accumulation of food crops, such as rice, sugarcane, maize and other cereals on irrigation schemes, and pockets of high population densities, encourage the build-up of rodents such as domestic rats (*Rattus rattus*) and *Praomys* (= *Mastomys*) *natalensis*, and this increases the risk of transmission of zoonotic infections such as leptospirosis, Lassa fever, plague, and various rickettsial diseases. There are of course other problems, such as loss of water. In Egypt, Lake Nubia has an annual evaporation loss of 2500 mm which leads to concentration of salts in the soil which if allowed to continue can pose agronomic problems. Waterlogging and degradation of the land can become a serious problem if due care is not taken. Many irrigation schemes are operating at less than 50 per cent efficiency, partly because of deterioration of maintenance and poor management; for example, more than 70 per cent of the 30 million hectares of irrigated land in Egypt, Iraq, Iran and Pakistan is affected by salinization and waterlogging.

Irrigation schemes often entail the resettlement into newly constructed villages of people who come from different areas and have dissimilar social structures and customs. Because of the diversity of their backgrounds there is frequently little community cohesion or spirit. Moreover, as the people do not own, but rent, their houses these often become delapidated and the villages tend to develop into shanty towns with the usual profusion of scattered litter, refuse, broken latrines and accumulations of polluted waters. This is then usually accompanied by large populations of *Culex quinquefasciatus* (= *C. fatigans*) breeding in the resettlement areas. This deterioration in sanitation, especially in the densely populated communities typical of many irrigation schemes, often results in the contamination of the land with faeces, thus enhancing the breeding of flies and the spread of a variety of enteric diseases. Resettlement may pose even more problems. Some immigrants are likely to be infected with malaria, filariasis or other mosquito-borne diseases, and due to the enormous populations of mosquitoes in the irrigated areas these diseases will soon spread through the community. Other people, however, may have come from areas where there is little malaria and will consequently have acquired little protective immunity and are therefore likely to suffer more than others if they become infected.

Another major disease associated with irrigation, especially of rice, is schistosomiasis. The snail intermediate hosts of this disease breed in both irrigated fields and the irrigation ditches and canals supplying them with water. In the massive Gezira irrigation scheme in the Sudan some 600 000 people are infected with schistosomiasis; on the Mwea irrigation scheme in Kenya 60 per cent of the children are now infected whereas in 1956 before the scheme started no cases were recorded. In Tanzania a survey in 1972 showed that 84 per cent of sugarcane cutters and irrigation workers have schistosomiasis. In Egypt farming in arid areas previously relied on annual flooding of the Nile Valley. This temporary inundation did not permit the permanent establishment of aquatic snails, whereas the now perennial irrigation of arid areas has resulted in the schistosomiasis rate rising from 5 per cent to 80 per cent in these areas, and about half the people of Egypt now have schistosomiasis.

Replacing earthen irrigation ditches and canals with concrete ones and maintaining them free of silt and weeds, and increasing water flow should make them unsuitable snail habitats, but in practice such high levels of sanitation are rarely maintained for long. Moreover, flowing water often constitutes suitable larval habitats for simuliid blackflies, certain species of which in Africa and the tropical Americas are vectors of river blindness (onchocerciasis), but this is a disease more usually associated with dams and spillways of hydro-electric schemes (see p. 338) than crop irrigation. It is more difficult to prevent snails breeding in the standing water of irrigated rice fields and water reservoirs than in irrigation ditches. Providing irrigation workers and their families with concrete paddling and swimming pools, a piped water supply and showers have sometimes reduced schistosomiasis transmission, but mainly among those not actually working in the irrigated fields.

It will be clear from the preceeding account that while crop irrigation increases food supply and hopefully improves living standards, at the same time it increases the transmission of a number of diseases. Moreover, problems may be aggravated by the frequent and widespread application of insecticides to combat agricultural pests. For example, it was observed in Kenya that some weeks after the application of organophosphate insecticides to control rice stemborer, populations of the *Anopheles gambiae* complex in the rice fields were substantially bigger than before spraying. The explanation was that whereas initially the insecticide killed most aquatic life, including mosquito larvae, the *An. gambiae* complex was able to recolonize the sprayed fields quicker than its predators. Consequently in the absence of predators which normally helped regulate population size mosquito populations increased explosively (see p. 40).

As already discussed (see p. 35) agricultural spraying of cotton and rice has several times caused the selection of insecticide resistance in populations of important disease vectors. At Hola in Kenya, *Culex quinquefasciatus* (= *C. fatigans*) breeding in polluted waters of the Bura irrigation settlements, is already highly resistant to DDT while *An. arabiensis* in the rice fields seems to be developing resistance, presumably as a result of cotton spraying.

As emphasized elsewhere in this book there is clearly an urgent need for closer collaboration between different disciplines, in this instance agriculturalists, economists and hygienists, to prevent the emergence of vector-borne diseases and other health problems on irrigated projects. A more comprehensive integrated approach to pest and vector management is required. In China, where there are about 36 million hectares under rice cultivation, the importance of developing pest management strategies which rely whenever possible on cultural and biological methods and minimize the use of pesticides, has been realized and put into practice. Insecticides, for example, are not used as routine measures against *possible* losses (i.e. insurance spraying), but only when pest population densities are approaching levels likely to cause economic losses.

Other agricultural practices

Other examples of man-made changes to the environment having a potentially adverse effect on human health include the policy in the 1970s of clearing forests in the Enugu region of Nigeria for planting the valuable timber trees teak and gmelina. Cutting down and removing the forest trees involved setting up forest camps for the workers and also resulted in new

Fig. 8.4 An example of a tree in Trinidad covered with bromeliad epiphytes. Water accumulates in the bases of the leaf axils and provides breeding places for *Anopheles bellator*, an important malaria vector. (M.W. Service.)

settlements becoming established at the edge of the forest; in addition people moved in to farm the newly cleared land. These activities greatly increased the people's exposure to bites of forest-breeding mosquitoes. This problem was aggravated because rot holes soon developed in tree stumps left behind after clearing operations and these filled with water and formed ideal larval habitats for *Aedes africanus*. This is a tree-hole mosquito that is known to transmit yellow fever among forest monkeys and in this area also to bite man readily. Thus development of the timber industry increased the risk of yellow fever transmission in the area, fortunately no human cases were recorded.

In Trinidad the principal vector of malaria is *Anopheles bellator*, a mosquito that breeds in the water-filled axils of bromeliads, plants that are epiphytes of a variety of trees (Fig. 8.4). During the 1940s it was the practice on the island to plant *Erythrina micropteryx* (which reaches a height of 20–25 m) as a shade tree on cocoa plantations. This tree is very susceptible to colonization by certain bromeliad species, with the result that within the cocoa ecosystem there were many more bromeliads than were present in the original forest before it was cleared for cocoa growing. Here then is another instance of environmental changes for agricultural purposes having unforeseen consequences, namely increasing the breeding of a mosquito vector.

Dams and man-made lakes

Some of the most spectacular, most complex and largest development projects in tropical countries have been the construction of dams, for instance in Angola, Zaire, Kampuchea, Egypt, Ghana, India, Ivory Coast, Mozambique, Nigeria, Pakistan, Zimbabwe, Senegal and Surinam, to supply hydro-electric power, and water for towns or for crops (Fig. 8.5). Ghana has the world's largest man-made lake, the Volta Lake, which is 8750 km^2; Lake Kainji in Nigeria is 1130 km^2; Lake Kariba on the borders of Zimbabwe and Zambia is 5250 km^2; and Lake Nasser in Egypt and the Sudan is about 7000 km^2. These massive artificial collections of water have created many problems, including reduced soil fertility due to build up of fertile silt behind the dams, deltaic soil erosion, high salinities, loss of water by evaporation, profound changes that have affected fisheries, as well as social and health problems. For example, the construction of the Lake Volta Dam displaced 80 000 people, Lake Nasser, 120 000 and the Kainji about 50 000 people. Dams provide important breeding sources for malaria vectors, such as the *Anopheles gambiae* complex in Africa, *An. hyrcanus* in Asia and *An. darlingi* in South America. Furthermore, if the waters become overgrown with rooted or floating vegetation they become colonized by *Mansonia* mosquitoes, which in Asia are important vectors of brugian filariasis (caused by the parasites *Brugia malayi*). Water

Fig. 8.5 Lake Volta in Ghana, the largest man-made lake. Both natural and man-made lakes have in many areas favoured the breeding of aquatic snails (*Bulinus* and *Biomphalaria* spp.) which are the intermediate hosts of schistosomiasis. In addition such lakes may increase the breeding of mosquitoes and thus enhance the spread of mosquito-borne diseases. (M.W. Service.)

pouring from lake outlets and over concrete spillways (Fig. 8.6) is colonized by larvae of simuliid blackflies, and in Africa this flowing water often constitutes larval habitats of the most important vectors of river blindness. Although not transmitting onchocerciasis hordes of biting simuliids have seriously disrupted work on the Inga dam in Zaire, which when completed will be the world's largest hydro-electric power plant. Although river blindness in Africa has occasionally posed problems for construction workers, it is a much greater threat to the more permanent lakeshore populations because it is a disease acquired after chronic exposure to bites.

Despite the dangers of mosquito-borne diseases and river blindness, the greatest public health hazard from the construction of artificial lakes is schistosomiasis. About 60 per cent of the fishermen behind the high dam of Lake Nasser are infected with schistosomiasis, as are 80 per cent of people living around the Kariba Lake and about 35 per cent of those living in the Kainji area. Schistosomiasis has reached epidemic proportions in people living near the Volta Dam in Ghana, and in some lakeside villages the prevalence rate has increased 20-fold.

Fig. 8.6 Dam spillways such as this one in Nigeria provide habitats for the immature stages of blackfly vectors of onchocerciasis, such as the *Simulium damnosum* complex. They also provide useful dosage points for applying insecticides to the rivers to kill the *Simulium* larvae. Here the insecticide is being dripped from two 4 gallon kerosene tins, and will be carried downstream by the flow of water. (Courtesy of R.W. Crosskey.)

A classic study on control of schistosomiasis was in St Lucia, a tropical island in the West Indies. Three separate methods of control were applied simultaneously to three separate valleys. During a 2 year period infection rates were reduced from about 19 per cent to 4 per cent by treating people with drugs; from about 23 per cent to 11 per cent by introducing a piped water supply, laundry facilities and swimming pools; and from 22 per cent to 10 per cent by snail control. Drugs provided the cheapest and quickest form of control and had the advantage of also reducing the reservoir of parasites in the community, but this required the co-operation and stability of the population; improved water supply also needed community co-operation, but not stability. Although exactly the same methods may not be applicable elsewhere they serve as an example of the type of approach needed. Another success story comes from China, a country with a population of about 1,000 million, and one that is increasing more rapidly than food production. Some 36 million hectares of rice are grown in China and schistosomiasis is understandably a problem, but the efforts of 20 million labourers armed with shovels, have largely succeeded in digging new irrigation canals while at the same time filling in the old ones and burying the snail hosts. This labour-intensive but attractively simple approach has apparently greatly reduced schistosòmiasis transmission in at least some areas. There are not many places, however, where the degree of community participation and discipline needed can be found. Also, the fact that the snails (*Oncomelania*) have separate sexes and are not hermaphrodite, as are the snails hosts, *Bulinus* and *Biomphalaria*, of schistosomiasis in Africa and the tropical Americas, makes them easier to control (see p. 322).

The Chinese approach was labour intensive and the new earthen drains will require considerable maintenance to prevent them becoming colonized by snails, and experience has shown that the failure of many control projects is caused by the inability to continue good maintenance. This approach contrasts greatly with the method in Japan of replacing earthen canals with concrete-lined ones, which although involving high initial cost require less maintenance.

A large ambitious irrigation scheme is presently being proposed in northern China to divert water from the Yangtze river to irrigate about 466 600 hectares of land for cultivation. Needless to say careful planning will be needed to minimize any detrimental ecological effects and increase in schistosomiasis and vector-borne diseases.

Conclusions

Dams, lakes, irrigation schemes and other development projects may be essential for the economic growth and development of many countries but unfortunately they invariably cause increases in schistosomiasis, malaria, filariasis, arboviruses and other health problems. As long ago as 1950 the World Health Organization cautioned against the dangers of increased schistosomiasis accompanying perennial irrigation, a warning that was largely ignored. Development and schistosomiasis seem irreversibly linked.

There will likely be increasing pressure to devise bigger and better irrigation schemes to reclaim deserts and semi-arid areas for cultivation. Such grandiose schemes will promote the spread of vector-borne diseases unless feasibility studies include serious considerations of the impact that development programmes are likely to have on disease transmission. Desk-studies and armchair-strategies will be of little practical use, however, unless the life-style of the people is understood and taken into consideration. For example, while community participation or enforced legislation may succeed in reducing schistosomiasis and vector-borne diseases in some countries, in others the life-style and social customs of the people may be so ingrained that they will hinder or prevent the implementation of ecologically sound but simple vector management systems. Although we have learnt a lot about disease vectors and the diseases themselves we have devoted surprisingly little attention to studying people and their behaviour relevant to the spread of such diseases. It seems that just as pest and vectors develop resistance so do people – resistance to change in life-style and habits. To overcome this we need to encourage a more understanding dialogue between 'the people' and those in authority – the politicians, administrators and scientists.

Recommended reading

Ackermann, W.C., White, G.F. and Worthington, E.B. (eds) (1973) *Man-made Lakes: Their Problems and Environmental Effects*, American Geophysical Union, Washington, DC, 847 pp.

Audy, J.R. (1972) 'Aspects of human behavior interfering with vector control'. In *Vector Control and the Recrudescence of Vector-borne Diseases*, Pan American Health Organization, Scientific Publication, No. **238**, 67–82.

Clegg, E.J. and Garlick J.P. (eds) (1980) *Disease and Urbanization,* Taylor and Francis, London, 171 pp.

Coumbaras, A.J. (1977) 'Travaux hydrauliques et problemès de santé dans les pay en voie de développement'. *Acta Tropica*, **34**, 229–48.

Gillett, J.D. (1975) 'Mosquito-borne disease: a strategy for the future'. *Science Progress, Oxford,* **62**, 395–414.

Gratz, N. (1973) 'Mosquito-borne disease problems in the urbanization of tropical countries'. *Critical Review in Environmental Control*, **3**, 455–95.

Holling, C.S. (ed.) (1978) *Adaptive Environmental Assessment and Management*, John Wiley, Chichester, 377 pp.

Obeng, L.E. (ed.) (1969) *Man-made Lakes: The Accra Symposium*, Ghana Univeristy Press, Accra, 398 pp.

Smith, C.E.G. (1972) 'Changing patterns of disease in the tropics'. In C.E.G. Smith (ed.) *Research in Diseases of the Tropics, Bulletin for Medical Research*, **28**, 3–9.

Stanley, N.F. and Alpers, M.P. (eds) (1975) *Man-made Lakes and Human Health*, Academic Press, London, 495 pp.

Waddy, B.B. (1975) 'Research into health problems of man-made lakes, with speical reference to Africa'. *Transactions of the Royal Society of Tropical Medicine*, **69**, 39–50.

Worthington, E.B. (ed.) (1977) *Arid Land Irrigation in Developing Countries: Environmental Problems and Effects*, Pergamon, Oxford, 463 pp.

9 Socio-economic considerations in the management of tropical pests and disease vectors

P. Rosenfield, A. Youdeowei and M. W. Service

In Chapters 6 and 7 we have presented a variety of components from which integrated management programmes for pests and vectors can be formulated and applied whenever possible. The methods outlined represent the technical aspects of pest and vector management. Considerable research and development (R and D) is being undertaken to accumulate relevant technical information which can be utilized to develop successful management programmes for major tropical pests and vectors. In many instances, however, experience has shown that these technical aspects alone do not necessarily provide entirely satisfactory solutions to pest and vector problems. In some extreme cases management proposals which seem technically sound in theory, fail woefully in practice. This is because the socio-economic implications of the management strategies were either poorly considered or entirely ignored. In effect, successful integrated pest and vector management must adequately consider both technical and socio-economic factors.

Recently, the economics of management programmes have been taken into consideration, particularly relating to resource balances in the ecosystem and resources for pest and vector control. G. Conway has declared: 'Thus in the final instance, pest control is a problem for the social sciences in particular for applied economics.' Nevertheless, decisions for action still usually omit the major actors – *the human population.* Cultural elements for the human environment which directly influence the ability to produce food and fibre and to reduce vector-borne diseases have until very recently been ignored. Culture is discussed usually only in the context of the local population's crop cultivation methods, and not in terms of people's beliefs and motivations. Generally speaking, people's attitudes, perceptions and behaviour have not been seriously taken into account in pest and vector management schemes. It is almost as if the over-riding objective leading to concern about pests and vectors has been forgotten. After all, the only justification for seeking to reduce pests and vectors is to improve human well-being.

Table 9.1 Social and economic activities associated with presence of pests and vectors. This list is illustrative but not exhaustive and may be modified to suit particular situations

Pests and vectors	Economic activities	Social activities
Crop pests	Cash crop production, subsistence crop production, storage of crops, marketing of crops, population movements	Population movements, siting of homes
Air-borne vectors (mosquitoes, tsetse flies, blackflies)	Farming, forestry, game hunting, fishing, livestock management, market attendance, population movements	Recreation (water and land), housing, waste disposal, fetching water, population movements, settlement patterns, leisure (sitting outside of houses)
Animal-borne vectors (ticks and mites)	Livestock management, game hunting,	Recreation, sanitation conditions
Water-borne 'vectors' (snails)	Fishing, irrigation, livestock management, market attendance, population movements	Siting of homes, bathing, washing, fetching water, recreation (water), waste disposal, population movements, settlement patterns

Upon closer examination, it becomes obvious that to ignore the human factor' in designing and implementing pest and vector management programmes must seriously hamper the chances for success. Furthermore, it is mainly because of human activities that problems with pests and vectors have become severe. Natural disruptions in the environment due to such forces as fires, floods or droughts, are more than matched by human activities involving shifting soil and water (such as irrigation and dams) for the sake of increasing food and power supplies (see Ch. 8). Rapid and massive human population increases characteristic of many tropical developing countries also influence demand for food and other resources. Population movements, on a permanent or temporary basis, can increase the risk of disease transmission. If the control of pests and vectors to prevent damage and disease is to be successful, it is clear that the plans must emphasize the relationship between the natural and human environments. Management strategies must explicitly take into account the social and economic activities associated with the presence of pests and vectors (Table 9.1) as well as the factors which influence the choice, implementation and maintenance of the technological control measures being proposed. This chapter focuses attention on social and economic factors associated with pest and vector management. Firstly, some specific examples are presented regarding the control of agricultural pests and disease vectors, then a more generalized account is given of the social and economic considerations which should be incorporated into control strategies. In this context a conceptual framework is suggested which relates a sequence of social and economic studies necessary for effective pest and vector management in addition to the traditional studies of the biological, chemical and physical aspects. Methods for carrying out social and economic studies at different stages of this sequence are briefly reviewed.

Crop pests

Pest control is needed to alleviate damage to crops which provide food and profit to the individual, and his or her neighbours, region, country and trading partners. The damages caused by crop pests can be considerable. Yet, like many issues in agricultural development, pest control is not only related to the required technical measures. As J.C. de Wilde states: 'It can succeed only with the co-operation of the farmer and on the basis of an understanding of the society of which the farmer is part.' Such an interdisciplinary approach requires an understanding of such social factors as population distribution and density, people's beliefs, patterns of settlement, land distribution systems, nature of the farming unit, sources of income, levels of output, migration patterns, levels of education, concern for food security, attitudes towards livestock, income aspirations, and response to prices.

Finance for pest management inputs

A large percentage, about 70 per cent, of the labour force in tropical developing countries is engaged in agriculture and most of this is on peasant farms of less than 2 hectares. In West Africa, for example, there is an estimated 12 million small-scale peasant farmers, 8 million of whom are in Nigeria. These farmers cultivate their small-holdings with minimum technical inputs. Their financial resources are very limited and so they are unable to pay for inputs such as expertise, labour, pesticides, fertilizers, biological control agents, and other contributions which could be introduced into their farms to enable them to increase their level of production. In fact, unless the governments of their countries fully undertake to provide such inputs entirely free of charge, or through heavily subsidized input schemes, it is unlikely that such farmers will utilize and benefit from the results of research into novel pest and vector management systems. The provision of such inputs places substantial financial burden on the governments of developing countries which usually find it very difficult to meet such demands.

Extension services

Many tropical countries have poorly developed extension services for agriculture and public health. An efficient extension service scheme guarantees the rapid transfer of pest and vector management technologies from research to the field, and conversely of farmers' problems to the research scientists. Governments in tropical developing countries do not seem to have established attractive training and career structures for extension service personnel (see Ch. 10) and therefore young people are not attracted to careers in this field. Consequently, results of scientific research undertaken in laboratories and experimental stations are not brought to the farmers. In addition, laboratory research is sometimes poorly undertaken and inappropriate to real field stations. This is in part due to lack of communication and understanding between laboratory scientists and those in the field. Nevertheless, where an efficient extension service has been developed it has proved to be a valuable mechanism for successful management of crop pests and disease vectors. Extension service personnel should be well trained and must be able to speak and understand the language of the farmers so as to build up the confidence of both the research scientists and farmers.

Acceptance of new concepts

For some time it has been recognized that there is a need for an overview of pest management strategies in agriculture. R.F. Smith and H.T. Reynolds realized this when they wrote in 1972: 'Pest control cannot be analysed or developed in isolation; rather, it must be considered and applied in the context of the agro-ecosystem in which the pest populations exist and in which control activities are taken.' The 1979 report from the Consultative Group on International Agricultural Research summarized an urgent need for pest management, which really also applies to vector control. The report states: 'Socio-economic research performs several roles. It is utilized in village level studies to examine existing farming systems and to reach understanding of why farmers behave as they do and thus indicate areas where changes might be induced. It is used to identify constraints to the adoption of new technology and to monitor the progress and consequences of the new technology. It is thus involved at both the beginning and the end of new technology.'

Building a conceptual framework to contain all these elements is an ambitious task. Yet planning within such a framework has the best chance of reaching the ultimate goal of improving human well-being. This framework (Fig. 9.1) shows the sequence of activities in developing socially and economically environmentally sound pest and vector management strategies. It must be recognized, however, that although this framework is specific to pest and vector management, the underlying social and economic factors relate also to other needs in the human environment. Moreover, measures broader than specific pest and vector control, such as improved housing, improved income and employment opportunities (from non-farm sources) and increased education, may influence the presence or absence of pests and vectors, and the reduction or prevention of the damage they cause.

Farmers are traditionally reluctant to accept and implement new farming methods unless they have been convinced through practical demonstration of the immediate benefits to them of the new approaches being presented (see pp. 349, 350). The integrated pest management (IPM) concept implies that pest populations be held in check at levels at which the damage they cause is well below the economic threshold (or action threshold, (see p. 75). This means that some degree of visible crop damage must be tolerated in the field. Experiments have shown that some crops like cassava and cotton can tolerate up to 50 per cent defoliation without any significant reduction in yields. Consequently, pest control measures for defoliating insects need not be applied if defoliation is lower than 50 per cent. This may be good scientific sense, but experience has shown that it is extremely difficult to convince those farmers in developing countries who can afford control measures to accept this concept of an economic threshold. They find it difficult to tolerate the presence of insect pest populations on their crops, let alone the idea of an economic threshold. This problem is aggravated by the regular presence of pesticide salesmen, who are experts in convincing the farmers of the immediate benefits of spraying against all insect pests causing any level of damage to farmer's crops. Unfortunately pesticide salesmen reach the farmers well before research and extension personnel, who therefore have a difficult time in convincing farmers about the IPM philosophy. An additional

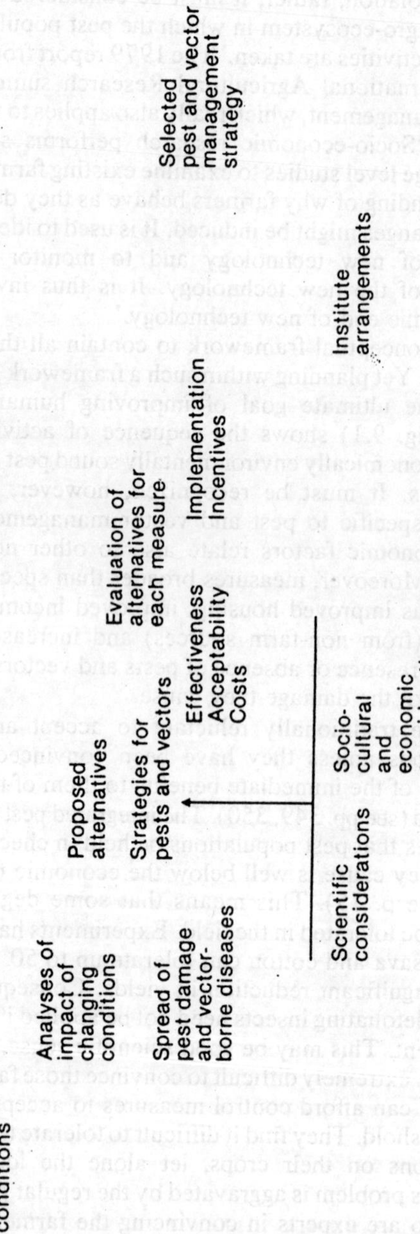

Fig. 9.1 Conceptual framework for consideration of social and economic aspects of pest and vector management.

problem is that most developing countries do not seem to have any efficient systems for regulating and controlling the activities of pesticide manufacturers and salesmen. Nevertheless, farmers have sometimes been persuaded that pesticides are not necessarily either the only or best solution for their pest problems. The case of cotton pests in the Cañete Valley in Peru demonstrates this; if farmers can recognize the threat of economic loss as being great enough then they may be prepared to change their behaviour. In the face of initial opposing recommendations by the government these farmers were able to recognize that cotton pests were still causing economic losses despite regular insecticidal spraying, and were thus persuaded to accept a more ecologically sound managagement strategy.

Some pest management practices look too simple and, unlike chemical control, are not dramatic and therefore do not yield immediate results. It is therefore difficult for farmers to accept their adoption. For example, W.W. Page and his co-workers in western Nigeria developed what seemed a simple and effective pest management system for the variegated grasshopper, *Zonocerus variegatus*. The approach involved digging up and destroying diapausing *Zonocerus* egg-pods from the soil where they are usually laid in dense aggregations. Although a few farmers in the area reacted favourably to this method of *Zonocerus* control, community participation by groups of farmers was essential for effective implementation of the scheme. Many farmers, however, were not prepared to allocate the time and expense of getting labour to identify the egg-laying sites and digging up the egg-pods. Besides it was not easy to convince them that this method could drastically reduce pest populations in the following cropping season, because they were used to the yearly arrival of this pest on their crops. Another problem was that there was a serious lack of agricultural extension personnel to propagate this idea to a large number of farmers. So, although this method of *Zonocerus* management was technically sound, the social and economic aspects could not be satisfied and implemented.

Sometimes a pest management practice interferes with people's traditional customs, and unless such customs are changed through education and provision of acceptable alternatives, the pest management strategy will never succeed. The cases of the control of cotton pink bollworms (*Pectinophora gossypiella*) in Egypt, and of sorghum stemborers in Nigeria illustrate this phenomenon. In Egypt, dried cotton bolls attached to còtton stalks are stored on the roofs of village houses adjacent to cotton fields to be used as domestic fuel. This collection of cotton bolls and stalks contains diapausing larvae of the bollworm. Adults emerge from these larvae in June – July and oviposit on new flower buds of the following year's neighbouring crop. This initial population gives rise to the new season's pest populations which spread rapidly into other cotton fields. Although entomologists have strongly advised that total destruction by burning of the dried cotton bolls and stalks would eliminate this initial source of bollworm infestation, villagers have refused to abandon their traditional practice of storing dried cotton stalks as domestic fuel. This is understandable because of the absence of any suitable alternative source of fuel of comparative cheapness. The pink bollwork therefore has to be controlled by more expensive chemical treatment of the cotton crop.

Stemborers (*Sesamia* spp. and *Busseola fusca*) are pests of maize and sorghum in Africa. Diapausing larvae and pupae of these pests are present in dried sorghum stalks which are piled around homes and used as a source of domestic fuel, as animal feed and as building material for construction of fences and houses. At the onset of the rains, adult pests emerge from the old dried stalks to lay eggs which produce larvae that attack the new crop. The original recommendation to deal with this problem was complete burning of the sorghum stalks after harvest in order to destroy the diapausing larvae and thus eliminate this source of pest infestation. Farmers in northern Nigeria rejected this recommendation in the absence of an alternative source of cheap fuel, animal feed and building materials. Work by A.A. Adesiyun and his colleagues at the Institute of Agricultural Research, Ahmadu Bello University in Zaria, Nigeria demonstrated that burning the dried leaves on partially dried sorghum stalks generated sufficient heat within the stalks to kill diapausing larvae. This practice does not destroy the stalks which can still be used as domestic fuel and as building material. Farmers have responded favourably to this modified pest management method which has partially taken their economic situation into consideration. Even with Adesiyun's recommendations, farmers still loose the sorghum leaves which serve as animal feed.

The control of cocoa swollen shoot disease in West Africa affords yet another example of the need to take into account socio-economic considerations. This virus disease is transmitted by mealybugs and causes a serious and widespread problem of cocoa cultivation in Ghana and Nigeria. Infected trees still bear fruit but yields are considerably reduced. The use of insecticides and natural enemies to control the vectors has failed, and the only feasible method of control is to remove all infected trees to eliminate the source of infection. Since 1946 when this cutting out scheme was introduced in Ghana over 160 million cocoa trees have been removed. Originally the Ghanaian government made cash payments to compensate farmers for the eliminated trees, but nowadays compensation is in the form of replanting schemes. Cocoa farmers in Ghana have found it difficult to accept the concept of management by cutting out cocoa trees because this crop takes up to 10 years to achieve appreciable yields, and therefore removal of mature trees means loss of a source of revenue for at least 8 years. Similar cutting out schemes of cocoa trees infested with larvae of the cocoa-pod borer moth, *Acrocercops cramerella*, in Indonesia have met with strong opposition from peasant farmers.

International pest control also faces considerable political and economic difficulties. In locust control in Africa for example, neighbouring countries which are affected by locust plagues only at irregular intervals are reluctant to contribute their financial quota to an international control organization, especially in years when there is no immediate threat of locust invasions. Thus, funds for surveys and control operations are seriously lacking and this leads to poor surveillance of the locust breeding sites and outbreak areas. Differences in political ideologies and border disputes between neighbouring countries can adversely affect the technical operations of international locust and armyworm control organizations. In fact in some cases it is very risky or impossible for pest control personnel to travel from one pest location to another in a neighbouring country. These are difficult problems which can be solved only through economic and political stability and understanding within and between the tropical developing countries concerned.

Another socio-economic consideration relating to crops is the effect of over-production of crops following successful pest management practices. In some instances when increased crop yields have been successfully achieved, farmers have considerable difficulty in disposing of the surplus. Storage facilities are usually inadequate (see p. 251), and transportation and marketing arrangements are often poor and unable to cope with the surpluses. Under such circumstances the market price of the product falls and the additional costs of storage and transportation of products to other markets are not economically justified.

These examples illustrate clearly that the successful implementation of a pest management strategy depends to a large extent on economic and social factors.

Disease vectors

People's attitudes

Whereas a farmer may well appreciate that the arrival of pests on his crops threatens an economic loss, he is less likely to understand that mosquitoes in his house may cause the death of his child from malaria. Firstly, he is unlikely to know that mosquitoes spread dangerous diseases (see p. 7), and secondly even if he has been told this he may find it difficult to accept, because he cannot see the insects actually causing ill health to his child, whereas he can see pests eating his crops. Even if he is concerned with protecting his family against mosquitoes there may be relatively little he can do. For although the use of mosquito nets will protect people against night-biting mosquitoes, good quality nets may be too costly, or unavailable. Moreover, in many houses there is simply just not sufficient room for nets to be placed over all the occupants. Another problem is that mosquito nets reduce ventilation and make the nights hotter. In Tanzania, trials showed that installing low white-painted ceilings in village huts reduced substantially the numbers of house-resting mosquitoes, but this inexpensive control method was unacceptable because it made the houses unbearably hot.

Curative medicine is likely to be accepted more readily than prophylactic administrations because it produces an almost immediate action in returning people to good health. In contrast the rationale behind preventive medicine may be more difficult to understand because it often implies radical changes in people's customs and attitudes without any obvious and immediate benefits. One form of change, however, that is accepted readily and which is universally asked for, is a *reliable* domestic supply of clean water (Fig. 9.2).

Fig. 9.2 Cartoon from the *Nairobi Times*, 19 July 1981, which well illustrates the credibility gap that can exist between administrators and the people. (Courtesy of the *Nairobi Times*.)

The free distribution of drugs to combat vector-borne diseases is frequently not feasible for a number of reasons, including high costs, administrative difficulties and the toxic side effects of many drugs, especially on people with poor general health and nutrition. However, even when the drugs are safe to administer (such as anti-malarial compounds) there may still remain problems. For example, many drugs are bitter and so people, especially children, may spit them out; there may also be totally unfounded rumours that the drugs make the women infertile or the men impotent – that is enough to halt any drug administration programme! What initially seemed a good idea to overcome some of these practical difficulties was the introduction of the anti-malarial drug chloroquin into domestic salt. The reasoning was that as we all take salt with our food everyone would be protected against malaria attacks. But this approach is feasible only in areas where there is a single supply of salt

Fig. 9.3 A typical gathering of women and children collecting water and washing clothes at a river bed during the dry season when water is scarce. (M.W. Service.)

and one that can be controlled, whereas in most communities crude salt can be purchased in local markets and this of course will not contain chloroquin. Moreover, not everyone takes salt, infants (0–9 months) frequently do not for example, and thus this vulnerable age group will remain unprotected. In addition there are several technical difficulties in incorporating chloroquin into salt.

People's customs

In any society people are resistant to change.

Schistosomiasis (bilharzia) is a disease that, in theory at least, should easily be prevented by stopping people coming into contact with water contaminated with human urine or faeces. But it is extremely difficult to educate or persuade people not to wash or swim in such dangerous waters. For example, it is a hygienic custom of some Moslems to wash their anus after defaecating, but the problem is that this ablution is often performed in a stream in which children bathe or play. In hot tropical countries it is understandably difficult to prevent children from playing in waters known to harbour infected snail intermediate hosts of schistosomiasis. As another example, women commonly wash clothes in streams or pools formed in the river-beds (Fig. 9.3). They are often accompanied by young children, who are consequently exposed to the risks of contracting schistosomiasis, and in Africa also to the possibility of being bitten by riverine tsetse fly vectors of sleeping sickness. The provision of alternative play sites has not usually been accepted, as this breaks a well established traditional and social custom.

The coir rope producers in Kerala, India and in Sri Lanka have not changed their behaviour despite the risks they encounter of contracting filariasis from being bitten by mosquitoes breeding in the coconut husk soaking-ponds, which traditionally are adjacent to their houses. Even though many of these people know their chances of becoming infected are high they prefer to seek medical treatment and use insecticides rather than change their pattern of activity or adopt more effective methods of getting rid of the breeding places. Similarly, throughout much of the tropics domestic water is stored in earthern pots or metal drums inside or outside houses and such man-made collections of water provide ideal larval habitats for dengue vectors, such as *Aedes aegypti* mosquitoes. It is almost impossible to persuade people to modify their water storage habits to reduce this type of peridomestic mosquito breeding.

One of the clinical symptoms of filariasis is the enlargement of the scrotum. In many East African communities the gradual increase in size of these organs is not regarded with alarm but with pride, as it is considered to show manliness and as such is a desired attribute! The fact that this condition is symptomatic of a disease that in later years may well result in gross swelling of the lower limbs and other discomforts is not appreciated. Consequently, the early signs of the disease fail to warn the people that treatment or preventive measures should be sought.

As has been stressed in several studies, human behaviour plays a major role in disease transmission and control and often derives from 'cultural idioms and themes'. The reasons for using certain water supplies, which may be inhabited by schistosome snails or used as mosquito or blackfly breeding places, may relate to convenience, long-standing tradition or even religious beliefs. Without fully understanding these underlying factors, control of diseases relying on modifying people's behaviour will most likely fail.

Economics of vector-borne diseases

Although monetary estimates of the damage caused by vector-borne diseases are difficult to make with any degree of reliability, the impacts of diseases such as river blindness (onchocerciasis), malaria, trypanosomiasis (sleeping sickness and nagana of cattle), schistosomiasis and elephantiasis are clearly evident through careful observation. As long ago in 1939 a study revealed that malaria cost India, at that time, some US$500 million a year. Other studies have shown that malaria increases absenteeism from schools, work and the number of patients in hospitals. It is of course one of the major causes of infant mortality in some tropical countries. What are the costs of introducing an effective malaria control programme? In the early 1970s, the cost in Asia and African countries was estimated at between US$0.30 and US$1.58/person/year. At that time wages accounted for 50 per cent of this cost, insecticides 33 per cent, transport, fuel, repairs to equipment etc. 10 per cent, and spraying equipment and laboratory facilities 7 per cent. This breakdown was based on the use of DDT, a cheap insecticides. Both the total amount and proportions will differ as labour costs rise and more expensive insecticides have to be used because of insects developing insecticide resistance. A malaria eradication programme would cost considerably more.

There are difficulties with trying to undertake cost-benefit analyses in health programmes if the only benefits taken into consideration are economic. Increasing life-expectancy for example, but not working life, is still highly desirable, but gives little direct economic benefit. It is generally agreed that interdisciplinary efforts are needed to demonstrate quantitatively all the consequences of ill health in the human environment (see p. 358). Thus the Onchocerciasis Control Programme (OCP) in West Africa is not aimed just at reducing the incidence of river blindness but takes into account the economic growth and improved social conditions of the community. As another example, a 1977 FAO report has stated: 'Active elimination of tsetse may be needed to make more land available for growing populations as well as to allow the development of newer forms of economic activity.'

In the past there have undoubtedly been some mistakes in planning and executing health and development programmes. For example, land needed for agricultural irrigation projects has sometimes been obtained by compulsory purchase from local farmers. As a result peasant farmers suddenly acquire new wealth, but since they no longer own land or cattle, their standing in their own community may decline. Many become tenant farmers and are accommodated in standard-sized houses with generally little regard to their family size. This sometimes leads to overcrowding, and because they do not own their houses there is little incentive to maintain them. As a consequence they fall into disrepair and slum conditions begin to develop; this is accompanied by deterioration in morale, community spirit, hygiene and general health.

Methods of analysing underlying social and economic conditions

From the outset it must be emphasized that the analysis of social and economic conditions which aims at presenting the choices of pest and vector management strategies must be linked with analyses of environmental conditions which lead to the presence of pests and vectors. The site-specific nature of pest and vector populations and peoples' attitudes have been described at length by many workers; and any field worker will protest strongly against generalizations in their subject drawn by others. Fortunately, however, the analytical methods available are similar for many situations, in fact, the main techniques are often talking and listening to people as well as observing their behaviour. The main tools of the social scientist or economist for collecting field data are informal or structured interviews, questionnaires, tests, participant or non-participant observations, and participatory research. Results of such intensive studies, although descriptive, provide the basis for the more extensive, analytical type of social or economic study by describing attitudes, perceptions, behaviour, families, education, needs, and philosophy – in other words whatever is relevant in the lives of the people at risk from damage or disease. Many textbooks exist on surveys and questionnaires but there are few publications on participant and non-participant observations and, practically none on participatory research.

The first-stage analysis, as outlined above, should provide the basis for developing socio-biological, epidemiological and economic tools of survey research. For one disease, schistosomiasis, these methods have been compared in detail. Table 9.2 lists the advantages and disadvantages, and each of the five methods outlined in the table can be used to study baseline conditions and also to assist in the evaluation of different control measures.

Table 9.2 Summary of advantages and disadvantages of social research methods

---

1 **Non-participant observations:**
(i) can provide reliable quantitative details about present behaviour;
(ii) can be standardized;
(iii) do not require local co-operation;
(iv) can use untrained personnel;
(v) have very low information yield per unit of time expended;
(vi) do not inform about past behaviour;
(vii) cannot reveal perceptions and attitudes.

2 **Participant observations:**
(i) can provide reliable quantitative details about present behaviour;
(ii) can inform about past behaviour;
(iii) can reveal perceptions and attitudes;
(iv) cannot be standardized;
(v) require some local co-operation and acceptance;
(vi) cannot usually use untrained personnel;
(vii) have very low information yield per unit of time expended.

3 **Questionnaires:**
(i) have high information yield per unit of time expended;
(ii) can yield some information about past behaviour (of questionable reliability);
(iii) can be standardized;
(iv) require local co-operation (may be difficult to achieve);
(v) can use untrained personnel.

4 **Testing:**
(i) can reveal perceptions and attitudes;
(ii) can be standardized;
(iii) requires local co-operation (may be difficult to achieve);
(iv) cannot use untrained personnel;
(v) has low yield of information per unit of time expended;
(vi) cannot inform about past behaviour.

5 **Structured interviews:**
(i) can inform about past behaviour;
(ii) can reveal perceptions and attitudes;
(iii) can be standardized;
(iv) require local co-operation (may be difficult to achieve);
(v) cannot use untrained personnel;
(vi) have low yield of information per unit of time expended.

Evaluating the choice of action

At this stage a different set of social and economic factors must be considered which relate to evaluating the decision to control pest and vector populations, and include such elements as acceptability, benefits, effectiveness, and costs. Although closely related to the specific choices to be evaluated these elements are also affected by social structures and customs of the society in which the project is being undertaken. Specifying which is the most acceptable and cost-effective pest and vector management strategy is not enough, the strategy has to be implemented. To achieve successful implementation, incentives pertinent to the specific strategy need to be applied and an institutional or organizational structure found for maintaining the use of the strategy. Implementation incentives can cover the range of economic, financial, informational and regulatory measures which can be applied at different stages in a management strategy. Organizational structures must be considered at the local, regional, national and international levels. Although they will of necessity also be country and culture specific, a range of possibilities is suggested.

Issues

In reviewing the social implications and economics of control measures it becomes clear that the choice of the most appropriate methods are not based purely on scientific knowledge of background, social, economic, biological, physical and chemical baseline conditions. More than local farmers' or individual acceptance is needed to obtain support for pest and vector control programmes; decision-makers at different levels of government must be convinced of the need to invest their scarce resources and efforts into this activity. Furthermore, in the choice of control methods, the differential impacts of pests and vectors influence the differing amounts of resources made available for their control. As already defined pest control relates directly to human survival and profit; people need crops for both food and income. Both the private and public sectors of the economy can justify the need for investment in cash crop protection; only slightly more difficult is the justification of investment in subsistence crop protection.

However, because of the difficulty in directly relating vector-borne disease control to a quantitative measure of human profit or pleasure, it is harder to justify these investments on purely monetary grounds (see p. 355). In this instance, the social elements associated with the consequences of the diseases need to be incorporated into the economic calculations to make those evaluations sensitive to the impact of disease on the quality of human life.

Acceptability of pest and vector control

The differing perception of the need for pest and vector control is reflected also in the acceptance of control technologies already referred to. Acceptibility of innovations relates closely to baseline conditions. Farming characteristics such as the tendency to risk-aversion and farm size play important, but not necessarily predictable, roles in acceptance of new technologies. Credit availability and market prices also affect the willingness to take risks on new technologies. The main issue that seems to tie together many of the findings is that of 'risk-aversion'.

If new pest management programmes are presented in a way that demonstrates their role in increasing the certainty of agricultural production, their acceptance may be higher (see p. 349). It has been demonstrated that many traditional farmers work in a cultural framework which emphasizes the practical and concrete; in this situation demonstration of success may increase the acceptability more than radios, posters or other means of transferring this information.

In the area of vector-borne diseases, the difficulties of acceptance may be even greater since the relationship, for example, between the bite of a mosquito and malaria may be unexplained, and more seriously it may be impossible to prevent except through measures which cause considerable inconvenience such as housespraying (see p. 275), regular taking of drugs, or expensive source-reduction methods. In several countries malaria is one of the major causes of infant mortality, its cure or prevention relies on specific measures, but when young children are dying due to intestinal infection, malnutrition and other causes it is difficult for a mother in the face of so many stresses to recognize the seriousness of a single disease such as malaria.

To encourage government interest, acceptance and financing of pest and vector management programmes, emphasis needs to be given to demonstrating the economic benefits of such activities.

Economic evaluations: benefits, effectiveness and costs

In setting out to evaluate the economics of pest or vector control, the first step is to estimate what the objectives and effects of the control measures will be. That is, the expected outcomes from such control need to be explicitly stated, although they may range from improving food supply or human health, to raising income, or to reducing populations of pests or vectors by a given percentage. With the emphasis on attempting to define the thresholds below which the damages due to the pests or vectors are tolerable (from an economic or public health perspective), it may be possible to estimate, based on empirical data, the critical disease or economic threshold. This threshold may, however, be difficult to define empirically partly because it is likely to be highly site-specific.

If the objectives are measured in monetary terms, then the techniques of cost-benefit analysis may be used. If, however, the objectives are expressed as physical units of output, such as case-years of infection prevented or area of leaf-crop protected, the techniques of cost-effectiveness analysis should be used.

If it is possible to assess the different levels of benefits resulting from use of specific measures or strategies then cost-benefit analysis can sometimes assist in choosing among competing demands for finite resources. Pest control programmes for cash crops can use cost-benefit analysis to provide the appropriate guidance for investment decisions. However, in an analysis of pest control programmes for improving subsistence crops or vector control for disease prevention, the estimation of benefits becomes more complex and confusing.

Social cost-benefit analysis attempts to account for the imperfections in the economic market-place. That is, usually the prices paid for pesticides at the store do not reflect the social cost of pesticide use which includes health hazards to humans, destruction of wildlife, environmental pollution and other adverse effects. Moreover, the costs may be felt by those who do not receive the direct economic benefits. On the other hand, the benefits gained by pesticides or implementation of integrated control methods may far exceed the income generated by the sale of the cash crop. Integrated pest management (IPM) may prevent the invasion of new pest species or may increase crop yields so that more choice is available to consumers. Nowadays in estimating social benefits more attention is increasingly paid to the distribution of the benefits, so that a project is evaluated according to its 'impact on society as a whole'.

Cost-effectiveness analysis is similar to cost-benefit analysis, except that it is used to assess the costs of reaching a certain target. Because of this, cost-effectiveness analysis may be easier to apply than cost-benefit analysis, but it too has its limitations. In 1978 N. Imboden wrote: 'Cost-effectiveness is used to evaluate projects for which outputs and inputs cannot be expressed in commensurable terms....Cost-effectiveness analysis defines the efficiency of different input combinations to achieve a goal or set of goals, if the goals are identified in an operational way so that the degree of goal achievement can be measured by a set of indicators.... While cost-effectiveness analysis provides the necessary criterion to choose between alternatives to achieve a specified goal or between different degrees of goal achievement for a specific input, it provides no information about the desirability of goals.'

In either cost-benefit or cost-effectiveness analysis a careful accounting framework must be established. Either technique can be applied to evaluate past decisions or to advise on future ones. It is most important, whether dealing with crop pests or disease vectors, that there is adequate background information on the ecology of the pests and vectors and the epidemiology of any disease the vectors spread. This knowledge is essential for making correct control decisions and for reliable predictions on the benefits or effects of future management strategies.

The selection and evaluation of the most beneficial or effective control programme in relationship to its costs is only the mid-point of a socially and economically sound management strategy. Once the choice is made it must be implemented and maintained, and therein lies the difficulty for many pest and vector management programmes.

Implementation and maintenance of pest and vector control programmes

No matter how detailed the baseline studies on ecology, sociology and economy are, and how carefully the choice has been made, incentives may be needed to ensure use of these measures. With carefully planned and culturally sensitive education programmes, pest and vector management techniques may possibly be accepted by the community. However, usually when new or untried techniques are introduced, the following factors have influenced the farmer (these criteria apply also to individuals at risk from vector-borne disease) in his use or non-use of the techniques: (i) fear of failure of conventional control methods; (ii) social pressure from other producers and other segments of the society; (iii) legislation concerning cultural practices, or legislation affecting availability or choice of chemicals; (iv) economics; and (v) education.

Thus, to ensure the use of different measures, a broad range of related implementation incentives is needed. According to P.L. Rosenfield and B. Bower (1979): 'Inducing action and changing behaviour requires both culturally sensitive implementation incentives – those instruments, techniques and procedures which induce individuals, groups, agencies to act – and institutional arrangements. The latter are the organizational structure, formal or informal, which has authority and responsibility for improving implementation incentives.' To induce people to use the advocated control measures, incentives may be financial, economic, informational, behavioural, regulatory, and evaluative. Incentives should be designed especially for the group responsible for maintenance. In Table 9.3 implementation incentives are linked to the institution appropriate for social and economically sound pest management activity. Such a table would have to be developed for each particular situation, but would likely contain some or all of these elements. In this table the household is shown to respond to certain incentives such as: bonuses for extra participation in spraying efforts; information presented in a culturally appropriate way about the hazards of pests and vectors; training; pressure from their peers to keep their houses sprayed (sometimes when people can afford to get treatment for malaria they refuse housespraying); and local laws (when enforced by fines). If the household or farmer is represented in the planning and evaluation processes they may develop more interest and commitment to the success of the control programme. This was clearly the case in Peru where the cotton farmers enthusiastically shifted to an integrated pest management system after learning first-hand about its comparative effectiveness (see p. 249). In much the same way, the local agency, be it the village council, agricultural extension worker or disease control team, is also likely to respond to incentives imposed by the national agencies, or indeed by the population at risk, if they are brought into the process.

National agencies can respond also in both directions, that is to constituent pressure from within the country and to donor agencies from outside. Moreover, the damages resulting from international migration of pests and vectors can lead to increased pressure from neighbouring countries to increase control efforts, as has happened for example in locust control and blackfly control campaigns (see also pp. 286, 350). However, it is usually the donor agencies which can try making loans conditional to environmental analysis of the impact of the control programme and ensuring that any resultant recommendations are implemented. Through loan agreements and technical assistance, donor agencies can also provide extra support for the use of appropriate pest management strategies, particularly when it will benefit not only the country but also its neighbours.

Table 9.3 Pest and vector management strategy (P and VMS) matrix relating implementation incentives and institutions. Each box in the table contains the incentives developed by the level in the next column to induce acceptance by the previous institution

Implementation incentive	Institution			
	Household	Local agency	National agency	External agency
Financial and economic	Bonuses Fines	Bonuses Fines	Foreign exchange earnings Loans contingent on P and VMS	Worldwide pressure
Information and behavioural	Training Participation in planning pest and vector education Peer group pressures	Training Participation in planning	Training programmes	Technical assistance
Regulatory	Local ordinances	Legislative Regulations		
Evaluation and monetary	Surveys of pest and vector control operations Review of behaviour Supervision	Surveys of pest and vector control by operations Supervision	Review of project objectives on a regular basis	

To ensure effective maintenance of control measures, appropriate institutional arrangements must exist. Maintenance involves not only ensuring that spare parts, pesticides and biological control agents are available but also the provision of support for on-going monitoring and evaluation of the control efforts. Such programmes need to be developed in advance of implementating any programme and should be included in the economic analysis. Moreover, there is a role for community participation, particularly in monitoring, because this could increase their feeling of partnership with the control programme. As has been proposed for some of the better irrigation projects, it would be useful to have a review committee for each project. This would consist of representatives from agencies responsible for pest and vector control, for agriculture, health, water resources, as well as representatives for informal and formal local organizations. They could be responsible for reviewing plans and monitoring progress. Such a review committee could become an effective mechanism for ensuring continued application of appropriate control measures.

To ensure effective management, it must also be emphasized that the teams within the pest and vector control agency should be interdisciplinary. This does not mean that anthropologists and economists are necessary specialists of every team, but it does mean that people having some training in these disciplines should be. Economic entomology has long been a legitimate profession. Now that increasing recognition is at last being given by pest and vector managers and research workers to the human factors, perhaps more training is needed in *social entomology* so that anthropologists and entomologists can learn together how to increase their contribution to pest and vector management.

Conclusions

The main theme in this chapter has stressed the importance of taking into consideration cultural and economic considerations in developing successful and realistic management strategies for tropical pests and vectors. Effective collaboration must be established between entomologists, economists, social scientists and of course the people themselves. In determining which set of pest and vector control measures to use, a variety of physical, chemical and biological measures are available which have been described throughout the book. The aim of most of these measures is to *reduce* pest and vector populations. However, given the recognition of the role of people in this system, measures need also be considered to modify people's activities which increase pest populations and lead to greater contact with disease vectors, and to increase people's acceptance of control methods. The choice of which combination of measures to use should be based on knowledge of the underlying social and economic conditions as well as on the availability of technology necessary to modify or prevent those conditions which lead to increased pests and vectors.

Recommended reading

Abel-Smith, B, and Leiserson, A. (1978) *Poverty, Development and Health Policy*, World Health Organization, Geneva, 109 pp.

Barducci, T.B. (1972) 'Ecological consequences of pesticides used for the control of cotton insects in Cañete Valley, Peru'. In M.T. Farvar and

J.P. Milton (eds) *The Careless Technology – Ecology and International Development*, National History Press, Garden City, New York, 423–38.

Consultative Group on International Agricultural Research (1979) '1979 Report on the Consultative Group and the International Agricultural Research System, An Integrative Report'. *CGIAR Secretariat*, 60 pp.

Conway, G. (1976) 'Man versus pests'. In K.M. May (ed.) *Theoretical Ecology: Principles and Applications,* Blackwell Scientific Publications, Oxford, 257–81.

Cvjetanovic, B, and Grab, B. (1976) 'Rough determination of the cost-benefit balance point of sanitation programmes'. *Bulletin of the World Health Organization*, **54**, 207–15.

Dunn, F.L. (1979) 'Behavioural aspects of the control of parasitic diseases'. *Bulletin of the World Health Organization*, **57**, 499–512.

Garlick, J.P. and Keay, R.W.Y. (eds) (1977) *Human Ecology in the Tropics*, Symposia of the society for the study of human biology, volume **16**, Taylor and Francis, London, 200 pp.

Haskell, P.T. (1972) 'Locust control: Ecological problems and international pests'. In M.T. Farvar and J.P. Milton (eds) *The Careless Technology – Ecology and International Development,* National History Press, Garden City, New York, 499–526.

Holling, C.S. (1978) *Adaptive Environmental Assessment and Management*, John Wiley, Chichester, 377 pp.

Hughs, C.C. and Hunter, J.M. (1972) 'The role of technological development in promoting disease in Africa. In M.T. Farvar and J.P. Milton (eds) *The Careless Technology – Ecology and International Development*, National History Press, Garden City, New York, 69–101.

Imboden, N. (1978) *A Management Approach to Project Appraisal and Evaluation*, Development Centre of the Organization for Economic Co-operation and Development, OECD, Paris, 172 pp.

Prescott, N.M. (1979) 'Schistosomiasis and development'. *World Development* **7**, 7–14.

Rosenfield, P.L. and Bower, B. (1979) 'Management strategies for mitigating adverse health impacts of water resources development projects'. *Progress in Water Technology*, **11**, 285–301.

Rosenfield, P.L., Wolman, M.G. and Smith, R.A. (1977) 'Development and verification of a schistosomiasis transmission model'. *American Journal of Tropical Medicine and Hygiene,* **26**, 505–16.

Smith, R.F. and Reynolds, H.T. (1972) 'Effects of manipulation of cotton agro-ecosystems on insect pest populations'. In M.T. Farvar and J.P. Milton (eds) *The Careless Technology – Ecology and International Development*, National History Press, Garden City, New York, 373–406.

Sugden, R. and Williams, A. (1978) *The Principles of Practical Cost-Benefit Analysis*, Oxford University Press, Oxford, 275 pp.

10 Training and policies of pest and vector management

The preceeding chapters of this book have outlined the problems of pests and vectors in the tropics and some examples of the ways in which important pests and vectors have been managed or attempted to be managed. The question of scientific expertise and research capabilities needed to develop management strategies for tropical pests and vectors are not considered. It is appreciated that institutional facilities for scientific research already exist in some countries, but frequently the main constraint for effective research and the implementation of pest and vector management recommendations is the shortage of funds and trained manpower. Some developing countries have funds, considerable expertise and well established research and development institutions, yet these countries have failed to develop satisfactory integrated, environmentally sound pest and vector control strategies for their major pests and vectors. Thus, the challenge of pests and vectors, the availability of finance, institutional facilities, and trained manpower have not been sufficient to stimulate the successful development and implementation of integrated management protocols. Clearly then other factors must be operating as serious constraints to progress in this field. Some socio-economic factors are considered in Chapter 9. This final chapter deliberately focuses attention on some specific issues associated with research, training and national policies relevant to the adoption of appropriate pest and vector control measures.

Expected trends

Pest and vector management is a dynamic exercise. Every year new methods and technologies are developed for dealing with specific pests and vectors. Research for more effective and ecologically sound, long-lasting management methods against troublesome organisms must be continued because such endeavours are critical to the success of integrated pest and vector management. Traditionally, research into the biology, ecology and control of tropical insect pests and disease vectors has been isolated and conducted individually with very little attempt to seek solutions to the problems in an integrated way. This has led to failure of the control methods developed because many gaps existed in the knowledge of the biology and ecology of the pests and vectors concerned. Effective pest and vector management involves the intelligent integration of several approaches into a harmonious system. This multi-component system is based not only on scientific knowledge but also on socio-economic and socio-cultural factors. Future trends in pest and vector management research must therefore be focused on the collaborative efforts of applied entomologists, agronomists, ecologists, economists and sociologists. It is only with the combined efforts of these and other relevant experts that a clear conception of the diversity of pest and vector problems can be reasonably achieved.

National policies

The results of scientific research can only be fully utilized for the development of a nation, if a definite science policy is formulated, adopted and vigorously pursued by the government. Such a science policy must, among other things, include the education and training of indigenous scientists who should be prepared to devote their working life to scientific research for national development.

Many developing countries lack the necessary scientific manpower capabilities to meet research and development needs. Quite apart from the mere shortage of trained scientists, two other aspects of the manpower situation seem to compound the problem. In many tropical developing countries, particularly in Africa, young scientists and technologists who should be actively practising research soon become saddled with administrative duties. Sometimes they are forced, because of lack of sufficient trained people, to be appointed to jobs for which they were not originally trained. Thus, their expertise is lost to scientific research and training. This is not to say that their contributions to administration are not valuable, but there is a kind of local 'scientific and technological brain drain' away from the country's scientific and technological problems.

Another aspect is that of national recognition. Pest and vector management specialists, like other scientists and technologists in developing countries, are not encouraged by the societies in which they live, to develop a real interest in solving the scientific problems which will benefit their countries, especially those problems affecting poor rural populations. Unlike priests, lawyers and physicians, they do not yet enjoy a high social status in the community and therefore they are forced to depend upon international sources for scientific recognition. (It may well be that developing societies do not yet fully appreciate the role that scientists and technologists could and should play in their community.) In frustration, scientists fashion their science after western patterns and pursue scientific research which will become published in international scientific journals. This approach may well result in international recognition, but to the neglect and detriment of research into the local pressing problems of hunger, malnutrition, poverty and disease.

Very low priority is given to funding and developing the relevant basic research that is needed to identify and develop the technology that the community needs for rapid industrial and socio-economic development. Because of the diversity and magnitude of the 'development problems' in tropical countries, demand for well-trained pest and vector management specialists, as well as for other scientists, is extremently high. Such specialists either have to develop their own science and technology relevant to the problems of their countries or, like Japan during the 1950s, acquire the latest and most productive technologies of industralized countries and modify them for introduction into their own countries. They may even do both. Experience has shown that successful development and adoption of appropriate technologies for any country can be best achieved by indigenous scientists and technologists. This is because such people are best equipped to appreciate and understand the socio-cultural background and economic structure of their own communities, and have the mental freedom to explore and elucidate their own scientific problems'. The late Homi Bhabha of India believed that the developing countries must have indigenous scientific capacities equivalent to that of industrialized countries for successful acquisition and transfer of technology. The poor countries of the tropics must therefore develop deliberate policies and programmes to train large numbers of indigenous scientists and technologists for the management of insect pests and disease vectors which constitute impediments to rapid socio-economic development.

Many tropical developing countries have not yet evolved a policy which makes applied entomology a registerable profession like human or veterinary medicine. Persons who advise and prescribe methods for pest and vector management should be those who have had appropriate training and proven expertise in the field. This trend towards professionalism in applied entomology has been developed in a few countries, such as the USA (especially in California and Mississippi). Such a policy ensures that the right persons have the monopoly of taking the decisions in integrated pest and vector management practice. Once a government has established a policy of professionalism in applied entomology, the next step is to create a scientific climate conducive for the entomologists to practise their trade successfully.

Research strategies

Successful pest and vector management can only be built on a sound foundation of knowledge acquired through research over many years. Although considerable research on tropical insect pests and vectors has been undertaken in many national and international research institutions throughout the tropics, there has been a very poor attempt to deliberately organize this effort towards integrated pest and vector management schemes. Research strategies must therefore be changed so as to concentrate more on applied problems, such as pest and vector management. There is need for entomologists to work in closer collaboration with other scientists, technologists and sociologists and to develop other areas of pest biology. For instance, more attention needs to be focused on insect genetics, the development of physiological strains which are poorly adapted to their environments and more susceptible to their natural enemies, and the effective use of chemicals based on the insect's own products such as pheromones, and insect growth regulators (IGRs). Integrated management tactics should be evaluated in the field soon after they have been developed.

International aid and co-operation

Research is expensive and some developing countries in the tropics lack adequate funds to allocate to research. Such poor countries therefore need considerable assistance from international foundations, United Nations agencies and other donor bodies, to effectively develop research and training programmes as well as to enable them to achieve scientific self-reliance. While such funding and expertise are essential, the philosophy behind the aid is even more important, and this is where problems have arisen in the past. We believe that one reason for the failure of many development aid programmes in developing countries is the erroneous assumption that the donors invariably know what is best for the recipient country. They then proceed to provide expertise and devise control tactics only to discover, too late, that the wrong assumptions have been made. It has been stressed in this chapter that indigenous scientists and technologists are the people who are best equipped to evaluate and determine their country's needs. International development aid should therefore adopt a system of working in partnership with the personnel in the poor countries.

The network of CGIAR (Consultative Group in International Agricultural Research) is a system of international co-operation in agricultural research in the tropics. Within this system it should be possible to intensify collaborative research efforts in pest and vector management.

National and regional support

Apart from providing adequate financial support the governments of developing countries need to recognize pest and vector management as a matter of utmost priority and commit the necessary resources of funds, manpower for training, and materials for the successful implementation of management programmes. It is their responsibility to determine the priority areas of training needed and to make accurate plans and projections of their manpower requirements for socio-economic development. Adequate recognition, and satisfactory job and career opportunities must be created for scientists and technologists so that these specialists can be retained to perform the jobs for which they will be trained. A number of pests and vectors such as grasshoppers, armyworm moths, tsetse flies and simuliid blackflies can be adequately controlled only with regional collaboration. Therefore resources must be made available by all the neighbouring countries concerned, and effective research and training organized at the regional level.

Support from industry

Research activities and development projects will from time to time require the production of materials for field application against pests and vectors. This is where industry inputs become relevant especially in integrated rural development schemes. It is therefore essential for industry to become involved at the research stage in order to hasten the build up of appropriate technologies.

Training programmes and training needs

National institutions exist in most developing countries for training professional staff in various aspects of the biology and control of insect pests and disease vectors. These range from undergraduate and postgraduate programmes in universities to diploma and certificate courses in polytechnics and advanced colleges of agriculture and animal health schools. The first two years of most 4 year university undergraduate programmes are usually devoted to courses in the basic sciences such as botany, zoology, biology, chemistry, physics, mathematics, climatology and biochemistry. This solid science background is highly desirable for an understanding of advanced courses in agricultural, biological or veterinary sciences. The last two years are usually devoted to general agriculture or veterinary sciences, in some cases with specializations in applied entomology, plant pathology, ecology, or plant breeding. From the faculties of science, graduates in biology, zoology, or botany are produced.

At higher educational levels such as for masters and doctorate degrees, further specialization occurs. After such specialization and training the former students may be employed by research institutes to undertake full-time research, or remain at university to teach and research. A large number are also absorbed into government bodies such as the Ministries of Agriculture and Health, while others enter 'agro-based' industries of the public and private sectors.

Examination of these programmes shows that at the undergraduate level a generalist is produced, while a specialist is the product of the higher degree programme. Very few institutions provide any kind of multi-disciplinary training in Integrated Insect Pest and Vector Management, which is a real need of developing countries. Unfortunately most institutions of higher education in developing countries are poorly staffed especially with trained indigenous professionals who have expertise in pest management. Often departments are poorly equipped and lack the necessary logistic support for teaching and research. As a result departments are unable to admit and train a sufficient number of students to meet the needs of their countries for pest and vector management. Furthermore, the few professional staff present are often over-burdened with teaching and administrative duties, and therefore have very limited time to carry out research on important pest and vector problems in their areas.

One consequence of these training inadequacies is that scientists and technologists from developing countries come to the industralized countries, like Britain, North America, Germany and France, to study for various degrees. In such countries they become involved in sophisticated research and acquire expensive scientific and technological tastes, which may be largely irrelevant to the specific problems of their home countries. Back home they grapple with readjustment difficulties, become frustrated by grossly inadequate organizational arrangements which hinder them from orientating their newly gained expertise in training towards national problems. Admittedly, there is merit in scientists and technologists from developing countries being regularly exposed to recent scientific advances and discoveries and learning new techniques which can be adapted to their countries' needs. However, the entire exercise should not degenerate, as often happens, into exact replication, in developing countries of what has been seen in the industralized countries.

Training for insect pest and vector management can be conveniently grouped into three major categories.

1 Professional training

This category includes training scientists and technologists at university level. Trainees undergo the usual undergraduate and postgraduate programmes and after graduation undertake fundamental research to provide a sound basis for developing long-term management systems for insect pests and disease vectors. This should result in providing a variety of high-level scientific and technological manpower which is so much needed by tropical countries. This is an important area of training in which ICIPE (International Centre of Insect Physiology and Ecology) in Nairobi has developed expertise for many years.

Besides these traditional undergraduate and postgraduate programmes, there is a further need to introduce programmes providing training in integrated pest and vector management specially for the tropics. This type of programme in crop protection, has already been designed in 1978 by A. Youdeowei and M.O. Adeniji with particular reference to pest problems of the peasant small-scale African farmer. This project was supported by FAO Plant Protection Service and the Rockefeller Foundation under the auspices of the Association of Faculties of Agriculture in Africa (AFAA). In this programme, African institutions of higher agricultural education are encouraged to introduce a 2 year postgraduate degree course leading to a master of science in crop protection. This course was designed specially for the tropical African environment, with special emphasis on the broad range of pest problems encountered by small-scale farmers having mixed cropping systems. The ultimate goal is to produce a cadre of well-trained pest management specialists in the area of integrated pest management who can function as agricultural extension specialists, pest management consultants, research scientists and technologists. Thus professional training programmes will build up the manpower who will not only be concerned with research and development of appropriate technologies for pest and vector management, but who will also participate fully in the field application and evaluation of the various management systems developed.

This flexible model course structure together with its philosophy was accepted by the AFAA general assembly in December 1978 in Rabat, Morocco. Later, the course was distributed to African institutions of higher agricultural education for necessary modification to suit particular local conditions. The adoption of this programme implies the introduction of a new degree, MSc (Crop Protection); perhaps this degree should be appropriately called the Master of Crop Protection (MCP). In the field of vectors of medical importance, MSc courses in medical entomology and parasitology lasting 1–2 years have been established recently in several developing countries enabling them to train their own scientists instead of having to send them to the developed countries. Hopefully, the curriculum will emphasize the practicalities of the subject, such as surveys and control operations. The World Health Organization is supporting some of these university MSc courses in various countries in Latin America, Africa and South-east Asia.

2 In-service training

In-service training is the system by which professionals and other personnel who are already in employment periodically receive instruction to improve their competence and skills.

Seminars, workshops, field courses and post-doctoral research programmes are included in this category, and all levels of staff can benefit from such in-service training. However, with regard to integrated pest and vector management this aspect of training has received very limited attention in the tropics. There is consequently a need to organize international in-service training programmes which will be multi-disciplinary in concept and approach, and also take into consideration strategies for the management of both agricultural pests and vectors of disease – because these aspects are frequently inter-related in many aspects. An initial attempt at this kind of training was made in 1977. In that year the International Centre of Insect Physiology and Ecology (ICIPE) (Fig. 10.1) in collaboration with the United Nations Environment Programme (UNEP), both in Nairobi, initiated an 'International Group-Training Course on the Components Essential for Ecologically Sound Pest and Vector Management Systems'. The objectives of this course were defined as follows:

(a) To provide a sound basis for scientific, modern and clear understanding of pest and vector management problems taking into account the preservation of environmental quality;
(b) To establish a foundation for field scientists who are concerned with, or who will be assigned, the responsibility for pest and vector management in developing countries;
(c) To provide trainees with information on modern trends and approaches in integrated pest and vector management stressing all the components essential for ecologically sound management systems;
(d) To provide an opportunity for exchange of ideas and experiences between pest and vector control operators from different geographical and ecological zones in the developing world;
(e) To develop personal and international co-operation between scientists in the area of practical pest and vector management and to encourage further consultation and advice regarding specific national and international pest and vector management programmes.

This in-service course was designed for young scientists in the field of agriculture, public health or livestock health who are either actively engaged in pest and vector management, or who are starting their professional careers in these fields. Trainees also include research scientists and university lecturers. Between 1977 and 1981 a total of 132 scientists from 30 developing countries have been successfully trained. They are in turn expected to train middle-level personnel in their home countries on new and up-to-date pest and vector management methods, and thereby help to build up the pest and vector management capabilities of their countries.

Fig. 10.1a

Fig. 10.1b (a) The main laboratories and administrative centre of the International Centre of Insect Physiology and Ecology (ICIPE) in Nairobi, Kenya. This is an important research and training centre for scientists, especially from tropical developing countries, concerned with arthropod pests and disease vectors. (b) The 'Duduville' ('Insect village') international guest house of ICIPE, where both trainees and lecturers are accommodated. (Courtesy of ICIPE.)

One area of in-service training which has not been sufficiently recognized and explored are the Post-doctoral Scholar and the Research Associate Schemes. The Post-doctoral Scholar Scheme provides young scientists who have just completed their doctorate programmes with an opportunity to expand and strengthen their research experience in a different tropical environment for 1–3 years. The Research Associateship Scheme enables mature scientists from a developing country to undertake research in a research centre in another developing country for short periods (3–6 months) every year for up to 4 years. This scheme promotes intellectual intercourse between research scientists in developing countries.

The ICIPE Research Centre in Nairobi has been in the forefront of pioneering efforts in these areas of training. Further development and expansion of such schemes is highly desirable.

3 Field operators

All the money and effort spent on research and development in pest and vector management will be wasted if the results of these efforts are not put into practical use. This requires well-trained personnel and efficient extension services to operate at field level. Surveillance personnel, spraymen, technical support personnel and other field staff are included in the category of field operators that need adequate training. Field operators are extension specialists who according to O.C. Onazi 'must be mature and experienced persons who are not only well trained in the professional area (technical subject) but also have knowledge and experience in extension methodology'. The extension services in most developing countries suffer the greatest shortage of trained manpower at all levels. The recently established UNDP/FAO Tsetse Applied Research and Training Project in Lusaka, Zambia hopes to train field operators in tsetse fly control in efforts to control human and animal trypanosomiasis.

Administrators and politicians

Another aspect of training, often overlooked, is the education of administrators and politicians on the level of priority which should be given to research and training in pest and vector management in developing countries. As funding for research and training depends on the goodwill of the politicians and administrators, it is essential that these people are constantly made aware of the dimensions of local pest and vector problems, and the financial implications and level of commitment required to tackle them. Seminars and workshops on pest and vector problems organized for politicians, administrators and government policy-makers can greatly influence political decisions. It is essential therefore that such people are fully aware of the adverse economic effects of insect pests and disease vectors, and the relevance of new approaches for their management. In other words the pest and vector problems must be made political issues in developing countries so as to give them the priority rating which they deserve.

Conclusions

Clearly there is an urgent need for the establishment of some form of effective institutional arrangement within tropical countries to provide relevant research and training in pest and vector management. One practical way to achieve this arrangement is to utilize a few specialized regional research and training centres located within tropical developing countries. The functional organization of such a centre is presented in Fig. 10.2.

The centres would be funded and fully supported by national governments, industry, international foundations, United Nations agencies and foreign donor organizations. Two distinct functions would be carried out, namely fundamental research on insect pests and disease vectors, and training. Research would be specifically developmental-orientated for application to pest and vector management. For both functions, the centre should collaborate with national universities and other higher educational institutions which have the authority to award diplomas and degrees; while the centre itself should provide the research base for professional training of scientists and technologists. This can be achieved readily by means of collaborative agreements between the centre and the educational institutions. Apart from professional training, the centre should train field operators and organize in-service training in the form of formal lectures, seminars, study workshops, symposia, group-training and field courses for all levels of staff, and involve participants from neighbouring countries.

To foster community participation, which is especially important for vector management programmes, the centre should aim to educate and enlighten the public. In addition, special seminars for administrators, politicians and policy-makers in governments should be arranged. These functions should lead to the development, application and evaluation of insect pest and vector management systems in both agriculture and public health, and eventually result in increased food production and a more healthy population. A feed-back mechanism should be established for referring problems of pest and vector management in the field to the regional research and training centre for investigation and the application of appropriate solutions. The efficient execution of all these activities would help to achieve sound socio-economic development.

The outcome of the successful operation of a few well-organized regional research and training centres is international and regional co-operation, and the exchange of information and ideas on pest and vector

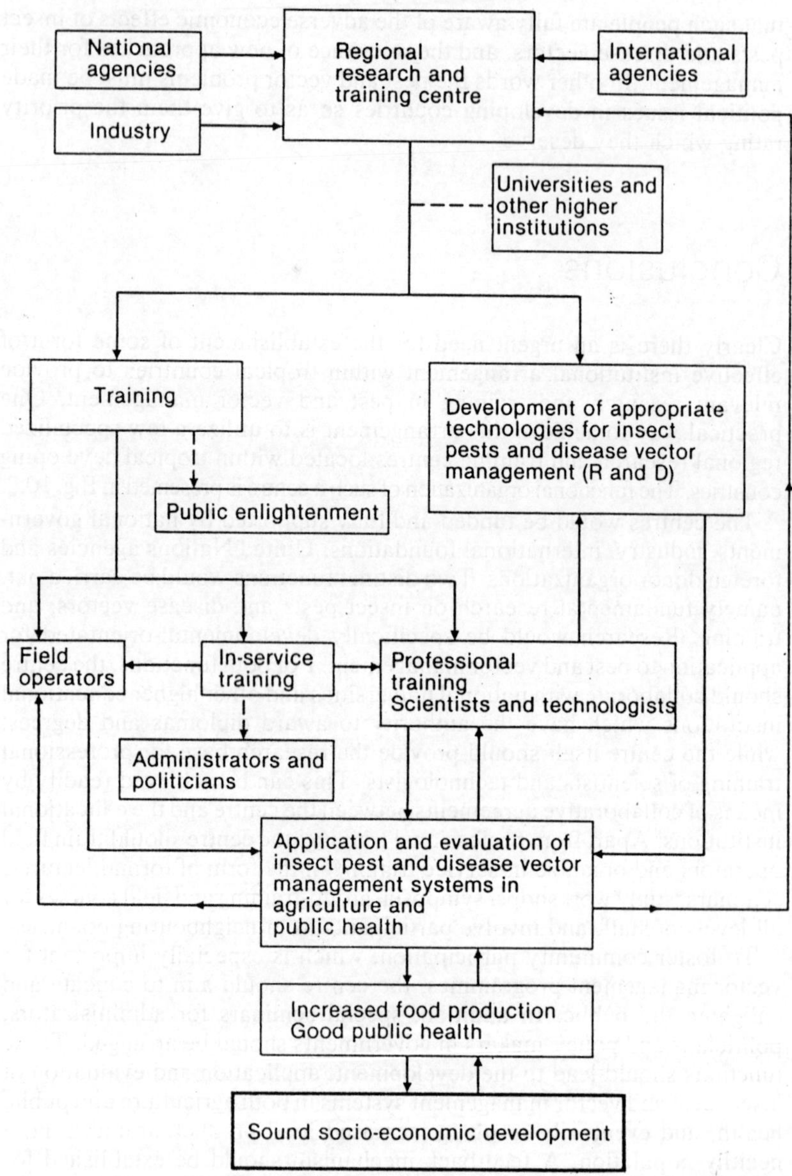

Fig. 10.2 Proposed functional organization of a model Regional Research and Training Centre.

management. The International Centre of Insect Physiology and Ecology (ICIPE) in Kenya is already vigorously pursuing such activities. The institution typifies the model of an 'activity centre' for research and training on insect pests and disease vectors in the tropical developing world.

Species index

General index